吴文忠　编著

Cracking the AP Calculus

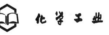
·北京·

《AP 微积分辅导手册》融汇众多成功案例，直击中国学生的薄弱点，解构整门考试的知识点、考点，为参加 AP 微积分考试的中国学生提供一套应对 AP 微积分（AB&BC）考试的完备方案。希望考生学完本书内容，可以顺利通过考试。

《AP 微积分辅导手册》一书的内容有：函数、极限和连续性、导数、微分、不定积分和定积分、积分的应用、微分方程和级数，涵盖了 AP 微积分 AB 和 AP 微积分 BC 考试大纲中要求的全部考点，并且有相关的例题演示，在理论讲解上兼顾实战性。

本书适合准备前往海外读大学的高中生，准备参加 AP 考试的考生学习使用，同时可用作相关培训和辅导机构的参考教材。

图书在版编目（CIP）数据

AP 微积分辅导手册/吴文忠编著. —北京：化学工业出版社，2018.8（2024.4重印）
ISBN 978-7-122-32488-7

Ⅰ.①A… Ⅱ.①吴… Ⅲ.①微积分-高等学校-入学考试-美国-自学参考资料 Ⅳ.①O172

中国版本图书馆 CIP 数据核字（2018）第 136537 号

责任编辑：耍利娜　　　　　　　　　文字编辑：吴开亮
责任校对：王素芹　　　　　　　　　装帧设计：刘丽华

出版发行：化学工业出版社（北京市东城区青年湖南街 13 号　邮政编码 100011）
印　　装：北京天宇星印刷厂
787mm×1092mm　1/16　印张 19½　字数 466 千字　2024 年 4 月北京第 1 版第 8 次印刷

购书咨询：010-64518888（传真：010-64519686）　售后服务：010-64518899
网　　址：http://www.cip.com.cn
凡购买本书，如有缺损质量问题，本社销售中心负责调换。

定　价：59.80 元　　　　　　　　　　　　　　　　　　　　　版权所有　违者必究

前言
Foreword

自 2012 年开始，不经意间竟攒下教学总结数百页．同时，伴随着学员的好成绩，越来越多家长找到我，希望我辅导孩子学习微积分，也在很多家长朋友建议下，我开始萌生编撰本书的想法．

2014 年本书部分内容线上发布，截止到 2017 年元旦，累计读者 12 万多人次，许多学生考出 5 分（满分）的优异成绩，我感到很自豪！

置身教学第一线多年，我深切体会到学生在微积分学习上的难点、易错点和薄弱点．微积分很抽象，学生在学习过程中，可能遭遇抽象性和数学语言符号化等困难，也可能遭遇有限运算到无限运算转变的困难，还可能遭遇基础不够等困难．

正是因为笔者知道如何才能培养和提高学生的能力，了解后继课程的需要以及沉淀教学经验，本书考虑到多数中国学生所遇到的问题和需求，并完全适应最新的 AP 微积分考纲（2017 年），确保学生能从这本教材中真正地受益，取得好成绩．

如何阅读本书

本书在编排方式上循序渐进，便于读者随时参阅并配合练习，合理使用本书，将大有裨益．

首先，充分了解 AP 微积分并及早制订学习计划．第 1 章，详细介绍了 AP 微积分及考试大纲、常见问题、试卷结构和考试时间等，帮你打"有准备的战"，帮你制订出适合自身的学习计划．

其次，全面掌握本书每一个知识点．第 2 章，集中回顾了 Pre-Calculus 知识点，用来确保学习微积分良好的数学基础；第 3~11 章，每章都有思维导图、英文思维的核心概念、精选的 Example、方法总结和 Note 标注，多层次地帮你全面掌握每一个必考知识点．

最后，注重练习与分享．演算典型例题和习题，真题实战训练，是保证考试成绩不可或缺的一个环节．并且，以分享者的角色来阅读，效果大不一样．

分享的阅读方式可以增强记忆、加深体会、扩大视野．开诚布公地与人分享读书心得，可以为你带来一群学习好伙伴．也欢迎通过微信与我取得联系，加入官方阅读群．

你将有何收益（本书的特点）

我想借用美国作家 Marilyn Ferguson 的一段话：

"谁也无法说服他人改变．我们每个人都守着一扇只能从内开启的改变之门，不论动之以情或晓之以理，我们都不能替别人开门．"

当你决定对微积分打开"接纳之门"，接纳本书，你将得到以下收获．

首先，一书在手，AP 微积分考试无忧．

微积分在近代数学的地位绝无仅有，本书聚焦 AP 微积分考试，共计 11 章，有效覆盖 AP 微积分考试所有考点，专注于中国学生的需求．

- 详细介绍了 AP 微积分及其考试，精心汇编了专业词汇；
- 以美国本土广为流行的 TI-89 Titanium * 为例，讲解图形计算器使用；
- 解构知识点、考点，精心配备了 380 余道典型例题，100 余道课后习题；
- 直击考生薄弱点，函数图像、图表 200 多幅，Note 标注近 80 处，方法总结超过 30 个．

作为一本内容翔实的备考教材，本书还针对考生 Free-Response 薄弱的状况，习题"分解式"紧密呼应考点并附详细解答，全面对接国内高中数学，衔接 Pre-Calculus，让基础薄弱的同学无忧，还总结了考试常用公式定理，实现"一书在手，AP 微积分考试无忧"．

其次，希望本书能弥补预定不上我们课程的遗憾．

本书一开始即为 AP 微积分辅导课程编撰，参加我的课程无疑是一个好选择，然而，由于笔者时间不足，一部分同学可能预定不上，希望本书弥补笔者的部分遗憾，以另一种形式为你的备考出力．

- 例题选取上，如本书 Example 5.42，在常规求导（Derivative）外，示例避开繁杂运算，Example 5.47 再示例简化方法，颇费心思；
- 梯度设置上，如本书 5.4.1 节 Chain Rule 的实际应用，分解举例，先结合 Quotient Rule，再结合三角函数，循序渐进，步步深入考生的难点；
- 类似本书 7.3.2 节和 7.3.3 节，点出学生容易掉入的"坑"，并不仅仅停留于理论分析，还分解具体题型（如本书 7.3 节、7.4 节和 7.5 节），帮考生"跳出坑"；
- 准确把握知识程度，如本书 3.1 节，明确指出极限的考试范围是函数极限，不包含数列极限，4.1 节讲解连续，不用一般性定义，选用 3 个判断条件；
- 力求帮助同学们"知其然，更知其所以然"，如以 Example 形式给出导数基本公式证明，同学们可以掌握导数基础公式，还可以跟随练习．

通过研究真题发现，只给函数图像，不给解析式，中国考生很不习惯，特别在 Free-Response 部分．于是，借助我们过去开发的针对性训练，在本书各个章节分布重点函数图像．

笔者清楚，选修 AP 微积分的学生的情况多种多样，基础参差不齐，部分学生对数学兴趣不大，对微积分的重要性也不了解．本书入门门槛不高，突出重点，注重知识点内在脉络，使学生能逐步学好微积分．

最后，中英文结合，不止于通过 AP 微积分考试．

本书定位：做更适合中国考生的 AP 微积分辅导教材．一些英文原版教材对中国考生而言，或讲解过于空泛，或例题与知识点呼应不够紧密；国内不乏优秀的微积分教材，而专门针对 AP 微积分的却比较少．

本书汲取多本知名英语原著教材之精华，参考国内知名高等数学教材（见参考文献），中英文结合，英文思维为主，辅以母语的优势：

- 专业词汇中英文对照；
- 基本概念、定义和性质，典型例题、习题和解答，英文思维，英文表达；
- 过程讲解发挥母语优势，脉络清晰，重点内容标注专业英语词汇；
- 让你轻松应对 AP 微积分考试，更适应未来国外的大学数学课堂．

一本 AP 微积分辅导教材，若能尽可能快地向学生介绍微积分的基本概念、方法和应用，鼓励学生"数形结合""特值代入"等直观形象地分析、演算及思考，重点加强学生代数运算和抽象思维，兼顾中英文数学差异，这才是更适合中国考生的．

读者对象

本书可以作为一些学校和老师的 AP 微积分课程参考教材，也适合很多刚进入 AP 微积分辅导行业老师入门参考．本书适合如下读者：
- 在初中和高中阶段选修 AP 微积分的同学；
- 有出国意愿和规划的国内同学；
- 就读于国内国际班或国际学校的同学；
- 就读于国外高中，计划申请一流大学的同学；
- 已经获得国外大学 offer，而又想有所提高的同学；
- 微积分或数学爱好者．

勘误和支持

由于笔者的水平有限，书中难免出现一些不妥之处，在此恳请读者批评指正．笔者个人微信号（709645945），将努力回答所有与本书内容相关的问题，虚心接受您的宝贵意见和各种批评建议．另外，微信订阅号"AP 微积分"（微信号：apcalculus）用于及时更新读者反馈和最新的研究成果．

致谢

感谢我的学生们，让笔者更清楚学生的薄弱点在哪里，更需要什么，以及如何更高效并更有保障地获取 5 分（满分）的成绩．部分学生已经步入社会的各领域，看着学生的成长，笔者觉得自己的工作很有意义．

感谢众多家长朋友的信任和支持，有些凭借十几分钟的电话，素未谋面，便把孩子的学习托付到我手上，这份信任一直鼓舞着我．

感谢化学工业出版社的编辑，始终支持我的写作，让我顺利完成本书全部书稿．

谨以此书献给我最亲爱的家人和朋友们！

编著者

目录 Content

第1章 AP 微积分简介 Introduction of AP Calculus — 001

- 1.1 课程及考试 The Courses and Examinations — 001
- 1.2 AP 微积分 AB 和 BC 大纲要求 The Examination Outline of AP Calculus AB & BC — 004
- 1.3 AP 微积分参考词汇表 Reference Vocabulary of AP Calculus — 006
- 1.4 图形计算器的使用 Use of Graphing Calculators — 013

第2章 函数 Functions — 019

- 2.1 函数的定义 Definition of Functions — 020
- 2.2 函数的基本性质 Function Basic Properties — 022
- 2.3 基本初等函数 Basic Elementary Functions — 023
- 2.4 反函数 & 复合函数 Inverse Functions & Composite Functions — 033
- 2.5 函数变换 Transforming of Functions — 035
- 2.6# 参数方程 & 向量函数 Parametric Equations & Vector Functions — 037
- 2.7# 极坐标函数 Polar Functions — 039
- 2.8 习题 Practice Exercises — 041

第3章 极限 Limit — 043

- 3.1 极限的定义 Definition of the Limit — 044
- 3.2 极限存在的判定 The Limit does Exist or Not — 045
- 3.3 极限的运算 Operations of Limit — 047
- 3.4 极限的应用 Applications of Limit — 052
- 3.5 习题 Practice Exercises — 053

第4章 连续 Continuity — 055

- 4.1 连续性的定义 Definition of the Continuity — 056

4.2 间断点的分类 Kinds of Discontinuities ⋯⋯⋯⋯⋯⋯⋯⋯⋯⋯⋯⋯⋯⋯ 059
4.3 连续函数定理 The Continuous Functions Theorem ⋯⋯⋯⋯⋯⋯⋯⋯ 061
4.4 习题 Practice Exercises ⋯⋯⋯⋯⋯⋯⋯⋯⋯⋯⋯⋯⋯⋯⋯⋯⋯⋯⋯⋯ 063

第 5 章　导数和微分 Derivative and Differential　　　*065*

5.1　导数的定义 Definition of the Derivative ⋯⋯⋯⋯⋯⋯⋯⋯⋯⋯⋯⋯⋯ 066
5.2　可导性和连续性 Derivability and Continuity ⋯⋯⋯⋯⋯⋯⋯⋯⋯⋯⋯ 072
5.3　导数的基本公式和法则 Basic Differentiation Formulas and Rules ⋯⋯ 075
5.4　链式法则和反函数求导 The Chain Rule & Derivative of an Inverse
　　　Function ⋯⋯⋯⋯⋯⋯⋯⋯⋯⋯⋯⋯⋯⋯⋯⋯⋯⋯⋯⋯⋯⋯⋯⋯⋯⋯⋯ 077
5.5　隐函数求导和二阶导数 Implicit Differentiation & Second Derivatives ⋯ 082
5.6# 参数方程求导 Derivatives of Parametric Equations ⋯⋯⋯⋯⋯⋯⋯⋯ 088
5.7# 向量函数和极坐标函数求导 Derivatives of Vector Functions and
　　　Polar Functions ⋯⋯⋯⋯⋯⋯⋯⋯⋯⋯⋯⋯⋯⋯⋯⋯⋯⋯⋯⋯⋯⋯⋯⋯ 090
5.8　微分 Differential ⋯⋯⋯⋯⋯⋯⋯⋯⋯⋯⋯⋯⋯⋯⋯⋯⋯⋯⋯⋯⋯⋯⋯ 093
5.9　习题 Practice Exercises ⋯⋯⋯⋯⋯⋯⋯⋯⋯⋯⋯⋯⋯⋯⋯⋯⋯⋯⋯⋯ 096

第 6 章　微分的应用 Applications of Differential Calculus　　　*098*

6.1　切线方程和法线方程 Equations of Tangent and Normal ⋯⋯⋯⋯⋯⋯ 099
6.2　最值问题 The Problems of Maxima and Minima ⋯⋯⋯⋯⋯⋯⋯⋯⋯ 101
6.3　运动问题 The Problems of Motion ⋯⋯⋯⋯⋯⋯⋯⋯⋯⋯⋯⋯⋯⋯⋯ 112
6.4　微分中值定理 The Mean Value Theorem for Derivatives ⋯⋯⋯⋯⋯ 118
6.5　洛必达法则 L'Hôpital's Rule ⋯⋯⋯⋯⋯⋯⋯⋯⋯⋯⋯⋯⋯⋯⋯⋯⋯⋯ 120
6.6　估算问题 The Problems of Estimate ⋯⋯⋯⋯⋯⋯⋯⋯⋯⋯⋯⋯⋯⋯ 125
6.7# 欧拉方法 Euler's Method ⋯⋯⋯⋯⋯⋯⋯⋯⋯⋯⋯⋯⋯⋯⋯⋯⋯⋯⋯ 129
6.8　习题 Practice Exercises ⋯⋯⋯⋯⋯⋯⋯⋯⋯⋯⋯⋯⋯⋯⋯⋯⋯⋯⋯⋯ 130

第 7 章　不定积分 The Indefinite Integral　　　*132*

7.1　不定积分的定义 Definition of The Indefinite Integral ⋯⋯⋯⋯⋯⋯⋯ 133
7.2　不定积分公式 Formulas of The Indefinite Integral ⋯⋯⋯⋯⋯⋯⋯⋯ 135
7.3　U-替换法 U-Substitution ⋯⋯⋯⋯⋯⋯⋯⋯⋯⋯⋯⋯⋯⋯⋯⋯⋯⋯⋯ 138
7.4# 分部积分法 Integration by Parts ⋯⋯⋯⋯⋯⋯⋯⋯⋯⋯⋯⋯⋯⋯⋯⋯ 148
7.5# 有理函数的积分 Integration of Rational Functions ⋯⋯⋯⋯⋯⋯⋯⋯ 153
7.6　不定积分的应用 Applications of Indefinite Integral ⋯⋯⋯⋯⋯⋯⋯⋯ 156
7.7　习题 Practice Exercises ⋯⋯⋯⋯⋯⋯⋯⋯⋯⋯⋯⋯⋯⋯⋯⋯⋯⋯⋯⋯ 157

第 8 章 定积分 The Definite Integral — *159*

- 8.1 黎曼和与梯形法则 Riemann Sums and Trapezoid Rule ⋯⋯ 160
- 8.2 定积分的定义 Definition of the Definite Integral ⋯⋯ 165
- 8.3 微积分基本定理 The Fundamental Theorem of Calculus ⋯⋯ 169
- 8.4 定积分的性质 Properties of Definite Integral ⋯⋯ 174
- 8.5 积分中值定理 The Mean Value Theorem for Integrals ⋯⋯ 176
- 8.6 定积分的计算 The Operations of Definite Integrate ⋯⋯ 178
- 8.7# 广义积分 Improper Integrals ⋯⋯ 180
- 8.8 习题 Practice Exercises ⋯⋯ 185

第 9 章 积分的应用 Applications of Integral — *186*

- 9.1 面积 Area ⋯⋯ 187
- 9.2 体积 Volume ⋯⋯ 195
- 9.3# 弧长 Arc Length ⋯⋯ 204
- 9.4 位移和距离 Displacement and Distance ⋯⋯ 206
- 9.5 习题 Practice Exercises ⋯⋯ 207

第 10 章 微分方程 Differential Equations — *209*

- 10.1 一阶微分方程 First-Order Differential Equations ⋯⋯ 210
- 10.2 求解可分离变量微分方程 Solving Separable D. E. ⋯⋯ 211
- 10.3 斜率场 Slope Fields ⋯⋯ 213
- 10.4 指数增长与衰减 Exponential Growth and Decay ⋯⋯ 216
- 10.5 约束增长与衰减 Restricted Growth and Decay ⋯⋯ 219
- 10.6# 逻辑斯谛微分方程 Logistic Differential Equation ⋯⋯ 222
- 10.7 习题 Practice Exercises ⋯⋯ 225

第 11 章 无穷级数 Infinite Series — *226*

- 11.1 数列的极限 The Limit of The Sequence ⋯⋯ 227
- 11.2 无穷级数 Infinite Series ⋯⋯ 228
- 11.3 四类重要级数 Four Important Series ⋯⋯ 232
- 11.4 正项级数的四大判别法 Four Tests of Nonnegative Series ⋯⋯ 235
- 11.5 绝对收敛和条件收敛 Absolute and Conditional Convergence ⋯⋯ 240
- 11.6 幂级数 Power Series ⋯⋯ 242
- 11.7 泰勒级数和麦克劳林级数 Taylor and Maclaurin Series ⋯⋯ 245
- 11.8 幂级数的计算 Computations with Power Series ⋯⋯ 251

11.9 习题 Practice Exercises ·················· 254

习题答案 Practice Answer   255

附录 Appendix   287

A.1 常用公式和定理 Common Formulas and Theorems ·················· 287
A.2 AP 微积分公式总结 Summary AP Calculus Formula ·················· 291
A.3 VIP 服务及网站 ·················· 298

参考文献 References   299

第 1 章

AP 微积分简介
Introduction of AP Calculus

微积分是近代数学中最伟大的成就,对它的重要性无论怎么评价都不过分.
The calculus was the first achievement of modern mathematics and it is difficult to overestimate its importance.

——【美】冯·诺依曼（Von Neumann）

1.1 课程及考试 The Courses and Examinations

恩格斯指出,"在一切理论成就中,未必再有什么像 17 世纪下半叶微积分的发现那样被看作人类精神的最高胜利了.如果在某个地方我们看到人类精神的纯粹和唯一的功绩,那就正是在微积分这里."

微积分是数学的一个基础学科；在物理、化学、生物、工程、医学和经济学等很多学科上,微积分也都有广泛应用,其价值难以估量.

这里以笔者微信的个性签名与大家共勉,一起开启这门课程的学习：

仰望莱布尼茨、牛顿肖像；

拉格朗日,傅里叶旁,追逐黎曼最初梦想……

1.1.1 为什么要参加 AP 微积分考试 Why would you like to join the examination of AP Calculus

其一,微积分的学习是世界各国高等教育的重要内容之一,被绝大多数大学专业列为必修课程,AP 微积分能使同学们提前接触到这门重要的大学课程.

其二,AP 微积分考试成绩达 4 分或 5 分,美国大多数的大学可以置换相应的学分,有些大学也接受 3 分的成绩.这意味着,虽然还没进大学,你已经拿到大学的学分,为你大学教育节约时间成本并省下一笔高昂的学费.

其三,AP 微积分成绩标志着一个学生的数学学术能力,对申请高水平大学具有重要的参考价值.即便最后 AP 微积分考试成绩不能置换学分,也足以证明你在中学已经学习了高等数学的课程,给大学数学学习打下坚实的基础.

另外,微积分知识对于理解 AP 经济学中边际、弹性等概念有很好的帮助,学习 AP 物理需要用微积分知识作为工具,所以,AP 微积分成为很多同学的首选科目.

1.1.2 AP 微积分 AB 和 BC 有什么区别 What are the differences between AP Calculus AB and BC

AP 微积分课程包括微积分 AB 和微积分 BC 两门课．微积分 AB 大致相当于国内文科高等数学第 1 学期的内容和难度．微积分 BC 大致相当于国内普通工科第 1 学年（2 学期）微积分课程的内容和难度．内容上，微积分 BC 涵盖了 AB 的所有内容，具体请参考本书 1.2 节（AP 微积分 AB 和 BC 大纲要求）．本书加 "♯" 内容为 BC 要求而 AB 不要求的内容．

因为考试时间重合，一年之内，AB 和 BC 只能选择一科，但并不要求一定得先考 AB 再考 BC. 假如同学们参加微积分 BC 的考试，微积分 BC 的成绩单上面同时会提供微积分 AB 的成绩．

对于未来专业选择生物化学、社会科学、普通商科及管理类的同学而言，BC 基本涵盖了大学本科所有的教学内容，拿到 5 分并兑换 AP 学分，意味着大学不需要再修习微积分课程．对于专业方向是数学、计算机、物理、工程或经济、金融等专业的同学，BC 能相当好地满足后续学习要求．

部分同学纠结于选择 AB 还是 BC，多数情况下，建议结合自身数学基础、专业选择以及所申请大学的要求，酌情选择即可．特殊情况下，欢迎与我联系，定制最优方案．

1.1.3 AP 微积分考题结构和考试时间 The test form and schedule of AP Calculus

改革后，最新版（2017 年）的 AP 微积分，考题结构和考试时长不变：总时长 195 分钟，共 108 分．选择题，选项由 5 个减少为 4 个，无计算器部分增加 2 道，有计算器部分减少 2 道，对于考试时长调整了 5 分钟．见 Table 1.1.3.

Table 1.1.3　AP 微积分考题结构和考试时间（最新）

多项选择题 Multiple-Choice Section	总共 45 题/105 分钟	答对得 1 分，不答或答错均不得分不扣分，卷面分乘以系数 1.2 为最后得分．答对全部得 54 分，占总分 50%
Part A：无计算器	30 题/60 分钟	
Part B：有计算器	15 题/45 分钟	
自由问答题 Free-Response Section	总计 6 题/90 分钟	每题 9 分，共 54 分，占总分 50%
Part A：无计算器	2 题/30 分钟	
Part B：有计算器	4 题/60 分钟	

每年具体考试日期可在微信订阅号 "AP 微积分"（微信号：apcalculus）中回复 "考试日期" 查看．

1.1.4 制订你的学习计划 Make your plan

因为申请一般都在秋季，对计划出国的高中生而言，三年时间减少为两年半．为此，建议同学们在高一、高二合理安排 AP 微积分课程计划．高一至少完成微积分 AB 的课程学习并拿到理想成绩，至少获得 3 方面优势：

① 由于专业方向可能未确定，主要考虑因素是考试成绩，数学基础好的同学，直接报 BC，基础一般的同学，先报 AB，再报 BC，策略取胜，都能拿到理想的微积分考试成绩，为高水平大学申请"攒下"优势.

② 微积分知识对于理解 AP 经济学很多概念有帮助，是学习 AP 物理的工具，优先选报微积分，为选报其他 AP 课程打下知识基础.

③ 若是到后面，只剩下一年多一点的时间，要完成托福、SAT 或者 ACT 的考试，再加上 SAT2 和 AP 的压力，时间真的不够. 所以，高一提前准备不止能获得更多考试机会，还能减轻压力.

1.1.5 自学或参加辅导 Learn By Oneself or Attend The After-school Class

理论上，任何课程都可以自学，本书十分适合中国学生用来自学 AP 微积分，在此，提供 4 点建议：

① 确保良好的数学基础，本书第 2 章重点回顾了微积分的预备知识（Pre-Calculus），希望同学们巩固掌握.

② 本书完全按照最新 AP 微积分考纲编写，突出重点，分章节编排，注重知识点内在脉络和联系，同学们应该把控进度，合理安排学习计划.

③ 依据本书各章给出的典型例题，一步一个脚印地掌握所有知识点，完成各章对应的练习题，做到举一反三，深入理解.

④ 结合 College Board 发布的 AP 微积分真题，贴近实战经验，保障考试发挥.

另外，AP 微积分是一门准大学水平课程，具有一定的挑战性，除了自学，本书配套 VIP（辅导）服务体系，可通过"AP 微积分·考试网"（calculus.apexams.net）或微信订阅号"AP 微积分"（ID：apcalculus）与我取得联系. 参加辅导将为你带来三方面的价值：

① 节约宝贵的时间成本（划重点），减轻压力. 面临 TOEFL、SAT 或 ACT 考试，再有 SAT2 和 AP 的压力，时间真心不够，完善的课程体系和课程配套、资深的教研和师资给你支持，部分同学还能获得本人一对一亲自辅导.

② 高分，有保障. 完善的课程体系，《AP 微积分真题导练》训练体系，对 AP 微积分考试的研究、评分规则的把握以及微积分的教学积淀，你将获得每一步推理根据、数学语言规范性和关键步骤（步骤分）等非常细节化的指导.

③ 启发思维. 我认为这一点特别重要，教育是鼓励、引领和启发，上述两点是教学，这一点接近教育，除去考试，启发思维，你还能获得更多.

1.1.6 AP 微积分等级分和原始分转换 AP Score Conversion Chart Calculus

AP 微积分采取 5 分制，分别是：5 分非常优秀（Extremely Well Qualified）；4 分优秀（Well Qualified）；3 分及格（Qualified）；2 分可能及格（Possibly Qualified）；1 分无建议（No Recommendation）. 考生原始成绩都转换成 5 分制，每年略有变化，但总体趋势稳定. Table 1.1.6.1 是 2011 年 AP 微积分等级分和原始分转换表，并且历年成绩分布也非常接近，没有显著变化，Table 1.1.6.2 给出 2017 年微积分 AB 和 BC 的成绩分布.

Table 1.1.6.1　AP 微积分等级分和原始分转换表（2011 年）

微积分 AB		微积分 BC	
考试实际分数	5 分制	考试实际分数	5 分制
68～108	5	69～108	5
52～67	4	59～68	4
39～51	3	44～58	3
27～38	2	36～43	2
0～26	1	0～35	1

Table 1.1.6.2　AP 微积分成绩分布（2017 年）

等级分	AB		BC	
	人数	百分比	人数	百分比
5	59250	18.7%	56422	42.6%
4	56775	18.0%	23987	18.1%
3	65851	20.8%	26341	19.9%
2	69631	22.0%	18694	14.1%
1	64592	20.4%	7070	5.3%
全球平均分	2.93		3.78	

1.2　AP 微积分 AB 和 BC 大纲要求 The Examination Outline of AP Calculus AB & BC

最新改革后的 AP 微积分（2017 年），微积分 AB 增加考点 L'Hopital's Rule，BC 在 Series 部分增加考点 Limit Comparison test，Absolute and Conditional Convergence 以及 The Alternating Series Error Bound，本书完全匹配支持，除此之外，因为 L'Hopital's Rule 很好用，本书 BC 部分扩了其他不定式类型（见 6.5.3 节），提升学生能力．

内容上，BC 在 AB 基础上，主要增加参数方程（Parametric Equations）、向量函数（Vector Functions）和极坐标函数（Polar Functions），广义积分（Improper Integrals）以及无穷级数（Infinite Series）等．下面根据 College Board 发布的考试大纲详细说明．

1.2.1　基础知识 Basic Facts

1）函数（Functions）和函数图像（Graphs）分析；

2）基本初等函数的导数（Derivatives）和不定积分（Antiderivatives）；

3）导数的积（Product）、商（Quotient）和链式法则（Chain Rules）；

4）使用矩形（Rectangle）的左边（Left）、中点（Midpoint）或右边（Right）估算定积分，使用梯形法则（Trapezoid Rules）估算定积分（Definite Integrals）；

5）重要的定理：罗尔定理（Rolle's Theorem）、中值定理（The Mean Value Theorem）和微积分基本定理（The Fundamental Theorem of Calculus）；

6）#掌握参数方程（Parametric Equations）、向量函数（Vector Functions）和极坐标

函数（Polar Functions）并应用于微积分.

1.2.2 极限和连续 Limits and Continuity

1）掌握函数的极限，包括单侧极限（One-sided Limits）；

2）理解渐近性（Asymptotic）和无界性（Unbounded）并掌握渐近线（Asymptote）；

3）理解函数的连续性（Continuity）并判断其连续性.

1.2.3 导数及其应用 Derivative and Its Applications

1）理解导数是一个瞬时的变化比率（Instantaneous Rate of Change），并且能够应用这个概念.

2）求解某点处的导数，以及求方程的切线（Tangent Line）和法线（Normal Line）.

3）运用导数、二阶导数研究函数在上升/下降（Increasing/Decreasing），上凹/下凹（Concave Up/Down），以及最大值/最小值（Maxima/Minima）和拐点（Inflection）.

4）在运动中分析一个物体的瞬时速度（Speed）、速度（Velocity）和加速度（Acceleration）；必要时，运用隐函数求导（Implicit Differentiation）的方式解决变化率（Rates）问题.

5）使用洛必达法则（L'Hopital's Rule）计算不定型的极限（Limits of Indeterminate Forms）.

6)# 掌握两个维度的运动分析. 掌握如何分析位置（Position）、速度（Velocity）、瞬时速度（Speed）、加速度（Acceleration）、物体运动的距离（Distance），并使用向量（Vectors）微积分的概念分析在两个维度上的运动.

1.2.4 积分及其应用 Integral and Its Applications

1）理解积分是代表着基于不定积分的累加函数（Accumulation Functions），并能够运用这个概念；

2）理解积分中值定理（The Mean Value Theorem for Integrals）并会求函数的平均值（Average Value）；

3）求面积（Area）和体积（Volume）；

4）求运动对象的位置（Position）和运动距离（Distance）；

5）给定一个比例（Accumulation），考虑累积的总量（Total Amount）；

6)# 使用分步积分法（Integration by Parts）和部分分数法（Partial Fractions）求解积分（Finding Antiderivatives）；

7)# 使用极限分析和求解广义积分（Improper Integrals）；

8)# 求解弧线长（Arc Lengths）.

1.2.5 微分方程 Differential Equations

1）求解微分方程（Differential Equations）和斜率场（Slope Fields）；

2）结合微分的应用，使用欧拉方法（Euler's Method）估算分析；

3)# 理解掌握指数增长（Exponential Growth）、约束增长（Restricted Growth）和

logistic 增长等微分方程，并简单运用解决实际问题．

1.2.6# 无穷级数 Infinite Series

1)# 熟悉重要的级数：几何级数（Geometric Series）、P 级数、调和级数（Harmonic Series）、交错级数（Alternating Series）；

2)# 使用比值判别法（Ratio Test）、极限比较判别法（Limit Comparison test）等判别常数项级数（Series of Constants）收敛/发散（Converges/Diverges）；

3)# 理解绝对收敛（Absolute Convergence）和条件收敛（Conditional Convergence）并能求交错级数的误差界（Alternating Series Error Bound）；

4)# 确定幂级数（Power Series）的收敛半径和收敛区间（The Radius and Interval of Convergence）；

5)# 使用泰勒级数（Taylor Series）和麦克劳林级数（Maclaurin Series）表示函数的幂级数；

6)# 使用拉格朗日误差界（Lagrange Error Bound）估算级数的误差边界．

1.3 AP 微积分参考词汇表 Reference Vocabulary of AP Calculus

①按知识体系分类整理；②意思相近的词汇尽量排列在一起；③标注词汇音标（部分过长词条标注主要音标）．使用本词汇表，更加高效、稳固地掌握课程专业词汇．

1.3.1 算数和代数 Arithmetic and Algebra

add	加上
subtract	减去
multiply	乘以
divide	除以
sum	和
difference	差
product	积
quotient	商
reciprocal	倒数
radians	弧度
degrees	角度
trigonometric identity	三角恒等式
Pythagorean theorem	勾股定理
double-angle formula	倍角公式
sum formula	和角公式
difference formula	差角公式
numerator	分子

续表

denominator	分母
polynomial	多项式
radical	根式
root	根
corollary	推论
figure out	计算
plug … into …	把…代入…
divide A by B	把 A 除以 B
multiply A by B	用 A 乘以 B
cross-multiply	交叉相乘
take the square root	求平方根
expand and simplify	展开化简
factor …out of	约去

1.3.2 几何 Geometry

perpendicular	垂直的
horizontal	水平的
parallel to	平行于
vertex	顶点
dimension	维度
length	长度
width	宽度
diameter	直径
perimeter	周长
area	面积
surface area	表面积
volume	体积
rectangle	矩形
trapezoid	梯形
semicircle	半圆
ellipse	椭圆
parabola	抛物线
cubic	立方体的
cylindrical	圆柱形的
hypotenuse	斜边
right triangle	直角三角形
equilateral triangle	等边三角形

isosceles triangle	等腰三角形
sphere	球体
cone	圆锥体
right circular cone	正圆锥

1.3.3 函数 Functions

interval	区间
open interval	开区间
closed interval	闭区间
length of an interval	区间长度
domain	定义域
range	值域
variable	变量；变量的
independent variable	自变量
dependent variable	因变量
intermediate variable	中间变量
function	函数
power function	幂函数
exponential function	指数函数
logarithmic function	对数函数
trigonometric function	三角函数
sign function	符号函数
piecewise function	分段函数
composite function	复合函数
elementary function	初等函数
function property	函数性质
Increasing function	递增函数
decreasing function	递减函数
even function	偶函数
odd function	奇函数
periodic function	周期函数
monotone function	单调函数
inverse function	反函数
vector function	向量函数
transforming function	函数变换
shifting transform	平移变换

		续表
stretching transform		拉伸变换
compressing transform		压缩变换
symmetric transform		对称变换
operation of function		函数运算
rectangular coordinates		直角坐标系
polar coordinate		极坐标
polar equation		极坐标方程
parametric equation		参数方程
sketch the function		画函数图像

1.3.4 极限和连续 Limit and Continuit

limit	极限
one-sided limit	单侧极限
left-hand limit	左极限
right-hand limit	右极限
approach	趋近
approximation	近似值
infinity	无穷
rational function	有理函数
positive infinity	正无穷
negative infinity	负无穷
natural logarithm	自然对数
the highest power of x	x 最高次项
boundedness	有界性
error	误差
horizontal asymptote	水平渐近线
vertical asymptote	垂直渐近线
continuity	连续性
continuous function	连续函数
discontinuity	间断点
jump discontinuity	跳跃间断点
removable discontinuity	可去间断点
infinite discontinuity	无穷间断点

1.3.5 导数和微分 Derivative and Differential

derivative	导数
differentiate	计算导数
take/find derivative	求导
differentiation	求导数[名词]
derivability	可导性
differential	微分,微分的
differentiability	可微性
differentiable	可微的
first derivative	一阶导数
second derivative	二阶导数
n-th derivative	n 阶导数
higher order derivative	高阶导数
difference quotient	差商
average rate of change	平均变化率
instantaneous rate of change	瞬时变化率
slope	斜率
steepness	倾斜度
secant line	割线
equation of tangent line	切线方程
equation of normal line	法线方程
x-intercept	x 轴的截距
implicit function	隐函数
chain rule	链式法则
y terms of x	用 x 表示 y
with respect to	关于,相对于
absolute/ relative	绝对的/相对的
local maximum/minimum	极大值/极小值
extreme value	极值
critical value	临界值
inflection	拐点
concavity	凹向
concave up/down	上凹的/下凹的
particle	质点
position	位置
velocity	速度(矢量)
speed	速率(标量)
acceleration	加速度

distance	距离
displacement	位移(矢量)
linearization	线性化
related rate	相关变率
linear approximation	线性估算

1.3.6　积分和微分方程 Integral and Differential Equation

antiderivative	原函数,不定积分
indefinite integral	不定积分
definite integral	定积分
integrand	被积函数
integration	积分法
U-substitution	U 代换法
integration by parts	分部积分法
linear factor	线性因子
quadratic factor	二次因子
inscribed	内接
circumscribed	外接
Riemann sums	黎曼和
trapezoidal rule	梯形法则
accumulation function	累积函数
washer method	"垫圈"法
disk method	圆盘法
vertical slices	垂直切片
horizontal slices	水平切片
infinitely thin strips	无限薄条
cylindrical shell	圆柱壳
cross-section	横截面
arc length	弧长
converge	收敛
diverge	发散
differential equation	微分方程
initial value	初始值
original equation	原始方程
general solution	通解
particular solution	特解

separation of variable	变量分离
slope fields	斜率场
be proportional to	成比例
exponential growth	指数增长
decay	衰变
restricted growth	约束增长

1.3.7　无穷级数 Infinite Series

sequence	数列
arithmetic sequence	等差数列
geometric sequence	等比数列
infinite series	无穷级数
coefficient	系数
subscript	下标
general term	通项
partial sum	部分和
the nth term	第 n 项
harmonic series	调和级数
geometric series	几何级数
alternating series	交错级数
nonnegative series	正项级数
power series	幂级数
Taylor series	泰勒级数
Maclaurin series	麦克劳林级数
ratio test	比值判别法
comparison test	比较判别法
integral test	积分判别法
convergence	收敛
absolute convergence	绝对收敛
conditional convergence	条件收敛
radius of convergence	收敛半径
interval of convergence	收敛区间
Taylor polynomial	泰勒多项式
error bound	误差界
Lagrange error	拉格朗日误差

1.4 图形计算器的使用 Use of Graphing Calculators

AP 微积分考生必须持用 College Board 规定型号的图形计算器参加考试，AB 和 BC 在计算器操作要求上基本相同．本节以美国本土广为流行的 TI-89 Titanium* 为例，讲解图形计算器的使用．

TI-89 功能强大，处理速度快，其他计算器型号的操作可参考 TI-89．VIP（辅导）服务体系含有图形计算器培训课程．

1.4.1 官方指定型号 List of Approved Graphing Calculators

Table 1.4.1.1～Table 1.4.1.3 为 AP 微积分图形计算器型号．

Table 1.4.1.1 AP 微积分图形计算器型号（Casio）

FX-6000 Series	FX-6200 Series	FX-6300 Series
FX-6500 Series	FX-7000 Series	FX-7300 Series
FX-7400 Series	FX-7500 Series	FX-7700 Series
FX-7800 Series	FX-8000 Series	FX-8500 Series
FX-8700 Series	FX-8800 Series	Graph25 Series
FX-9700 Series *	FX-9750 Series *	FX-9860 Series *
CFX-9800 Series *	CFX-9850 Series *	CFX-9950 Series *
CFX-9970 Series *	FX 1.0 Series *	FX-CG-10 *
Algebra FX 2.0 Series *	FX-CG-20 Series *	FX-CG-50 *
Graph35 Series *	Graph75 Series *	Graph95 Series *
Graph100 Series *	FX-CG500 *（The use of the stylus is not permitted）	

Table 1.4.1.2 AP 微积分图形计算器型号（Texas Instruments）

TI-73	TI-80	TI-81
TI-82 *	TI-83 *	TI-83 Plus *
TI-83 Plus Silver *	TI-84 Plus *	TI-84 Plus CE *
TI-84 Plus Silver *	TI-84 Plus C Silver *	TI-84 Plus T
TI-84 Plus CE-T	TI-85 *	TI-86 *
TI-89	TI-89 Titanium *	TI-Nspire
TI-Nspire CX *	TI-Nspire CAS *	TI-Nspire CX CAS *
TI-Nspire CM-C *	TI-Nspire CM-C CAS *	TI-Nspire CX-C CAS *

Table 1.4.1.3 AP 微积分图形计算器型号（其他）

Hewlett-Packard		
HP-9G	HP-28 Series *	HP-38G *
HP-39 Series *	HP-40 Series *	HP-48 Series *
HP-49 Series *	HP-50 Series *	HP Prime *

续表

Sharp		
EL-5200	EL-9200 Series *	EL-9300 Series *
EL-9900 Series *	EL-9600 Series * （The use of the stylus is not permitted）	
Radio Shack		
EC-4033	EC-4034	EC-4037
Other		
Datexx DS-883	Micronta	Smart[2]

Note：本表依据 College Board（2017 年）最新发布的权威信息更新．订阅号 "AP 微积分"（微信 ID：apcalculus）将持续更新．

1.4.2 计算函数导数 Compute The Derivative of Function

掌握图形计算器求解一个函数的导函数和在某一点处的导数值的操作方法．

Example 1.1　$\dfrac{d}{dx}(e^x + x\cos x) = \cos x - x\sin x + e^x$

计算器主屏幕（Home）是数学运算的基本屏幕，要显示计算器主屏幕，按下 ◆ HOME 或 APPS 桌面上选择 Home 图标，并按下 ENTER 键显示计算器主屏幕．

在每个例子执行之前，可按下键 F1 并选择 8：Clear Home 清空屏幕，使之仅显示该例的键击结果．按键顺序：

2nd [d] ◆ [e^x] X) + X × 2nd [cos] X) , X) ENTER

显示，见 Figure 1.4.2(a)．

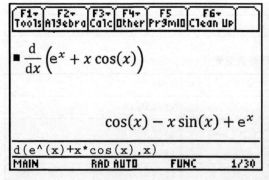

Figure 1.4.2(a)

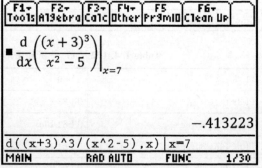

Figure 1.4.2(b)

Example 1.2　$\dfrac{d}{dx}\left(\dfrac{(x+3)^3}{x^2-5}\right)\bigg|_{x=7} = -0.413223\cdots$

计算器主屏幕（Home）下，按键顺序：

2nd [d] (X + 3) ^ 3 ÷ (X ^ 2 − 5) , X) | X = 7 ◆ [≈]

显示，见 Figure 1.4.2 (b)．

1.4.3 计算积分 Compute Integrals

掌握图形计算器求解不定积分和定积分的操作方法．

Example 1.3 $\int \dfrac{\arctan x}{1+x^2} dx = \dfrac{(\arctan x)^2}{2}$

计算器主屏幕（Home）下，按键顺序：

$\boxed{\text{2nd}}$ $[\int]$ ◆ $[\tan^{-1}]$ X $)$ $\boxed{\div}$ $\boxed{(}$ 1 $\boxed{+}$ X $\boxed{\wedge}$ 2 $\boxed{)}$ $\boxed{,}$ X $\boxed{)}$ $\boxed{\text{ENTER}}$

显示，见 Figure 1.4.3(a)。

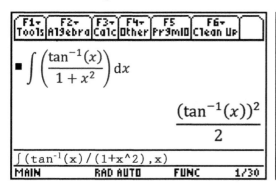

Figure 1.4.3(a)　　　　　　Figure 1.4.3(b)

Example 1.4 $\int_2^5 (x^3 + \sin x) dx = 151.55\cdots$

计算器主屏幕（Home）下，按键顺序：

$\boxed{\text{2nd}}$ $[\int]$ X $\boxed{\wedge}$ 3 $\boxed{+}$ $\boxed{\text{2nd}}$ $[\sin]$ X $\boxed{)}$ $\boxed{,}$ X $\boxed{,}$ 2 $\boxed{,}$ 5 $\boxed{)}$ ◆ $[\approx]$

显示，见 Figure 1.4.3(b)。

1.4.4　求解一个方程 Solve an Equation

掌握图形计算器求解方程的根以及函数零点值的操作方法。

Example 1.5　求解关于 x 的方程 $x^3 - 5x^2 + 9 = 0$。

使用 Ti-84 求根往往需要猜测所求根的所在范围，遭遇反复输入；Ti-89 一步求得所有根，这是 Ti-89 的强大之处。计算器主屏幕（Home）下，按键顺序：

$\boxed{\text{F2}}$ 1 X $\boxed{\wedge}$ 3 $\boxed{-}$ 5 X $\boxed{\wedge}$ 2 $\boxed{+}$ 9 $\boxed{=}$ 0 $\boxed{,}$ X $\boxed{)}$ $\boxed{\text{ENTER}}$

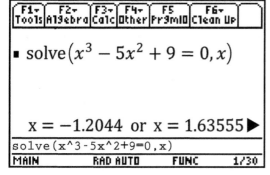　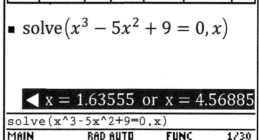

Figure 1.4.4(a)　　　　　　Figure 1.4.4(b)

显示，见 Figure 1.4.4(a)，显示结果右侧的小箭头（▶）提示该方程还有其他根。

移动光标键，选中显示结果查看，见 Figure 1.4.4(b)。

1.4.5 绘制函数图像 Produce The Graph of Function

使用图形计算器绘制函数图像并使用函数图像，分 3 类情况：①直角坐标系，见 Example 1.6；②参数方程，见 Example 1.7；③极坐标，见 Example 1.8。

Example 1.6 绘制图像 $y=x^3-5x^2+9$ 并使用图像求最大值。

函数作图使用 Y＝Editor，按下 ◆[Y＝] 或 APPS 桌面上选择 Y＝ 图标，并按下 ENTER 键进入输入行，见 Figure 1.4.5(a)。

输入表达式：

X ^ 3 − 5X ^ 2 + 9 ENTER，见 Figure 1.4.5(b)。

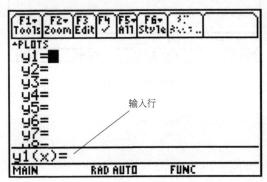

Figure 1.4.5(a)

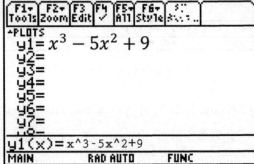

Figure 1.4.5(b)

显示函数图像：按下 F2，将光标移动到 6：ZoomStd 后按下 ENTER 来选取 6：ZoomStd 显示，见 Figure 1.4.5(c)。

开启 Trace 功能：按下 F3，屏幕上出现显迹光标、x 轴和 y 轴的坐标，见 Figure 1.4.5(d)。

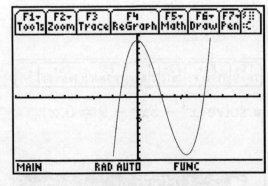

Figure 1.4.5(c)

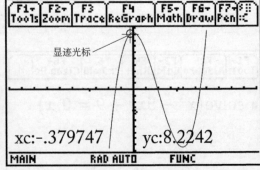

Figure 1.4.5(d)

求 $[a,b]$ 上最大值（最小值）：

按下 F5，选取 4：Maximum，见 Figure 1.4.5(e)，使用光标键（◀/▶）拖动显迹光标到 $x=a$，按下 ENTER 设定下界（Lower Bound）；再使用右光标键▶拖动显迹光标到

$x=b$，按下 ENTER 设定上界（Upper Bound）并获得最小值，见 Figure 1.4.5(f)。

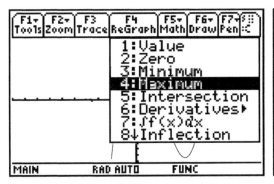
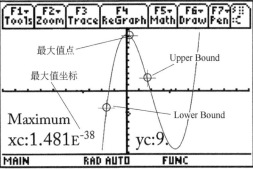

Figure 1.4.5(e)　　　　　　　　　Figure 1.4.5(f)

Example 1.7　绘制图像 $\begin{cases} x=6\cos t \\ y=3\sin t \end{cases}$

参数方程函数作图首先要把 Graph 模式调整到 PARAMETRIC，按键操作：
MODE ▶ 2 ENTER，见 Figure 1.4.5(g)。其余操作和直角坐标系下函数作图类似。

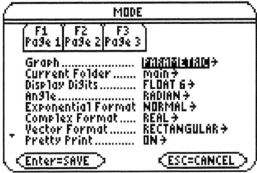
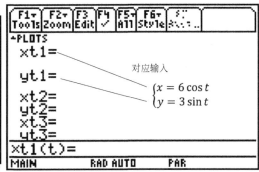

Figure 1.4.5(g)

进入 Y=Editor 并清屏：◆[Y=] F1 8 ENTER，见 Figure 1.4.5(h)。

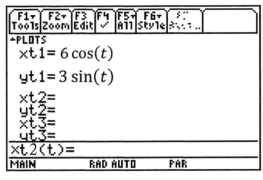
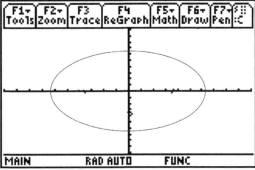

Figure 1.4.5(h)　　　　　　　　　Figure 1.4.5(i)

第 1 章　AP 微积分简介 Introduction of AP Calculus

输入水平分量 $xt1(t)=6\cos t$ 表达式：6 [2nd] [cos] T) [ENTER]

输入垂直分量 $yt1(t)=3\sin t$ 表达式：3 [2nd] [sin] T) [ENTER]

显示函数图像：按下 [F2] 6，见 Figure 1.4.5(i).

Example 1.8　绘制图像 $r=1+\sin\theta$.

极坐标函数作图首先要把 Graph 模式调整到 POLAR，Angle 模式调整到 RADIAN. 按键操作：

[MODE] ▷ 3 ▽ ▽ ▽ ▷ 1 [ENTER]，见 Figure 1.4.5(j). 其余操作和直角坐标系下函数作图类似.

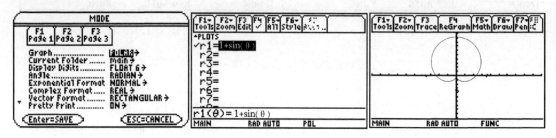

Figure 1.4.5(j)　　　　Figure 1.4.5(k)　　　　Figure 1.4.5(l)

进入 Y=Editor 并清屏：◆ [Y=] [F1] 8 [ENTER]，输入表达式：

1 [+] [2nd] [sin] ◆ [θ]) [ENTER]，见 Figure 1.4.5(k).

显示函数图形：按下 [F2] 6，见 Figure 1.4.5(l).

【方法总结】图形计算器的使用是 AP 微积分考试不可缺失的，考场允许考生携带两台图形计算器，但不允许在考场上借用其他同学的图形计算器，并且图形计算器上内存存储的数据和程序进入考场前须清空.

第 2 章

函数
Functions

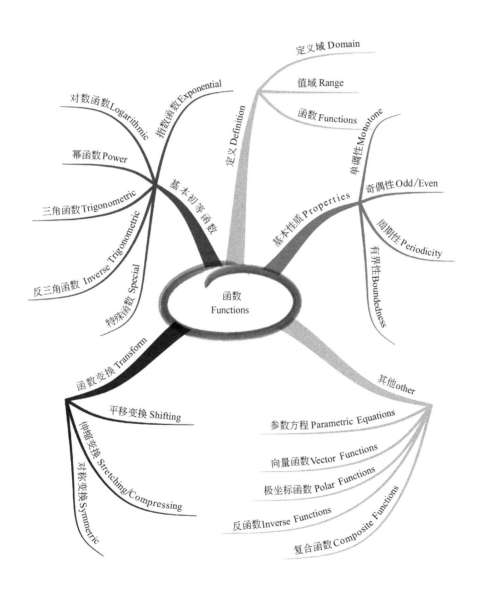

2.1 函数的定义 Definition of Functions

在被誉为数学七大名著之一的《无穷小分析引论》中，大数学家欧拉（Euler）指出："整个无穷分析所讨论的都是变量及其函数"（注：无穷分析即微积分）．函数（Function）是本书所有章节的基础．

2.1.1 定义域 Domain

A function $f(x)$ from a set D to a set E is a correspondence that assigns to each element x of D exactly one element y of E.

集合 D 是非空实数集，如果有某个对应法则 $f(x)$，使得在集合 D 上的每一个元素 x，都有一个确定的实数 y 与之对应，$f(x)$ 称为定义在集合 D 上的一个函数．即 $y=f(x)$，集合 D 是函数 $f(x)$ 的定义域，集合 E 为函数 $f(x)$ 的值域 $E=\{y=f(x)|x\in D\}$．

Example 2.1 Find the domains of：

a) $f(x)=\dfrac{8}{x-3}$； b) $f(x)=\dfrac{x}{x^2-16}$； c) $f(x)=\dfrac{\sqrt{5-x}}{x}$；

d) $f(x)=\log_a(3+2x-x^2)$

Solution：

a) The domain of $f(x)$ is the set of all reals except $x=3$．

b) The domain of $f(x)$ is $\{x|x\neq\pm 4\}$．

c) The domain of $f(x)$ is $\{x|x\neq 0, x\leqslant 5\}$．

d) The domain of $f(x)$ is $\{x|-1<x<3\}$．

【方法总结】求定义域（Domain）的常见方法：

1）若 $f(x)$ 为整式，定义域是全体实数（Real Number）；

2）若 $f(x)$ 为分式，定义域是使分母（Denominator）不为零的全体实数；

3）若 $f(x)$ 为偶次根式，定义域是使被开方式大于等于零的全体实数；

4）若 $f(x)$ 为对数函数（Logarithmic Function），定义域是使真数大于零，底数大于零且不为 1 的全体实数；

5）若 $f(x)$ 为复合函数（Composite Function），定义域由各个简单函数的定义域组成的不等式组确定；

6）实际问题列出的函数式，定义域由自变量的实际意义确定．

2.1.2 值域 Range

函数的值域（Range）取决于定义域和对应法则，不管采用什么方法求函数的值域都必须考虑函数的定义域．

Example 2.2 If $f(x)=\ln x$ for all positive x and $g(x)=16-x^2$ for all real x．Find the range of $y=f(g(x))$．

Solution：

The domain of $f(x)$ is the set of positive reals．We get

$$g(x) > 0 \text{ and } -4 < x < 4$$

Since the domain of $f(g(x))$ is $(-4, 4)$, $\ln(16-x^2)$ takes on every real value less than or equal to ln16.

Then, the range of $f(g(x))$ is $\{y \mid y \leqslant \ln 16\}$.

2.1.3 函数 Functions

定义域（Domain）、值域（Range）和对应法则（Corresponding Rule）统称为函数的三要素．在教学过程中发现，许多同学存在的问题都与对函数的理解不足有关．

函数表达一个变量对另外一个变量的依赖关系．函数的定义域和值域常用区间表达，Table 2.1.3 梳理了常见区间的定义、符号和几何表示．

Table 2.1.3　常用区间定义、符号及几何表示

定义/Definition	符号/Notation	几何表示/Graph
$\{x \mid a \leqslant x \leqslant b\}$	$[a, b]$	
$\{x \mid a < x < b\}$	(a, b)	
$\{x \mid a \leqslant x < b\}$	$[a, b)$	
$\{x \mid a < x \leqslant b\}$	$(a, b]$	
$\{x \mid \geqslant a\}$	$[a, +\infty)$	
$\{x \mid > a\}$	$(a, +\infty)$	
$\{x \mid \leqslant b\}$	$(-\infty, b]$	
$\{x \mid < b\}$	$(-\infty, b)$	
R	$(-\infty, +\infty)$	

Example 2.3　If $f(x) = 3x - 1$ and $g(x) = x^2$, then does $f(g(x)) = g(f(x))$?

Solution: Since
$$f(g(x)) = 3(x^2) - 1 = 3x^2 - 1$$

And
$$g(f(x)) = (3x-1)^2 = 9x^2 - 6x + 1$$

In general, $f(g(x)) \neq g(f(x))$.

Note: 两个函数的定义域和对应法则相同，则两个函数相同．

2.2 函数的基本性质 Function Basic Properties

函数具有单调性（Monotone）、奇偶性（Odd or Even）、周期性（Periodicity）和有界性（Boundedness）. 特别是函数的单调性，在微积分中应用十分广泛.

2.2.1 单调性 Monotone

A function $f(x)$ is called increasing on an interval I if $f(x_1) < f(x_2)$ whenever $x_1 < x_2$ in I；

It is called decreasing on an interval I if $f(x_1) > f(x_2)$ whenever $x_1 < x_2$ in I.

证明函数的单调性（Monotone）常用"作差法"，研究单调性要结合函数图像，发挥数形结合的优势. AP 微积分中，我们还将运用"导数"的观点讨论函数的单调性，见本书 6.2.1.

Example 2.4 Let $f(x) = \sqrt{16-x^2}$. Find the intervals on which $f(x)$ is increasing or decreasing.

Solution：

The domain of function $f(x)$ is the closed interval $[-4, 4]$, and the range of $f(x)$ is the interval $[0, 4]$. Sketch the graph of $f(x)$, see Figure 2.2.1.

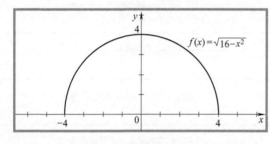

Figure 2.2.1

The graph rises as x increases from -4 to 0, so $f(x)$ is increasing on the closed interval $[-4, 0]$. Thus, as shown in the preceding chart, if $x_1 < x_2$ in $[-4, 0]$, then $f(x_1) < f(x_2)$.

The graph falls as x increases from 0 to 4, so $f(x)$ is decreasing on the closed interval $[0, 4]$. In this case, the preceding chart, if $x_1 < x_2$ in $[0, 4]$, then $f(x_1) > f(x_2)$.

2.2.2 奇偶性 Odd or Even

A function $f(x)$ is called odd if $f(-x) = -f(x)$ for every x in its domain. It is called even if $f(-x) = f(x)$ for every x in its domain.

奇函数关于原点对称（With Respect to The Origin）；偶函数关于 y 轴对称（With Respect to The Y-axis）. 举例：$f(x) = x^2$ 是偶函数，关于 y 轴对称；$f(x) = x^3$ 是奇函数，关于原点对称.

Example 2.5 Determine whether $f(x)$ is even, odd, or neither even nor odd.

a) $f(x) = x^4 - 2x^2 + 3$; b) $f(x) = x^5 - 3x^3 + x$; c) $f(x) = x^3 - x^2$

Solution：

a) $f(-x) = (-x)^4 - 2(-x)^2 + 3 = x^4 - 2x^2 + 3 = f(x)$

Since $f(-x) = f(x)$, $f(x)$ is an even function.

b) $f(-x)=(-x)^5-3(-x)^3+(-x)=-(x^5-3x^3+x)=-f(x)$

Since $f(-x)=-f(x)$, $f(x)$ is an odd function.

c) $f(-x)=(-x)^3-(-x)^2=-x^3+x^2$

Since $f(-x)\neq f(x)$, and $f(-x)\neq -f(x)$, the function $f(x)$ is neither even nor odd.

2.2.3 周期性 Periodicity

A function $f(x)$ is periodic, if there exists a positive real number T, such that, $f(x+T)=f(x)$ for every x in the domain of $f(x)$.

The least such positive real number T, if it exists, is the period of $f(x)$.

周期函数的图像是"有规律地重复着",典型的代表是三角函数,见 Table 2.2.3.

Table 2.2.3 各三角函数最小正周期（Period）

最小正周期(Period)$T=2\pi$			
$f(x)=\sin x$	$f(x)=\cos x$	$f(x)=\sec x$	$f(x)=\csc x$
最小正周期(Period)$T=\pi$			
$f(x)=\tan x$		$f(x)=\cot x$	

根据三角函数的周期性,可得到下述 6 道公式：

周期 π：　　$\tan(x+\pi)=\tan x$　　　$\cot(x+\pi)=\cot x$

周期 2π：　　$\sin(x+2\pi)=\sin x$　　　$\cos(x+2\pi)=\cos x$

　　　　　　$\sec(x+2\pi)=\sec x$　　　$\csc(x+2\pi)=\csc x$

2.2.4 有界性 Boundedness

A function $f(x)$ is Boundedness, if there exists a positive real number M, such that, $|f(x)|\leq M$ for every x in the interval I.

一般来说,连续函数在闭区间上具备有界性.极限、零点定理、极值定理和介值定理等,都牵涉到函数的有界性,有界性是微积分中很多基本概念和基础定理的先决条件.

举例：$f(x)=\sin x$、$f(x)=\cos x$、$f(x)=\sec x$、$f(x)=\csc x$、$f(x)=\arcsin x$、$f(x)=\arccos x$、$f(x)=\arctan x$ 和 $f(x)=\text{arccot} x$ 都是常见的有界函数.

2.3 基本初等函数 Basic Elementary Functions

基本初等函数包括：指数函数（Exponential）、对数函数（Logarithmic）、幂函数（Power）、三角函数（Trigonometric）和反三角函数（Inverse Trigonometric）.此外,AP 微积分考试经常出现绝对值函数、取整函数等其他常用函数,要求掌握各类函数的定义域、值域和函数图像等.

2.3.1 指数函数 Exponential Functions

$$f(x)=a^x, a>0 \text{ and } a\neq 1.$$

1) 定义域（Domain）和值域（Range）：R & $(0,+\infty)$.

The exponential function $f(x)$ whose domain is R and range is the set of positive real numbers.

2) 过定点 (Fixed Point): $(0,1)$.

Since $a^0=1$, the y-intercept is 1 for every a.

3) 单调性 (Monotone):

If $a>1$, then the exponential function $f(x)$ is increasing on R, see Figure 2.3.1(a);

If $0<a<1$, then the exponential function $f(x)$ is decreasing on R, see Figure 2.3.1(b).

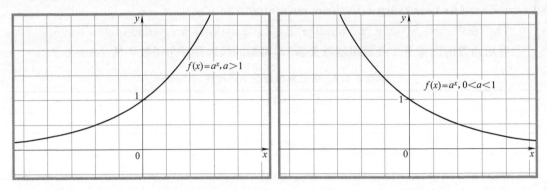

Figure 2.3.1(a)　　　　　　　　　　Figure 2.3.1(b)

4) 指数运算法则 (Laws of Exponents):

$a^m a^n = a^{m+n}$	$(a^m)^n = a^{mn}$
$(ab)^m = a^m b^m$	$\sqrt[n]{a^m} = a^{\frac{m}{n}}$
Real numbers a and b; Integers m and n	

指数函数以底数 $a>1$ 或 $0<a<1$ 分为两类，过定点 $(0,1)$，既不是奇函数，也不是偶函数. 注意掌握指数运算公式.

Example 2.6 Solve the equation $3^{5x-7} = 9^{x+1}$

Solution:

Express both sides with the same base, we get
$$3^{5x-7} = (3^2)^{x+1}$$
$$3^{5x-7} = 3^{2x+2}$$

Since exponential function are one-to-one, then
$$5x-7 = 2x+2$$
$$x = 3$$

2.3.2 对数函数 Logarithmic Functions

$$f(x) = \log_a x, \ a>0 \text{ and } a \neq 1.$$

1) 定义域 (Domain) 和值域 (Range): $(0, +\infty)$ & R.

The logarithmic function $f(x)$ whose domain is the set of positive real numbers and range is R.

2) 过定点（Fixed Point）：$(1,0)$.

Since $\log_a 1 = 0$, the y-intercept is 0 for every a.

3) 单调性（Monotone）：

If $a > 1$, then the logarithmic function $f(x)$ is increasing on $(0, +\infty)$, see Figure 2.3.2(a);

If $0 < a < 1$, then the logarithmic function $f(x)$ is decreasing on $(0, +\infty)$, see Figure 2.3.2(b);

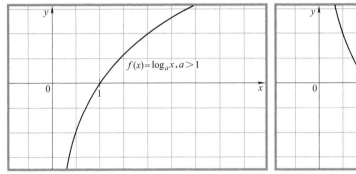

Figure 2.3.2(a)　　　　　　　　Figure 2.3.2(b)

4) 对数运算法则（Laws of Logarithms）：

$\log_a(MN) = \log_a M + \log_a N$	$\log_a \dfrac{M}{N} = \log_a M - \log_a N$
$\log_a b = \dfrac{\log_c b}{\log_c a}, c > 0$ and $c \neq 1$	$\log_{a^n} b^m = \dfrac{m}{n} \log_a b, n \neq 0$
$a^{\log_a b} = b$	$e^{\ln b} = b$
Positive Real Numbers M and N	

指数函数和对数函数在知识点上有雷同之处，都要求掌握各自的定义、定义域和值域，单调性和函数图像．同时，它们都既不是奇函数，也不是偶函数．另外，对数函数和指数函数互为反函数，是反函数的典型范例．

Example 2.7　Solve the equation $\lg\sqrt{x} = \sqrt{\lg x}$ for x.

Solution：Since law of exponents, we get
$$\lg x^{\frac{1}{2}} = \sqrt{\lg x}$$

Then,
$$\frac{1}{2}\lg x = \sqrt{\lg x}$$

Square both sides,
$$\frac{1}{4}(\lg x)^2 = \lg x$$

Make one side 0,
$$(\lg x)^2 - 4\lg x = 0$$

Factor out $\lg x$,
$$\lg x(\lg x - 4) = 0$$
$$\lg x = 0 \text{ or } \lg x - 4 = 0$$
$$x = 10^0 = 1, \text{ or } x = 10^4$$

2.3.3 幂函数 Power Functions

$$f(x) = x^\alpha, \alpha \text{ is a constant}$$

幂函数是五大初等函数中相对复杂的一类. 总体考察幂函数的定义域和值域、单调性和奇偶性，并非办不到，但容易显得生硬、不自然. 以我的教学经验出发，建议从五个特殊的幂函数图像 $\left(\alpha = -1, \dfrac{1}{2}, 1, 2, 3\right)$ 入手.

1) $f(x) = \dfrac{1}{x}$ and $f(x) = \sqrt{x}$，see Figure 2.3.3(a).

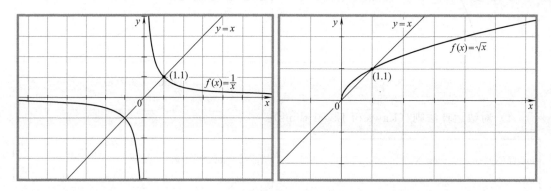

Figure 2.3.3(a)

2) $f(x) = x^2$ and $f(x) = x^3$，see Figure 2.3.3(b).

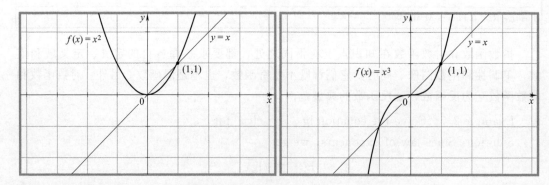

Figure 2.3.3(b)

3) 幂函数图像规律：①必定经过第一象限；②必定不经过第四象限；③是否经过第二、三象限，取决于函数的奇偶性；④最多只能同时经过两个象限；⑤若与坐标轴相交，交点一定是原点.

4) 化繁为简，各个击破：结合五个特殊的幂函数图像 $\left(\alpha = -1, \dfrac{1}{2}, 1, 2, 3\right)$ 和幂函数图像规律，可以逐一考查各个幂函数的定义域和值域，单调性和奇偶性等情况.

5）多项式函数（Polynomial Functions）：由多个幂函数加减乘运算得到.
$f(x)=a_0x^n+a_1x^{n-1}+\cdots+a_{n-1}x+a_n$, n is a positive integer or zero.

$n=1$，一次函数（Linear Function），$f(x)=kx+b(k\neq 0)$；

$n=2$，二次函数（Quadratic Function），$f(x)=ax^2+bx+c(a\neq 0)$；

$n=3$，三次函数（Cubic Function），$f(x)=ax^3+bx^2+cx+d(a\neq 0)$.

6）有理函数（Rational Functions）：由两个多项式函数除法运算得到.
$f(x)=\dfrac{P(x)}{Q(x)}$, $Q(x)\neq 0$, $P(x)$ and $Q(x)$ are polynomial functions.

Example 2.8 Sketching the graph of a rational function $f(x)=\dfrac{x^4}{x^4+1}$.

Solution：

Since $f(-x)=f(x)$, the function is even, and hence the graph is symmetric with respect to the y-axis. The graph intersects the x-axis at $(0,0)$. The numerator and denominator of $f(x)$ have the same degree. Since the equation $f(x)=1$ has no real solution, the graph does not cross the line $y=1$.

Plotting the point $(0,0)$ and $\left(2,\dfrac{16}{17}\right)$ and making use of symmetry leads to the sketch, see Figure 2.3.3(c).

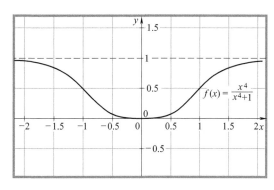

Figure 2.3.3(c)

2.3.4 三角函数 Trigonometric Functions

1）正弦函数（Sine Function）
$$f(x)=\sin x, x\in R, \text{ see Figure 2.3.4(a)}.$$

① Domain & Range：R & $[-1, 1]$.

The sine functions whose domain is R and range is $[-1, 1]$.

② The sine function is odd：$\sin(-x)=-\sin x$.

③ The period of the sine functions is 2π：$\sin(x+2k\pi)=\sin x$, $k\in Z$.

④ The sine function is increasing on $x\in\left(2k\pi-\dfrac{\pi}{2}, 2k\pi+\dfrac{\pi}{2}\right)$ and it is decreasing on $x\in\left(2k\pi+\dfrac{\pi}{2}, 2k\pi+\dfrac{3\pi}{2}\right)$, $k\in Z$.

2）余弦函数（Cosine Function）
$$f(x)=\cos x, x\in R, \text{ see Figure 2.3.4(b)}.$$

① Domain & Range：R & $[-1, 1]$

The cosine functions whose domain is R and range is $[-1, 1]$.

② The sine function is even：$\cos(-x)=\cos x$.

③ The period of the cosine functions is 2π：$\cos(x+2k\pi)=\cos x$, $k\in Z$.

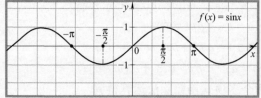
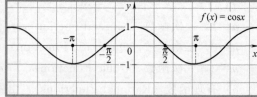

Figure 2.3.4(a)   Figure 2.3.4(b)

④ The cosine function is increasing on $x \in (2k\pi-\pi, 2k\pi)$ and it is decreasing on $x \in (2k\pi, 2k\pi+\pi)$, $k \in Z$.

3) 正切函数 (Tangent Function)

$$f(x) = \tan x, \ x \neq (2k+1)\frac{\pi}{2}, \ k \in Z, \text{ see Figure 2.3.4(c)}.$$

① Domain & Range: $\left\{x \mid x \neq (2k+1)\frac{\pi}{2}, \ k \in Z\right\}$ & R.

The tangent functions whose domain is $\left\{x \mid x \neq (2k+1)\frac{\pi}{2}, \ k \in Z\right\}$ and range is R.

② The tangent function is odd: $\tan(-x) = -\tan x$.

③ The period of the tangent functions is π: $\tan(x+k\pi) = \tan x$, $k \in Z$.

④ The tangent function is increasing on $x \in \left(k\pi - \frac{\pi}{2}, \ k\pi + \frac{\pi}{2}\right)$, $k \in Z$.

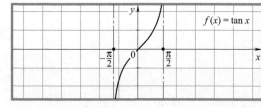

Figure 2.3.4(c)   Figure 2.3.4(d)

4) 余切函数 (Cotangent Function)

$$f(x) = \cot x, \ x \neq k\pi, \ k \in Z, \text{ see Figure 2.3.4(d)}.$$

① Domain & Range: $\{x \mid x \neq k\pi, k \in Z\}$ & R.

The cotangent functions whose domain is $\{x \mid x \neq k\pi, \ k \in Z\}$ and range is R.

② The cotangent function is odd: $\cot(-x) = -\cot x$.

③ The period of the cotangent functions is π: $\cot(x+k\pi) = \cot x$, $k \in Z$.

④ The cotangent function is decreasing on $x \in (k\pi, \ k\pi+\pi)$, $k \in Z$.

5) 正割函数 (Secant Function)

$$f(x) = \sec x, \ x \neq (2k+1)\frac{\pi}{2}, \ k \in Z, \text{ see Figure 2.3.4(e)}.$$

① Domain & Range: $\left\{x \mid x \neq (2k+1)\frac{\pi}{2}, k \in Z\right\}$ & $\{y \mid |y| \geq 1\}$.

The secant functions whose domain is $\{x \mid x \neq (2k+1)\frac{\pi}{2}, k \in Z\}$ and range is $\{y \mid |y| \geq 1\}$.

② The secant function is even: $\sec(-x) = \sec x$.

③ The period of the secant functions is 2π: $\sec(x + 2k\pi) = \sec x$, $k \in Z$.

④ The secant function is increasing on $x \in \left(2k\pi, 2k\pi + \frac{\pi}{2}\right) \cup \left(2k\pi + \frac{\pi}{2}, 2k\pi + \pi\right)$, and it is decreasing on $x \in \left(2k\pi + \pi, 2k\pi + \frac{3\pi}{2}\right) \cup \left(2k\pi + \frac{3\pi}{2}, 2k\pi + 2\pi\right)$, $k \in Z$.

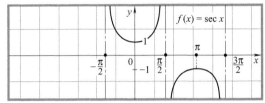
Figure 2.3.4(e)

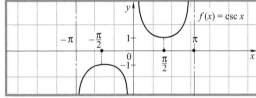
Figure 2.3.4(f)

6) 余割函数 (Cosecant Function)

$$f(x) = \csc x, \quad x \neq k\pi, \quad k \in Z, \text{ see Figure 2.3.4(f)}.$$

① Domain & Range: $\{x \mid x \neq k\pi, k \in Z\}$ & $\{y \mid |y| \geq 1\}$.

The cosecant functions whose domain is $\{x \mid x \neq k\pi, k \in Z\}$ and range is $\{y \mid |y| \geq 1\}$.

② The cosecant functions is odd: $\csc(-x) = -\csc x$.

③ The period of the cosecant function is 2π: $\csc(x + 2k\pi) = \csc x$, $k \in Z$.

④ The cosecant function is increasing on $x \in \left(2k\pi + \frac{\pi}{2}, 2k\pi + \pi\right) \cup \left(2k\pi + \pi, 2k\pi + \frac{3\pi}{2}\right)$, and it is decreasing on $x \in \left(2k\pi, 2k\pi + \frac{\pi}{2}\right) \cup \left(2k\pi + \frac{3\pi}{2}, 2k\pi + 2\pi\right)$, $k \in Z$.

国内的高考仅要求熟练 sin、cos 和 tan 三个函数，AP 微积分要求熟练掌握上述 6 个三角函数的定义域和值域、单调性和奇偶性，以及函数图像.

另外，三角基本恒等式、和差公式、诱导公式、倍角公式、半角公式和正弦余弦定理等是三角函数非常重要的组成部分，具体参考《附录：A.1 常用公式和定理 Common Formulas and Theorems》，此处不加赘述.

Example 2.9 Verify the following identity by transforming the left-hand side into the right-hand side, $\cos(-x) - \sin x \tan(-x) = \sec x$.

Solution:

Formulas for negatives,

$$\cos(-x) - \sin x \tan(-x) = \cos x + \sin x \tan x$$

Tangent identity,

$$\cos x + \sin x \tan x = \cos x + \sin x \cdot \frac{\sin x}{\cos x}$$

Multiply,

$$\cos x + \sin x \times \frac{\sin x}{\cos x} = \cos x + \frac{\sin^2 x}{\cos x} = \frac{\cos^2 x + \sin^2 x}{\cos x}$$

Pythagorean identity,

$$\frac{\cos^2 x + \sin^2 x}{\cos x} = \frac{1}{\cos x} = \sec x$$

Then,

$$\cos(-x) - \sin x \tan(-x) = \sec x$$

有同学觉得:"三角函数太难了,公式多,学不会". 事实上,在教学实践上,我们 1 次课就能帮助零基础的学生"秒杀"三角函数,并不存在"学不会的三角". 同学们要坚定信心,夯实基础很重要,Table 2.3.4 梳理了三角函数的常用特殊值.

Table 2.3.4　三角函数常用特殊值

x(Radians)	x(Degrees)	$\sin x$	$\cos x$	$\tan x$	$\cot x$	$\sec x$	$\csc x$
0	0°	0	1	0	—	1	—
$\frac{\pi}{6}$	30°	$\frac{1}{2}$	$\frac{\sqrt{3}}{2}$	$\frac{\sqrt{3}}{3}$	$\sqrt{3}$	$\frac{2\sqrt{3}}{3}$	2
$\frac{\pi}{4}$	45°	$\frac{\sqrt{2}}{2}$	$\frac{\sqrt{2}}{2}$	1	1	$\sqrt{2}$	$\sqrt{2}$
$\frac{\pi}{3}$	60°	$\frac{\sqrt{3}}{2}$	$\frac{1}{2}$	$\sqrt{3}$	$\frac{\sqrt{3}}{3}$	2	$\frac{2\sqrt{3}}{3}$
$\frac{\pi}{2}$	90°	1	0	—	0	—	1

2.3.5　反三角函数 Inverse Trigonometric Functions

1) 反正弦函数 (Arcsine Function)

$f(x) = \arcsin x$, $x \in [-1, 1]$, see Figure 2.3.5(a).

① Domain & Range: $[-1, 1]$ & $\left[-\frac{\pi}{2}, \frac{\pi}{2}\right]$.

The arcsine functions whose domain is $[-1, 1]$ and range is $\left[-\frac{\pi}{2}, \frac{\pi}{2}\right]$.

② The arcsine function is odd: $\arcsin(-x) = -\arcsin x$.

③ The arcsine function is increasing on $x \in [-1, 1]$.

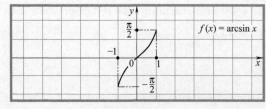

Figure 2.3.5(a)

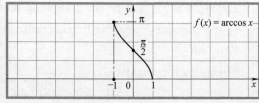

Figure 2.3.5(b)

2) 反余弦函数 (Arccosine Function)

$f(x) = \arccos x$, $x \in [-1, 1]$, see Figure 2.3.5(b).

① Domain & Range: $[-1, 1]$ & $[0, \pi]$.

The arccosine functions whose domain is $[-1, 1]$ and range is $[0, \pi]$.

② The arccosine function is decreasing on $x \in [-1, 1]$.

3) 反正切函数（Arctangent Function）

$$f(x) = \arctan x, x \in R, \text{ see Figure 2.3.5(c)}.$$

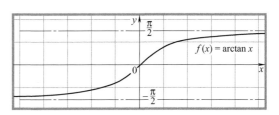

Figure 2.3.5(c)

① Domain & Range: R & $\left[-\dfrac{\pi}{2}, \dfrac{\pi}{2}\right]$.

The arctangent functions whose domain is R and range is $\left[-\dfrac{\pi}{2}, \dfrac{\pi}{2}\right]$.

② The arctangent function is odd: $\arctan(-x) = -\arctan x$.

③ The arctangent function is increasing on $x \in R$.

掌握反三角函数的定义域和值域、单调性和奇偶性，熟练函数图像，运用反三角函数一般要结合三角函数解决问题.

Example 2.10 Find the exact value, a) $\arcsin\left(\sin\dfrac{2\pi}{3}\right)$; b) $\arccos\left[\sin\left(-\dfrac{\pi}{6}\right)\right]$; c) $\tan(\arctan 100)$.

Solution:

a) $\arcsin\left(\sin\dfrac{2\pi}{3}\right) = \arcsin\left(\dfrac{\sqrt{3}}{2}\right) = \dfrac{\pi}{3}$

b) $\arccos\left[\sin\left(-\dfrac{\pi}{6}\right)\right] = \arccos\left(-\dfrac{1}{2}\right) = \dfrac{2\pi}{3}$

c) $\tan(\arctan 100) = 100$

2.3.6 特殊函数 Special Functions

1) 绝对值函数（The Absolute-Value Functions）: $f(x) = |x|$

Example 2.11 Let $f(x) = |x|$.

a) Determine whether $f(x)$ is even or odd.

b) Sketch the graph of $f(x)$.

c) Find the intervals on which $f(x)$ is increasing or is decreasing.

Solution:

a) The domain of $f(x)$ is R, since
$$f(-x) = |-x| = |x| = f(x)$$

Thus, $f(x)$ is an even function.

b) Since $f(x)$ is even, its graph is symmetric with respect to the y-axis. The first quadrant part of the graph coincides with the line $y = x$. Sketching this half-line and using sym-

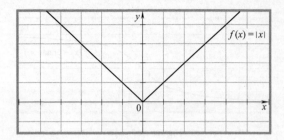

Figure 2.3.6(a)

metry gives us, see Figure 2.3.6(a).

c) Referring to the graph, we see that $f(x)$ is decreasing on $(-\infty, 0]$ and is increasing on $[0, +\infty)$.

2) 取整函数 (The Greatest-Integer Functions): $f(x) = [x]$

Example 2.12 Let $f(x) = |x|$, sketch the graph of $f(x)$.

Solution:

x	$f(x) = [x]$
⋮	⋮
$-1 \leq x < 0$	-1
$0 \leq x < 1$	0
$1 \leq x < 2$	1
⋮	⋮

Whenever x is between successive integers, the corresponding part of the graph is a segment of a horizontal line, see Figure 2.3.6(b).

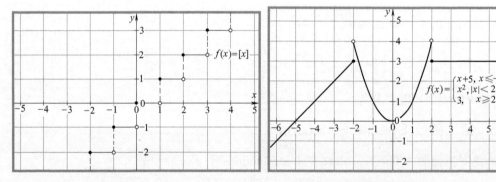

Figure 2.3.6(b) Figure 2.3.6(c)

3) 分段函数 The Piecewise Defined Functions.

The function is defined by different formulas in different interval of its domain.

Example 2.13 sketch the graph of $f(x) = \begin{cases} x+5, & x \leq -2 \\ x^2, & |x| < 2 \\ 3, & x \geq 2 \end{cases}$

Solution:

If $x \leq -2$, then $f(x) = x+2$ and the graph of $f(x)$ coincides with the line $y = x+2$ and is represented by the portion of the graph to the left of the line $x = -2$. The small dot indicates that the point $(-2, 3)$ is on the graph, see Figure 2.3.6(c).

If $|x|<2$, we use x^2 to find values of $f(x)$, and therefore this part of the graph of $f(x)$ coincides with the parabola $y=x^2$. Note that the points $(-2,4)$ and $(2,4)$ are not on the graph.

Finally, if $x \geq 2$, the values of $f(x)$ are always 3.

2.4 反函数 & 复合函数 Inverse Functions & Composite Functions

在基本初等函数基础上，求反函数和函数"复合"运算成为创造新函数的手段，也成为分解形式相对"复杂"函数的有力武器，是学习微积分必备技能之一．

2.4.1 一一对应函数 One-to-One Function

A function $f(x)$ with domain D and range R is a one-to-one function if either of the following equivalent conditions is satisfied：

1) Whenever $a \neq b$ in D, then $f(a) \neq f(b)$ in R；
2) Whenever $f(a) = f(b)$ in R, then $a = b$ in D.

反函数存在的前提条件是：原函数是一一对应函数，并且原函数在定义域上单调．

2.4.2 反函数 Inverse Functions

Let $f(x)$ be a one-to-one function with domain D and range R. A function $g(x)$ with domain R and range D is the inverse function of $f(x)$, provided the following condition is true for every x in D and every y in R：

$$y = f(x) \text{ if and only if } x = g(x).$$

Domain of $f^{-1}(x)$ = range of $f(x)$; Range of $f^{-1}(x)$ = domain of $f(x)$.

反函数的定义域是原函数的值域，反函数的值域是原函数的定义域．反函数图像与原函数图像关于 $y=x$ 对称．

Example 2.14 Let $f(x) = 4x - 5$. Find the inverse function of $f(x)$.

Solution：
Solve the equation $y = f(x)$ for x：

$$x = \frac{y+5}{4}$$

We now formally let $x = f^{-1}(y)$, that is,

$$f^{-1}(y) = \frac{y+5}{4}$$

Since the symbol used for the variable is immaterial, we may also write：

$$f^{-1}(x) = \frac{x+5}{4}$$

where x is in the domain $f^{-1}(x)$.

【方法总结】求解反函数（Inverse Function）的步骤：
1) 变形原函数，以 y 表示 x；

2）将 x 与 y 互换；

3）求出定义域（Domain）。

2.4.3 复合函数 Composite Functions

The composite function $f \circ g$ of two functions $f(x)$ and $g(x)$ is defined by $(f \circ g)(x) = f(g(x))$. The domain of $f \circ g$ is the set of all x in the domain of $g(x)$ such that $g(x)$ is in the domain of $f(x)$. A number x is in the domain of $(f \circ g)(x)$ if and only if both $g(x)$ and $f(g(x))$ are defined.

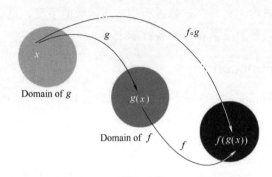

AP 微积分考试中，复合函数 (Composite Function) 是由两个或两个以上的基本初等函数经过复合运算得到的。复合运算不是简单的加减乘除运算，复合函数的生成过程是这样的：用一个基本初等函数去替换另一个基本初等函数的自变量，调整定义域，得到复合函数。

Example 2.15 Let $f(x) = x^2 - 9$ and $g(x) = \sqrt{x}$, finding composite functions.

a) Find $(f \circ g)(x)$ and the domain of $f \circ g$.

b) Find $(g \circ f)(x)$ and the domain of $g \circ f$.

Solution：

a) $(f \circ g)(x) = f(g(x)) = f(\sqrt{x}) = (\sqrt{x})^2 - 9 = x - 9$

The domain of $f \circ g$ is set of all x in $[0, +\infty]$.

b) $(g \circ f)(x) = g(f(x)) = g(x^2 - 9) = \sqrt{x^2 - 9}$

$x^2 - 9 \geqslant 0$, $x^2 \geqslant 9$, $|x| \geqslant 3$

The domain of $g \circ f$ is set of all x in $[-\infty, -3] \cup [3, +\infty]$.

Example 2.16 Which basic elementary functions compose the following composite functions?

a) $f(x) = \sqrt[3]{(-x)}$；b) $f(x) = \ln x^2$；c) $f(x) = 2^{\frac{1}{x-1}}$；d) $f(x) = \cos^2 \sqrt{3x^2 - 1}$

Solution：

Using the definition of composite function，we get：

a) we let $f(u) = \sqrt[3]{u}$ and $u(x) = -x$ which is a composite function from for $f(x) = \sqrt[3]{(-x)}$.

b) we let $f(u) = \ln u$ and $u(x) = x^2$ which is a composite function from for $f(x) = \ln x^2$.

c) we let $f(u) = 2^u$ and $u(x) = \dfrac{1}{x-1}$ which is a composite function from for $f(x) =$

$2^{\frac{1}{x-1}}$.

d) we let $f(u)=u^2$, $u(v)=\cos v$, $v(w)=\sqrt{w}$ and $w(x)=3x^2-1$ which is a composite function from for $f(x)=\cos^2\sqrt{3x^2-1}$.

Note：复合函数使得一些复杂的函数问题"化繁为简"，我们能根据需要，把一个复合函数拆分为两个或者多个基本初等函数去分析，化整为零，各个攻破，围而击之．

2.5 函数变换 Transforming of Functions

函数图像是函数变量之间关系的直观体现，掌握函数图像的基本变换，可以探索较为复杂的函数图像并了解它们的性质，这是学习微积分的必备基础技能．

2.5.1 平移变换 Shifting Transform

1）垂直平移 Vertically Shifting the Graph of $y=f(x)$

$y=f(x)+c$ with $c>0$：The graph of $f(x)$ is shifted vertically upward a distance c；

$y=f(x)-c$ with $c>0$：The graph of $f(x)$ is shifted vertically downward a distance c，see Figure 2.5.1(a).

2）水平平移 Horizontally Shifting the Graph of $y=f(x)$.

$y=f(x+c)$ with $c>0$：The graph of $f(x)$ is shifted horizontally the left a distance c；

$y=f(x-c)$ with $c>0$：The graph of $f(x)$ is shifted horizontally the right a distance c，see Figure 2.5.1(b).

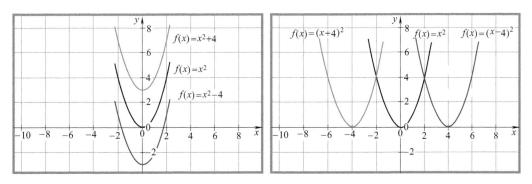

Figure 2.5.1(a)　　　　　　　　Figure 2.5.1(b)

【方法总结】函数图像平移变换技巧：左加右减，上加下减．

2.5.2 伸缩变换 Stretching or Compressing Transform

1）垂直伸缩 Vertically Stretching or Compressing the Graph of $y=f(x)$.

$y=cf(x)$ with $c>1$：The graph of $f(x)$ is stretched vertically by a factor c；

$y=cf(x)$ with $0<c<1$：The graph of $f(x)$ is compressed vertically by a factor $\frac{1}{c}$，see Figure 2.5.2（a）.

2) 水平伸缩 Horizontally Stretching or Compressing the Graph of $y=f(x)$.

$y=f(cx)$ with $c>1$: The graph of $f(x)$ is compressed horizontally by a factor c;

$y=f(cx)$ with $0<c<1$: The graph of $f(x)$ is stretched horizontally by a factor $\dfrac{1}{c}$, see Figure 2.5.2(b).

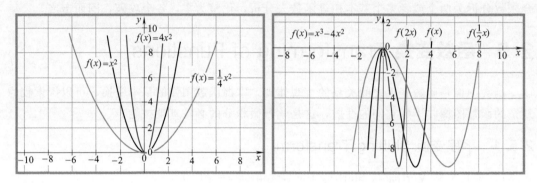

Figure 2.5.2(a)　　　　　　　　　　Figure 2.5.2(b)

2.5.3 对称变换 Symmetric Transform

1) 垂直翻折 Vertical Reflecting about x-axis

$y=|f(x)|$, Reflect points on $f(x)$ with negative y-values through the x-axis, see Figure 2.5.3(a).

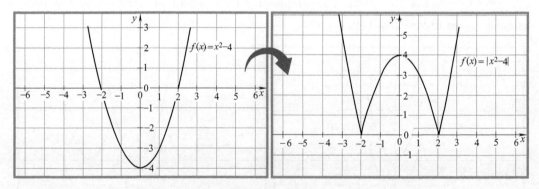

Figure 2.5.3(a)

2) 水平翻折 Horizontal Reflecting about y-axis.

$y=f(|x|)$, Reflect points on $f(x)$ with positive x-values through the y-axis, see Figure 2.5.3(b).

Example 2.17 Graph the indicated translations $f(x)=\left|\dfrac{3x+1}{x+1}\right|$

Solution: Separate variable

$$f(x)=\left|\dfrac{3x+1}{x+1}\right|=\left|\dfrac{3(x+1)-2}{x+1}\right|=\left|3-\dfrac{2}{x+1}\right|$$

Shift the graph $f(x)=-\dfrac{2}{x}$ a distance 1 unit to the left, then distance 3 units up-

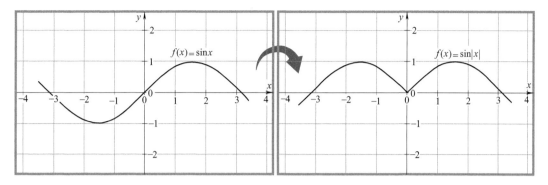

Figure 2.5.3(b)

ward, and then, reflect points on the graph with negative y-values through the x-axis, see Figure 2.5.3(c).

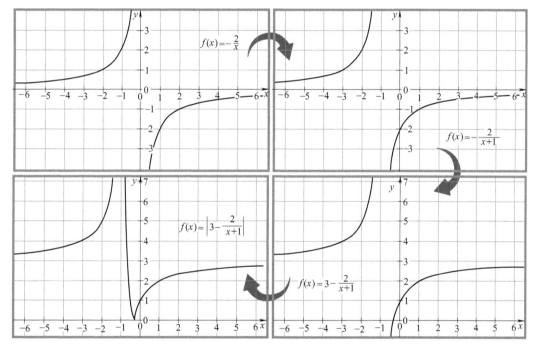

Figure 2.5.3(c)

2.6# 参数方程 & 向量函数 Parametric Equations & Vector Functions

微积分 BC 要求掌握参数方程和向量函数. 事实上, 含有向量的函数, 常常借助参数方程的形式解答, 向量函数可以借助参数方程相关知识得到解答.

2.6.1# 参数方程 Parametric Equations

Let C be the curve consisting of all ordered pairs $(f(t), g(t))$, where $f(t)$ and $g(t)$

are defined on an interval I. The equations $\begin{cases} x = f(t) \\ y = g(t) \end{cases}$, for t in I, are parametric equations for C with parameter t.

参数方程和函数相类似，参数等同于函数自变量，决定着因变量的结果．运动学问题中，参数通常是"时间"，参数方程的结果是位置、速度或加速度．

Example 2.18 Find the Cartesian equation represented by the parametric equations $\begin{cases} x = 3\cos t \\ y = 3\sin t \end{cases}$, $0 \leqslant t \leqslant 2\pi$.

Solution：

Using Trigonometric Identify $\sin^2 t + \cos^2 t = 1$：

$$\cos t = \frac{x}{3} \text{ and } \sin t = \frac{y}{3}$$

Then，

$$\left(\frac{x}{3}\right)^2 + \left(\frac{y}{3}\right)^2 = 1$$

Therefore，

$$x^2 + y^2 = 9$$

This is a circle, centered at the origin, with radius 3.

Note：利用三角恒等式 $\sin^2 x + \cos^2 x = 1$ 是参数方程和（笛卡尔的）普通方程互相转化的常用技巧．

Example 2.19 A point moves in a plane such that its position $P(x, y)$ at time t is given by $\begin{cases} x = a\cos t \\ y = b\sin t \end{cases}$, t in R, where $a \neq 0$, $b \neq 0$. Describe the motion of the point.

Solution：

First, simplify expression like this：

$$\begin{cases} \dfrac{x}{a} = \cos t \\ \dfrac{y}{b} = \sin t \end{cases}$$

Next, using Trigonometric Identify,

$$\frac{x^2}{a^2} + \frac{y^2}{b^2} = 1$$

If $a = b$, it represents a circle, see Figure 2.6.1(a); else if, $a \neq b$, it represents an ellipse, see Figure 2.6.1(b).

2.6.2[#] 向量函数 Vector Functions

Vector is a quantity that has both magnitude and direction.

Denote by：$\vec{a} = \overrightarrow{OP}$, $\vec{a} = \langle x, y \rangle$, $\vec{a} = x\vec{i} + y\vec{j}$, see Figure 2.6.2.

Note：国内教材中，向量用圆括号表示，国外用尖括号，区别细微，但请注意．

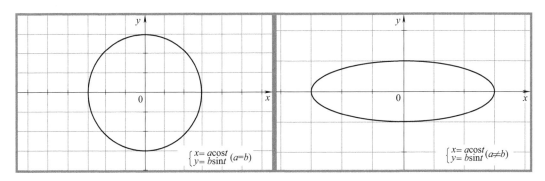

Figure 2.6.1(a) Figure 2.6.1(b)

The magnitude of the vector $\vec{a} = \langle x, y \rangle$, denoted by $|\vec{a}|$, is given by $|\vec{a}| = |\langle x, y \rangle| = \sqrt{x^2 + y^2}$.

向量的模表示该向量的大小. 在 AP 微积分中，常常把力、速度和加速度沿着 x 与 y 轴方向分解. 如：$\vec{v} = \langle v_x, v_y \rangle$.

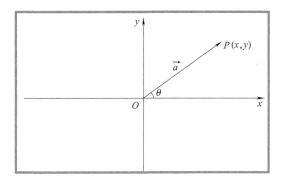

Example 2.20 Find a vector $\vec{b}$ in the opposite direction of $\vec{a} = \langle 3, -4 \rangle$ that has magnitude 3.

Figure 2.6.2

Solution：The magnitude of $\vec{a}$ is given by
$$|\vec{a}| = \sqrt{3^2 + (-4)^2} = 5$$

A unit vector $\vec{u}$ in the direction of $\vec{a}$ can be found by multiplying $\vec{a}$ by $\frac{1}{|\vec{a}|}$. Thus,
$$\vec{u} = \frac{1}{|\vec{a}|} \times \vec{a} = \frac{1}{5} \times \langle 3, -4 \rangle = \left\langle \frac{3}{5}, -\frac{4}{5} \right\rangle$$

Multiplying $\vec{u}$ by 3 gives us a vector of magnitude 3 in the direction of $\vec{a}$, so we'll multiply $\vec{u}$ by -3 to obtain the desired vector $\vec{b}$.
$$\vec{b} = -3\vec{u} = -3 \times \left\langle \frac{3}{5}, -\frac{4}{5} \right\rangle = \left\langle -\frac{9}{5}, \frac{12}{5} \right\rangle$$

2.7# 极坐标函数 Polar Functions

极坐标系中，点 P 的坐标 (r, θ)，用点 P 的极角 θ 和点 P 到极点 O 的距离 r 来确定. 极坐标系下描述的函数曲线称作极坐标函数（Polar Function）.

2.7.1# 极坐标 Polar Coordinates

We begin with the pole O and the polar axis with endpoint O. Next, we consider any point P in the plane different from O.

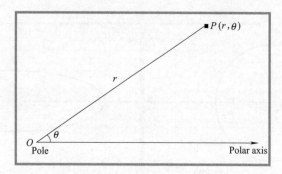

Figure 2.7.1

If, as illustrated in Figure 2.7.1, $r = d(O, P)$ and θ denotes the measure of any angle determined by the polar axis and OP, then r and θ are polar coordinates of P and the symbols (r, θ) or $P(r, \theta)$ are used to denote P.

极坐标中角的规定同三角函数中角的规定一致. 规定 r 可以为正的, 也可以为负的. 点 $P'(-r,\theta)$ 为 $P(r,\theta)$ 绕极点逆时针旋转 $180°$, 即:

$$P'(-r,\theta) = P(r, \theta + \pi)$$

Example 2.21 Plot the points in the polar coordinate

a) $A = \left(3, \dfrac{\pi}{2}\right)$ b) $B = \left(3, -\dfrac{3\pi}{4}\right)$ c) $C = \left(-3, \dfrac{\pi}{2}\right)$

Solution:

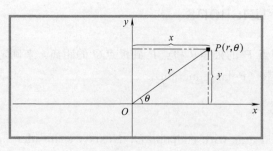

2.7.2# 直角坐标系和极坐标系的关系 Relationships Between Rectangular and Polar Coordinates

The rectangular coordinates (x, y) and polar coordinates (r, θ) of a point P are related as follows, (see Figure 2.7.2):

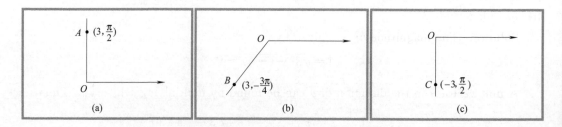

Figure 2.7.2

$$\begin{cases} r^2 = x^2 + y^2 \\ \tan\theta = \dfrac{y}{x}, x \neq 0 \end{cases}, \quad \begin{cases} x = r\cos\theta \\ y = r\sin\theta \end{cases}$$

Example 2.22 If $(r, \theta) = \left(2, \dfrac{7\pi}{6}\right)$ are polar coordinates of a point P, find the rectangular coordinates of P.

Solution:

Substituting $r=2$ and $\theta=\dfrac{7\pi}{6}$, we obtain the following:

$$x = r\cos\theta = 2\cos\dfrac{7\pi}{6} = 2\times\left(-\dfrac{\sqrt{3}}{2}\right) = -\sqrt{3}$$

$$y = r\sin\theta = 2\sin\dfrac{7\pi}{6} = 2\times\left(-\dfrac{1}{2}\right) = -1$$

Hence, the rectangular coordinates of P are $(x,y)=(-\sqrt{3},-1)$.

2.7.3# 极坐标函数 Polar Functions

极坐标函数 $r=f(\theta)$，即：r 为因变量，θ 为自变量. 极坐标方程描绘的图形经常会表现出不同的对称形式.

Example 2.23 What curves are represented by the following equations?

a) $r=2$; b) $r=\cos\theta$; c) $r=1+\sin\theta$

Solution:

a) Since $r=2$, $r^2=4$, then, $x^2+y^2=4$.

It represents a circle with center $(0,0)$ and radius 2.

b) Since $x=r\cos\theta$ and $r=\cos\theta$, then, $r^2=x$, $x^2+y^2=x$, $\left(x-\dfrac{1}{2}\right)^2+y^2=\dfrac{1}{4}$.

It represents a circle with center $\left(\dfrac{1}{2},0\right)$ and radius $\dfrac{1}{2}$.

c) We get a table by the given θ and the calculation of r.

θ	0	$\dfrac{\pi}{4}$	$\dfrac{\pi}{2}$	$\dfrac{3\pi}{4}$	π	$\dfrac{5\pi}{4}$	$\dfrac{3\pi}{2}$	$\dfrac{7\pi}{4}$	2π
r	1	$1+\dfrac{\sqrt{2}}{2}$	2	$1+\dfrac{\sqrt{2}}{2}$	1	$1-\dfrac{\sqrt{2}}{2}$	0	$1-\dfrac{\sqrt{2}}{2}$	2

Sketch the graph, see Figure 2.7.3.

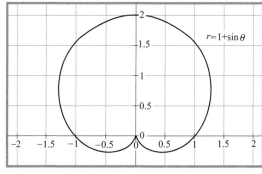

Figure 2.7.3

2.8 习题 Practice Exercises

(1) Find the domains of:

1) $f(x)=\dfrac{1}{1-x^2}+\sqrt{x+1}$; 2) $f(x)=\dfrac{\ln(2-x)}{\sqrt{|x|-1}}$; 3) $f(x)=\begin{cases}\cos\dfrac{1}{x},&x\neq 0\\ 1,&x=0\end{cases}$

(2) If $f(x)=\ln x$ and $g(x)=25-x^2$. Find the range of $y=f(g(x))$.

(3) Determine whether $f(x)$ is even, odd, or neither even nor odd.

1) $f(x)=x^3-x$; 2) $f(x)=x^6-x^2+7$; 3) $f(x)=x+x^2$

(4) Solve the equation $2^{5x-7}=8^{x+1}$.

(5) Solve the equation $\log_2(4+x)=4$.

(6) Find the period of the function $f(x)=\dfrac{1}{2}\cos^2 x$.

(7) Let $f(x)=x^2-2$ for $x\geqslant 0$. Find the inverse function of $f(x)$.

(8) Verify the identity $\sec x=\sin x(\tan x+\cot x)$.

(9) Find the exact value, 1) $\arcsin\left(\sin\dfrac{\pi}{3}\right)$ 2) $\arccos\left[\sin\left(-\dfrac{\pi}{3}\right)\right]$.

(10) Let $f(x)=x^2-3$ and $g(x)=\sqrt{x}$, find $f\circ g$ and the domain of $f\circ g$.

(11)$^{\#}$ Find the Cartesian equation represented by the parametric equations $\begin{cases}x=2\cos t\\ y=2\sin t\end{cases}$, $0\leqslant t\leqslant 2\pi$.

第 3 章

极限
Limit

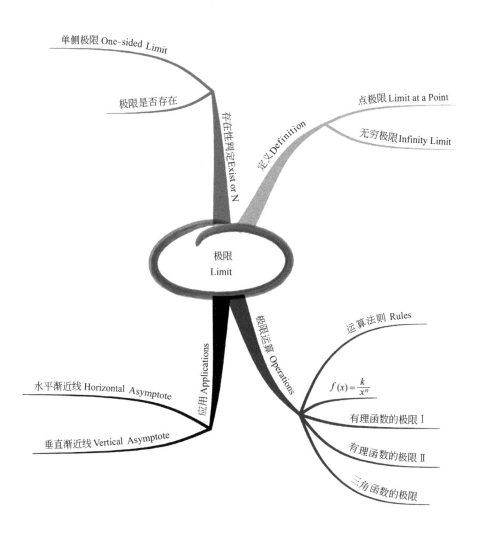

3.1 极限的定义 Definition of the Limit

使用"任意靠近"或者"充分靠近"（Approach）给予极限"形象化"的定义，在数学上虽然不够确切，但是在 AP 微积分中是足够的，它能使我们计算很多特定函数的极限．

3.1.1 x→x_0 时，函数 f(x) 的极限 The limit of f(x) when x approaches x_0

Let $f(x)$ be defined on an open interval about x_0, except possibly at x_0 itself. If $f(x)$ close to L for all x sufficiently close to x_0, we say that $f(x)$ approaches the limit L as x approaches x_0. Denoted by

$$\lim_{x \to x_0} f(x) = L$$

因为"靠近"（Approach）依赖于不同情况，使用"靠近"说法的极限定义在数学上并不确切．对于机械模具制造，"靠近"可能意味着小于一根头发丝的直径；对于天文学研究，"靠近"可能就意味着几千光年以内．但是，这仍然是极限学习很好的入手点．

Example 3.1 Find $\lim\limits_{x \to 2} x^2$.

Solution：Plug in 2 for x, and you get 4.

Example 3.2 Find $\lim\limits_{x \to 1}(32+4x)$.

Solution：Plug in 1 for x, get

$$\lim_{x \to 1}(32+4x) = 32+4 \times 1 = 36$$

Example 3.3 Find $\lim\limits_{x \to 0}\dfrac{10}{1+x}$.

Solution：Plug in 0 for x, get

$$\lim_{x \to 0}\frac{10}{1+x} = \frac{10}{1+0} = 10$$

3.1.2 x→∞ 时，函数 f(x) 的极限 The limit of f(x) when x approaches infinity

Suppose the values of $f(x)$ get close to the number L as x approaches infinity. Then we say that the limit of $f(x)$ is L as x approaches infinity. Denoted by

$$\lim_{x \to \infty} f(x) = L$$

假设 $|x|>1$，$n>0$，那么 $\lim\limits_{x \to \infty}\dfrac{k}{x^n}=0$（$n$ 和 k 为常数）.

Example 3.4 Find $\lim\limits_{x \to \infty}\dfrac{2}{x-1}$.

Solution：

$$\lim_{x \to \infty}\frac{2}{x-1} = 0$$

Example 3.5　Find $\lim\limits_{x\to\infty}\dfrac{2x+1}{6x-1}$.

Solution：

$$\lim_{x\to\infty}\frac{2x+1}{6x-1}=\lim_{x\to\infty}\frac{\frac{2x}{x}+\frac{1}{x}}{\frac{6x}{x}-\frac{1}{x}}=\lim_{x\to\infty}\frac{2+\frac{1}{x}}{6-\frac{1}{x}}=\frac{1}{3}$$

Example 3.6　If $f(x)=\begin{cases}3, x\neq 2\\ 1, x=2\end{cases}$, find $\lim\limits_{x\to\infty}f(x)$ and $\lim\limits_{x\to 2}f(x)$.

Solution：

$$\lim_{x\to\infty}f(x)=3 \text{ and } \lim_{x\to 2}f(x)=1$$

【方法总结】计算极限，当自变量（Independent Variable）x 趋于某定值（或无穷）时，$\lim\limits_{x\to\infty}f(x)$ 和 $\lim\limits_{x\to x_0}f(x)$ 与该点是否有定义域（Domain）及该点的函数值均没有必然的联系．

3.2　极限存在的判定 The Limit does Exist or Not

高等数学中，极限分为数列（Sequence）的极限和函数（Function）的极限．AP 微积分中，主要研究函数的极限，分为 $x\to x_0$ 和 $x\to\infty$ 两种情况．求解一个函数极限的前提，是这个函数存在极限．

3.2.1　单侧极限 One-sided limit

We write $\lim\limits_{x\to x_0^-}f(x)$ for the Left-hand limit as x approaches and increases x_0, and $\lim\limits_{x\to x_0^+}f(x)$ for the Right-hand limit as x approaches and decreases x_0.

求左（右）极限是判定极限是否存在的前提条件．

Example 3.7　Find $\lim\limits_{x\to 3^-}\left(\dfrac{x+2}{x^2-x-6}\right)$.

Solution：

Here we should think about what happens when we plug in a value that is very close to 3，but a little bit less. The top expression will approach 5. The bottom expression will approach 0，but will be a little bit less. Thus，the limit will be $\dfrac{5}{0^-}$，which is $-\infty$.

$$\lim_{x\to 3^-}\left(\frac{x+2}{x^2-x-6}\right)=-\infty$$

Example 3.8　Find $\lim\limits_{x\to 3^+}\dfrac{x+2}{x^2-9}$.

Solution：

$$\lim_{x\to 3^+}\frac{x+2}{x^2-9}=+\infty$$

Example 3.9　Find $\lim\limits_{x\to 3^+}\dfrac{x+2}{x^2-2}$.

Solution：

$$\lim_{x\to 3^+}\frac{x+2}{x^2-2}=\frac{3+2}{3^2-2}=\frac{5}{7}$$

3.2.2　极限是否存在 The Limit does Exist or Not

$\lim\limits_{x\to x_0}f(x)=L$ if and only if $\lim\limits_{x\to x_0^-}f(x)=\lim\limits_{x\to x_0^+}f(x)=L$.

如 Table 3.2.2 所示，当 x 趋近 2 时，函数 $f(x)=\dfrac{x^2-x-2}{x-2}=x+1(x\neq 2)$ 的极限存在并等于 3，当且仅当 x 趋近（Approach）2 时，$f(x)$ 的左极限（Left-hand Limit）和右极限（Right-hand limit）相等．

Table 3.2.2　当 x 趋近 2 时，函数 $f(x)$ 的极限值

x	$f(x)$	x	$f(x)$
1.9	2.9	2.1	3.1
1.99	2.99	2.01	3.01
1.999	2.999	2.001	3.001
1.9999	2.9999	2.0001	3.0001
...	...	...	...
$x\to 2^-$	$f(x)\to 3$	$x\to 2^+$	$f(x)\to 3$
$\lim\limits_{x\to 2^-}f(x)=3$		$\lim\limits_{x\to 2^+}f(x)=3$	
Since the left-hand limit equals the right-hand limit, we can write：$\lim\limits_{x\to 2}f(x)=3$			

Example 3.10　Find $\lim\limits_{x\to 0}\dfrac{1}{x}$.

Solution：

The left-hand limit：$\lim\limits_{x\to 0^-}\dfrac{1}{x}=-\infty$, and the right-hand limit：$\lim\limits_{x\to 0^+}\dfrac{1}{x}=+\infty$.

Because the left-hand limit is not equal to right-hand limit, the limit does not exist.

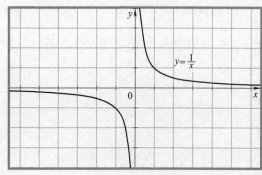

Figure 3.2.2

Look at the graph of $y=\dfrac{1}{x}$, see Figure 3.2.2.

Wu can see that on the left side of $x=0$, the curve approaches $-\infty$, and on the right side of $x=0$, the curve approaches $+\infty$.

Example 3.11　Find $\lim\limits_{x\to 1}\dfrac{1}{x}$.

Solution：

$$\lim_{x \to 1^-} \frac{1}{x} = 1 \text{ and } \lim_{x \to 1^+} \frac{1}{x} = 1$$

So,

$$\lim_{x \to 1} \frac{1}{x} = 1$$

Example 3.12 If $[x]$ is the greatest integer not greater then x, find $\lim_{x \to 1}[x]$.

Solution：

$$\lim_{x \to 1^-}[x] = 0 \text{ and } \lim_{x \to 1^+}[x] = 1$$

Because $\lim_{x \to 1^-}[x] \neq \lim_{x \to 1^+}[x]$, the limit $\lim_{x \to 1}[x]$ does not exist.

3.3 极限的运算 Operations of Limit

极限（Limit）是微积分（Calculus）中的基础概念，后面我们要讲的连续（Continuity）、导数（Derivative）、微分（Differential）和积分（Integral）等都建立在极限基础之上．所以，理解并掌握极限的概念和运算显得很重要．

3.3.1 极限的运算法则 Rules of Limit

If L, M, x_0 and k are real numbers, and $\lim_{x \to x_0} f(x) = L$, $\lim_{x \to x_0} g(x) = M$, Then

1) Constant Multiple Rule：$\lim_{x \to x_0} kf(x) = k \times \lim_{x \to x_0} f(x) = kL$

2) Sum/Difference Rule：$\lim_{x \to x_0}[f(x) \pm g(x)] = \lim_{x \to x_0} f(x) \pm \lim_{x \to x_0} g(x) = L \pm M$

3) Product Rule：$\lim_{x \to x_0}[f(x)g(x)] = \lim_{x \to x_0} f(x) \times \lim_{x \to x_0} g(x) = LM$

4) Quotient Rule：$\lim_{x \to x_0} \frac{f(x)}{g(x)} = \frac{\lim_{x \to x_0} f(x)}{\lim_{x \to x_0} g(x)} = \frac{L}{M} (M \neq 0)$

5) Power Rule：$\lim_{x \to x_0}[f(x)]^n = [\lim_{x \to x_0} f(x)]^n$ (n is rational number)

上述极限运算法则同样适用于 $x \to \infty$ 的情况，并且运用极限运算法则时，注意等号两边的极限都必须存在．

Example 3.13 Find $\lim_{x \to 3} \pi^2$.

Solution：

$$\lim_{x \to 3} \pi^2 = \pi^2$$

Note：依据极限的运算法则，求函数极限最直接的方法是代入法．当 x_0 在函数定义域内，求 $x \to x_0$ 的极限，只需把 $x = x_0$ 代入 $f(x)$．

Example 3.14 If $f(x) = \begin{cases} \dfrac{3}{x-1}, & x \neq 1 \\ 4, & x = 1 \end{cases}$, find $\lim_{x \to 1} f(x)$.

Solution：

$$\lim_{x\to 1^-}f(x)=-\infty \text{ and } \lim_{x\to 1^+}f(x)=+\infty$$

Because $\lim_{x\to 1^-}f(x)\neq \lim_{x\to 1^+}f(x)$, the limit $\lim_{x\to 1}f(x)$ does not exist.

Note：计算函数极限之前，注意先检验所求函数极限是否存在，切不可盲目求解．

Example 3.15 Find $\lim_{x\to 2}(x^4+3x^2-5)$.

Solution：
$$\lim_{x\to 2}(x^4+3x^2-5)=\lim_{x\to 2}x^4+\lim_{x\to 2}3x^2-\lim_{x\to 2}5=16+12-5=23$$

Example 3.16 Find $\lim_{x\to -2}\sqrt{4x^2-5}$.

Solution：
$$\lim_{x\to -2}\sqrt{4x^2-5}=\sqrt{\lim_{x\to -2}(4x^2-5)}=\sqrt{\lim_{x\to -2}4x^2-\lim_{x\to -2}5}=\sqrt{16-5}=\sqrt{11}$$

Example 3.17 Using the limit laws, evaluate $\lim_{x\to 3}\left(\dfrac{x^3+x^2-2}{x^2+3}\right)$.

Solution：
$$\lim_{x\to 3}\left(\frac{x^3+x^2-2}{x^2+3}\right)=\frac{\lim_{x\to 3}(x^3+x^2-2)}{\lim_{x\to 3}(x^2+3)}=\frac{\lim_{x\to 3}x^3+\lim_{x\to 3}x^2-\lim_{x\to 3}2}{\lim_{x\to 3}x^2+\lim_{x\to 3}3}$$
$$=\frac{27+9-2}{9+3}=\frac{17}{6}$$

3.3.2 函数 $f(x)=\dfrac{k}{x^n}$ 的极限 Limit of the Function $f(x)=\dfrac{k}{x^n}$

无穷小量（Infinitesimal）和无穷大量（Infinity）是极限理论的基础概念，AP 微积分中以函数 $f(x)=\dfrac{k}{x^n}$（k, n are natural number）的极限为代表．我们分两种情况：

1) $x\to 0$：When n is an odd integer, $\lim_{x\to 0}\dfrac{k}{x^n}$ does not exist；When n is an even integer, $\lim_{x\to 0}\dfrac{k}{x^n}=+\infty$.

2) $x\to \infty$：$\lim_{x\to \infty}\dfrac{k}{x^n}=0$.

Example 3.18 Evaluate $\lim_{x\to 0}\dfrac{1}{x^2}$ and $\lim_{x\to \infty}\dfrac{1}{x^2}$.

Solution：
$$\lim_{x\to 0^-}\frac{1}{x^2}=+\infty \text{ and } \lim_{x\to 0^+}\frac{1}{x^2}=+\infty$$

So,
$$\lim_{x\to 0}\frac{1}{x^2}=+\infty$$

Since $\lim_{x\to -\infty}\dfrac{1}{x^2}=0$ and $\lim_{x\to +\infty}\dfrac{1}{x^2}=0$, then $\lim_{x\to \infty}\dfrac{1}{x^2}=0$. Look at the graph of $y=\dfrac{1}{x^2}$, see Figure 3.3.2.

Example 3.19 Evaluate $\lim\limits_{x\to 0}\dfrac{1}{x^3}$.

Solution:

$\lim\limits_{x\to 0^-}\dfrac{1}{x^3}=-\infty$ and $\lim\limits_{x\to 0^+}\dfrac{1}{x^3}=+\infty$

Because $\lim\limits_{x\to 0^-}\dfrac{1}{x^3}\neq \lim\limits_{x\to 0^+}\dfrac{1}{x^3}$, the limit $\lim\limits_{x\to 0}\dfrac{1}{x^3}$ does not exist.

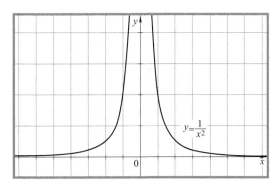

Figure 3.3.2

Note：类似地，如 Example 3.10，可自行画出函数图像观察，此处不赘述.

3.3.3 有理函数的极限 | Limit of a Rational Function |

If $P(x)$ and $Q(x)$ are polynomials and $Q(x_0)\neq 0$, then
$$\lim_{x\to x_0}\dfrac{P(x)}{Q(x)}=\dfrac{P(x)}{Q(x)}$$

有理函数（Rational Function）极限是另一类常见极限类型. 如果分母的极限不等于零，代入法求有理函数的极限；如果分母的极限等于零，消去分子和分母的公因式后，分母的极限不等于零，再用代入法求得极限.

Example 3.20 Find $\lim\limits_{x\to -1}\dfrac{x^3+3x^2+2}{x+5}$.

Solution:
$$\lim_{x\to -1}\dfrac{x^3+3x^2+2}{x+5}=\dfrac{(-1)^3+3(-1)^2+2}{(-1)+5}=1$$

Example 3.21 Find $\lim\limits_{x\to 2}\dfrac{x^2-x-2}{x^2-2x}$.

Solution:

We cannot substitute $x=2$ because it makes the denominator zero. We test the numerator to see if it, is zero at $x=2$. It is, so it has a factor of $(x-2)$ in common with the denominator.

$$\dfrac{x^2-x-2}{x^2-2x}=\dfrac{(x-2)(x+1)}{x(x-2)}=\dfrac{x+1}{x}(x\neq 2)$$

Using the simpler fraction, we find the limit of these values as $x\to 2$ by substitution:

$$\lim_{x\to 2}\dfrac{x^2-x-2}{x^2-2x}=\lim_{x\to 2}\dfrac{x+1}{x}=\dfrac{2+1}{2}=\dfrac{3}{2}$$

Note：因式分解（Factoring）是有理函数极限求解的常用技巧之一.

Example 3.22 Find $\lim\limits_{x\to 0}\dfrac{\sqrt{16+x}-4}{x}$.

Solution:

We cannot substitute $x=0$, and the numerator and denominator have no obvious common factors. We can create a common factor by multiplying both numerator and denominator by the expression $(\sqrt{16+x}+4)$.

$$\frac{\sqrt{16+x}-4}{x} = \frac{(\sqrt{16+x}-4)(\sqrt{16+x}+4)}{x(\sqrt{16+x}+4)} = \frac{(16+x)-4^2}{x(\sqrt{16+x}+4)}$$

$$= \frac{x}{x(\sqrt{16+x}+4)} = \frac{1}{\sqrt{16+x}+4}$$

Therefore,

$$\lim_{x \to 0} \frac{\sqrt{16+x}-4}{x} = \lim_{x \to 0} \frac{1}{\sqrt{16+x}+4} = \frac{1}{\sqrt{16+0}+4} = \frac{1}{8}$$

Note：如果有理函数分母的极限等于零，并且不能直接消去分子和分母的公因式，考虑创建公因式并消去公因式，再用代入法求得极限．

3.3.4 有理函数的极限 II Limit of a Rational Function II

If $P(x)$ and $Q(x)$ are polynomials and $Q(x_0) \neq 0$, $a_0 b_0 \neq 0$

$$\lim_{x \to \infty} \frac{P_n(x)}{Q_m(x)} = \lim_{x \to \infty} \frac{a_0 x^n + a_1 x^{n-1} + \cdots + a_{n-1} x + a_n}{b_0 x^m + b_1 x^{m-1} + \cdots + b_{m-1} x + b_m}$$

then,

$$\lim_{x \to \infty} \frac{P_n(x)}{Q_m(x)} = \begin{cases} \dfrac{a_0}{b_0}, & n=m \\ \infty, & n>m \\ 0, & n<m \end{cases}$$

当 $x \to \infty$ 时，求有理函数的极限，先把分子分母同时除以 x 的最高次数项，然后化简求得极限．利用上述结论求解更方便．

Example 3.23 Find $\lim\limits_{x \to \infty} \dfrac{3-4x+7x^3}{3x^3-5x+7}$.

Solution：Divide each term by x^2,

$$\lim_{x \to \infty} \frac{3-4x+7x^3}{3x^3-5x+7} = \lim_{x \to \infty} \frac{\frac{3-4x+7x^3}{x^3}}{\frac{3x^3-5x+7}{x^3}} = \lim_{x \to \infty} \frac{\frac{3}{x^3}-\frac{4}{x^2}+7}{3-\frac{5}{x^2}+\frac{7}{x^3}} = \frac{\lim\limits_{x \to \infty}\left(\frac{3}{x^3}-\frac{4}{x^2}+7\right)}{\lim\limits_{x \to \infty}\left(3-\frac{5}{x^2}+\frac{7}{x^3}\right)}$$

$$= \frac{\lim\limits_{x \to \infty}\frac{3}{x^3} - \lim\limits_{x \to \infty}\frac{4}{x^2} + \lim\limits_{x \to \infty}7}{\lim\limits_{x \to \infty}3 - \lim\limits_{x \to \infty}\frac{5}{x^2} + \lim\limits_{x \to \infty}\frac{7}{x^3}} = \frac{0-0+7}{3-0+0} = \frac{7}{3}$$

Example 3.24 Find $\lim\limits_{x \to \infty} \dfrac{3x^7-x^3+31x}{73x^6-3x^5+7}$.

Solution：Divide each term by x^7,

$$\lim_{x \to \infty} \frac{3x^7-x^3+31x}{73x^6-3x^5+7} = \lim_{x \to \infty} \frac{3-\frac{1}{x^4}+\frac{31}{x^6}}{\frac{73}{x}-\frac{3}{x^2}+\frac{7}{x^7}} = \infty$$

Example 3.25 Evaluate $\lim\limits_{x \to \infty} \dfrac{4x^5+3x^3+9}{3x^9-5x^4+x^2}$.

Solution: Divide each term by x^9,

$$\lim_{x \to \infty}\frac{4x^5+3x^3+9}{3x^9-5x^4+x^2}=\lim_{x \to \infty}\frac{\dfrac{4}{x^4}+\dfrac{3}{x^6}+\dfrac{9}{x^9}}{3-\dfrac{5}{x^5}+\dfrac{1}{x^7}}=\frac{0+0+0}{3-0+0}=0$$

3.3.5 三角函数的极限 Limit of Trigonometric Functions

$$\lim_{x \to 0}\frac{\sin x}{x}=1 \qquad (1)$$

$$\lim_{x \to \infty}\left(1+\frac{1}{x}\right)^x=e \qquad (2)$$

上述极限也称两个重要极限（Two Important Limits）。求解三角函数的极限，往往离不开极限（1）的帮助；极限（2）是自然底数 e，AP 微积分考查力度不大．

Example 3.26 Evaluate $\lim\limits_{x \to 0}\dfrac{\cos x-1}{x}$.

Solution:

$$\lim_{x \to 0}\frac{\cos x-1}{x}=\lim_{x \to 0}\frac{-2\sin^2\dfrac{x}{2}}{x}=\lim_{\frac{x}{2} \to 0}\left(-\frac{\sin\dfrac{x}{2}}{\dfrac{x}{2}}\times\sin\frac{x}{2}\right)$$

$$=-\lim_{\frac{x}{2}}\frac{\sin\dfrac{x}{2}}{\dfrac{x}{2}}\times\lim_{\frac{x}{2} \to 0}\sin\frac{x}{2}=-1\times 0=0$$

Example 3.27 Evaluate $\lim\limits_{x \to 0}\dfrac{\sin(ax)}{x}$, a is constant.

Solution:

$$\lim_{x \to 0}\frac{\sin(ax)}{x}=\lim_{x \to 0}a\times\frac{\sin(ax)}{ax}=\lim_{x \to 0}a\times\lim_{ax \to 0}\frac{\sin(ax)}{ax}=a\times 1=a$$

Example 3.28 Evaluate $\lim\limits_{x \to 0}\dfrac{\sin(ax)}{\sin(bx)}$, a,b is constant.

Solution:

$$\lim_{x \to 0}\frac{\sin(ax)}{\sin(bx)}=\lim_{x \to 0}\left(\frac{\dfrac{\sin ax}{ax}}{\dfrac{\sin bx}{bx}}\times\frac{a}{b}\right)=\frac{\lim\limits_{ax \to 0}\dfrac{\sin ax}{ax}}{\lim\limits_{bx \to 0}\dfrac{\sin bx}{bx}}\times\lim_{x \to 0}\frac{a}{b}=1\times\frac{a}{b}=\frac{a}{b}$$

Note: 上述结论可以看作极限（1）的推论．

Example 3.29 Find $\lim\limits_{x \to 0}\dfrac{\tan x-\sin x}{x}$.

Solution:

$$\lim_{x\to 0}\frac{\tan x - \sin x}{x} = \lim_{x\to 0}\frac{\frac{\sin x}{\cos x} - \sin x}{x} = \lim_{x\to 0}\left(\frac{\sin x}{x\cos x} - \frac{\sin x}{x}\right)$$

$$= \lim_{x\to 0}\left(\frac{\sin x}{x}\times\frac{1}{\cos x} - \frac{\sin x}{x}\right)$$

$$= \lim_{x\to 0}\frac{\sin x}{x}\times\lim_{x\to 0}\frac{1}{\cos x} - \lim_{x\to 0}\frac{\cos x}{x}$$

$$= 1\times 1 - 1 = 0$$

Example 3.30 Evaluate $\lim\limits_{x\to\infty}\left(1+\dfrac{5}{x}\right)^x$.

Solution:Let $\dfrac{5}{x}=\alpha$, if $x\to\infty$, then $\alpha\to 0$

$$\lim_{x\to\infty}\left(1+\frac{5}{x}\right)^x = \lim_{\alpha\to 0}(1+\alpha)^{\frac{5}{\alpha}} = \lim_{\alpha\to 0}[(1+\alpha)^{\frac{1}{\alpha}}]^5$$

$$= \lim_{\alpha\to 0}[(1+\alpha)^{\frac{1}{\alpha}}]^5 = \lim_{x\to\infty}\left[\left(1+\frac{5}{x}\right)^{\frac{x}{5}}\right]^5 = e^5$$

3.4 极限的应用 Applications of Limit

讨论一个函数的极限,得到函数的水平渐近线(Horizontal Asymptote)和垂直渐近线(Vertical Asymptote).

3.4.1 水平渐近线 Horizontal Asymptote

A line $y=b$ is a horizontal asymptote of the graph of a function $y=f(x)$ if either $\lim\limits_{x\to\infty}f(x)=b$ or $\lim\limits_{x\to-\infty}f(x)=b$.

水平渐近线意味着:x 趋近(Approach)于无穷大时,$f(x)$ 靠近某一平行 x 轴的直线. 注意到 $x\to-\infty$、$x\to+\infty$ 或 $x\to\infty$ 满足其一即可.

Example 3.31 Finding horizontal asymptotes of the function $f(x)=\dfrac{1-2x}{x+7}$.

Solution:

Using the definition, we obtain

$$\lim_{x\to-\infty}f(x) = \lim_{x\to-\infty}\frac{1-2x}{x+7} = -2 \text{ and } \lim_{x\to+\infty}f(x) = \lim_{x\to+\infty}\frac{1-2x}{x+7} = -2$$

So, the line $y=-2$ is a horizontal asymptote. Look at the graph of $y=\dfrac{1-2x}{x+7}$, see Figure 3.4.1.

3.4.2 垂直渐近线 Vertical Asymptote

A line $x=a$ is a vertical asymptote of the graph of a function $y=f(x)$ if either $\lim\limits_{x\to a^+}f(x)=\pm\infty$ or $\lim\limits_{x\to a^-}f(x)=\pm\infty$.

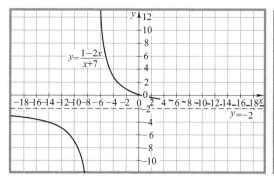
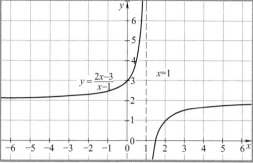

Figure 3.4.1   Figure 3.4.2

垂直渐近线意味着：函数在 $x=a$ 周围，$f(x)$ 不断靠近，直到无穷大．

Example 3.32 Finding vertical asymptotes of the function $f(x)=\dfrac{2x-3}{x-1}$.

Solution：

we know that the function has no definition at $x=1$.

$$\lim_{x\to 1^-}f(x)=\lim_{x\to 1^-}\frac{2x-3}{x-1}=+\infty \text{ and } \lim_{x\to 1^+}f(x)=\lim_{x\to 1^+}\frac{2x-3}{x-1}=-\infty$$

So，the line $x=1$ is a vertical asymptote.

Look at the graph of $y=\dfrac{2x-3}{x-1}$，see Figure 3.4.2.

3.5 习题 Practice Exercises

(1) $\lim\limits_{x\to 0}\sqrt{\pi}=$

(2) $\lim\limits_{x\to -1}(3x^2+9)=$

(3) $\lim\limits_{x\to 1}(2x^2-1)(\sqrt{8x^2+1})=$

(4) $\lim\limits_{x\to 2}\left(\dfrac{3x+1}{x-9}\right)=$

(5) $\lim\limits_{x\to\infty}\dfrac{x+1}{\sqrt{6x-1}}=$

(6) $\lim\limits_{x\to 0^-}\left(\dfrac{|x|}{x}\right)=$

(7) $\lim\limits_{x\to 0^+}\left(\dfrac{|x|}{x}\right)=$

(8) $\lim\limits_{x\to 4}\left(\dfrac{x+3}{x^2-x-12}\right)=$

(9) $\lim\limits_{x\to 0}\dfrac{7}{x^3}=$

(10) $\lim\limits_{x\to\infty}\dfrac{8x^4+64x+1}{x^5+4\sqrt{x}+9}=$

(11) $\lim\limits_{x\to\infty}\dfrac{x^4+64x+5x^5}{x^5+x^2+9}=$

(12) $\lim\limits_{x\to 0}\dfrac{\tan x}{x}=$

(13) $\lim\limits_{x\to 0}\dfrac{1-\cos x}{x^2}=$

(14) $\lim\limits_{x\to 0}\dfrac{(4+x)^2-16}{x}=$

(15) $\lim\limits_{x\to 0}\dfrac{\sqrt{x^2+16}-4}{x^2}=$

(16) Let $f(x)=\begin{cases}2\sqrt{x}-3, & x\leqslant 9\\ x-6, & x>9\end{cases}$.

Find: 1) $\lim\limits_{x\to 9^-}f(x)$; 2) $\lim\limits_{x\to 9^+}f(x)$; 3) $\lim\limits_{x\to 9}f(x)$

(17) Let $f(x)=\begin{cases}2\sqrt{x}-3, & x\leqslant 9\\ x-5, & x>9\end{cases}$.

Find: 1) $\lim\limits_{x\to 9^-}f(x)$; 2) $\lim\limits_{x\to 9^+}f(x)$; 3) $\lim\limits_{x\to 9}f(x)$

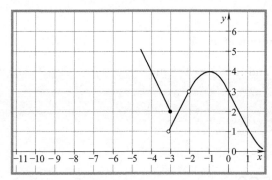

Figure 3.5

(18) The graph of the function $f(x)$ is shown in the Figure 3.5. Which of the following statements about $f(x)$ is true?

(A) $\lim\limits_{x\to -2}f(x)=\lim\limits_{x\to -3}f(x)$

(B) $\lim\limits_{x\to -2}f(x)=3$

(C) $\lim\limits_{x\to -3}f(x)=1$

(D) $\lim\limits_{x\to -3}f(x)=2$

(E) $\lim\limits_{x\to -2}f(x)$ does not exist.

(19) Which statement is true about the curve $f(x)=\dfrac{7-2x^3}{2-6x^2+4x^3}$?

(A) The line $y=\dfrac{1}{2}$ is a horizontal asymptote.

(B) The line $x=\dfrac{1}{2}$ is a vertical asymptote.

(C) The line $x=-\dfrac{1}{2}$ is a vertical asymptote.

(D) The graph has no vertical or horizontal asymptote.

(E) The line $y=1$ is a horizontal asymptote.

(20) Finding all asymptotes of the function $f(x)=\arctan x$.

第 4 章
连续
Continuity

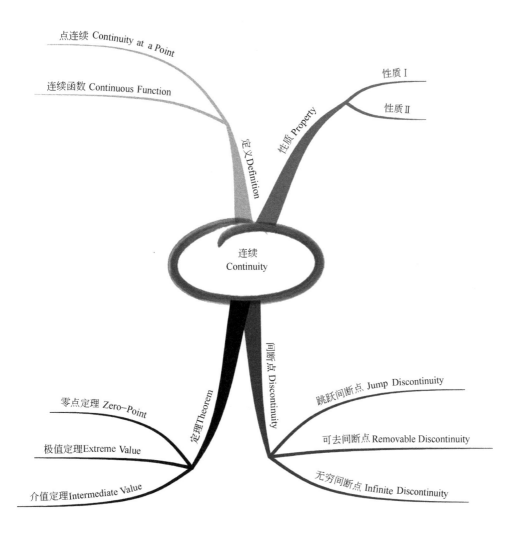

4.1 连续性的定义 Definition of the Continuity

函数的连续性（Continuity）是函数（Function）的一种重要属性，它是微积分中一个重要的概念．函数在 $x=x_0$ 处连续，意味着在 $x=x_0$ 处，当函数的自变量（Independent Variable）的变化足够小之时，因变量（Dependent Variable）的变化也会随之足够小．

4.1.1 点连续 Continuity at a Point

A function $f(x)$ to be continuous at a point $x=x_0$ must fulfill all three of the following conditions：

Condition 1：$f(x_0)$ exists；
Condition 2：$\lim\limits_{x \to x_0} f(x)$ exists；
Condition 3：$\lim\limits_{x \to x_0} f(x) = f(x_0)$．

以上是函数在点 $x=x_0$ 处连续的三个条件，严格而言，这不是点连续的定义，而是函数在点 $x=x_0$ 处连续的判定定理．

Example 4.1 If the function $f(x) = \begin{cases} \dfrac{5}{x-1}, & x \neq 1 \\ 2, & x=1 \end{cases}$ is continuous at $x=0$?

Solution：
Using the definition, we get
$$\begin{cases} f(0) = -5 \\ \lim\limits_{x \to 0^-} f(x) = \lim\limits_{x \to 0^+} f(x) = -5 \\ \lim\limits_{x \to 0} f(x) = f(0) = -5 \end{cases}$$

So, $f(x)$ is continuous at $x=0$.

Note：判别函数在点 $x=x_0$ 处是否连续，就是逐一验证是否连续的三个条件，全部满足则连续，否则不连续．

Example 4.2 If the function $f(x) = \begin{cases} \dfrac{5}{x-1}, & x \neq 1 \\ 2, & x=1 \end{cases}$ is continuous at $x=1$?

Solution：
$$\begin{cases} \lim\limits_{x \to 1^-} f(x) = \lim\limits_{x \to 1^-} \dfrac{5}{x-1} = -\infty \\ \lim\limits_{x \to 1^+} f(x) = \lim\limits_{x \to 1^+} \dfrac{5}{x-1} = +\infty \end{cases}$$

So, $\lim\limits_{x \to 1} f(x)$ not exists.

Therefor, $f(x)$ is not continuous at $x=1$. Look at the graph of $f(x)$, see Figure 4.1.1.

Note：函数 $f(x)$ 在 $x=x_0$ 处连续，直观上来说：x_0 在 $f(x)$ 的定义域上；当 $x \to x_0$ 时，$f(x)$ 有极限，并且，这个极限值等于 $f(x_0)$.

Example 4.3 If the function $f(x)=\begin{cases} x+2, x<1 \\ 3x+1, x=1 \\ 9x-6, x>1 \end{cases}$ is continuous at $x=1$?

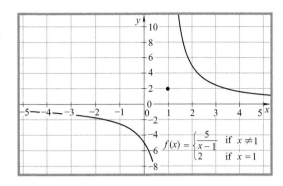

Figure 4.1.1

Solution：
$$\begin{cases} \lim_{x \to 1^-} f(x) = \lim_{x \to 1^-} (x+2) = 3 \\ \lim_{x \to 1^+} f(x) = \lim_{x \to 1^+} (9x-6) = 3 \end{cases}$$
Then, $\lim_{x \to 1} f(x) = 3$. But,
$$f(1) = 3 \times 1 + 1 = 4$$
Therefor, $\lim_{x \to 1} f(x) \neq f(1)$, $f(x)$ is not continuous at $x=1$.

4.1.2 连续函数 Continuous Functions

A function is continuous on the domain if and only if it is continuous at every point over the domain.

如果函数图像连续，那么这个函数的图像必定是一笔画. 幂函数、指数函数、对数函数、三角函数和反三角函数在定义域上都是连续函数.

Example 4.4 For what value of the constant k is the function
$$f(x) = \begin{cases} x+k & (x \leq 2) \\ \dfrac{kx^2+(3-2k)x-6}{x-2} & (x>2) \end{cases}$$
is continuous on $(-\infty, +\infty)$?

Solution：
The function $f(x)$ is continuous on $(-\infty, +\infty)$, if it is continuous at $x=2$. Then
$$\lim_{x \to 2^-} f(x) = \lim_{x \to 2^-} (x+k) = 2+k$$
$$\lim_{x \to 2^+} f(x) = \lim_{x \to 2^+} \frac{kx^2+(3-2k)x-6}{x-2} = \lim_{x \to 2^+} \frac{(kx+3)(x-2)}{x-2}$$
$$= \lim_{x \to 2^+} (kx+3) = 2k+3$$
Since
$$\lim_{x \to 2^-} f(x) = \lim_{x \to 2^+} f(x)$$
So, $2+k = 2k+3$, $k=-1$

Example 4.5 Is the function $f(x) = \dfrac{3x^2+5x+15}{x^2-x-20}$ continuous everywhere?

Solution：
No. It is discontinuous at $x=-4$ and $x=5$.

$$f(x)=\frac{3x^2+5x+15}{x^2-x-20}=\frac{3x^2+5x+15}{(x+4)(x-5)}$$

This means that $f(x)$ is undefined at $x=-4$ and $x=5$.

Therefore, $f(x)$ is discontinuous at $x=-4$ and $x=5$.

Note：有理函数 $f(x)=\dfrac{P(x)}{Q(x)}[Q(x)\neq 0]$ 在点 $Q(x)=0$ 处不连续（Discontinuous）.

4.1.3 连续函数的性质 I Properties of Continuous Function I

If the functions $f(x)$ and $g(x)$ are both continuous at $x=x_0$, then the following combinations are continuous at $x=x_0$:

1) $f(x)\pm g(x)$;
2) $kf(x)$, where k is a constant;
3) $f(x)g(x)$;
4) $\dfrac{f(x)}{g(x)}$, provided that $g(x)\neq 0$.

由函数在某点连续的定义和极限的四则运算法则，即可得到连续函数的和、差、积和商的连续性. 连续函数的和、差、积和商的连续性结合指数函数、对数函数、幂函数等常见函数的连续性，是判断函数连续性问题的常用方法.

Example 4.6 $f(x)$ is continuous at $x=0$. If the function $h(x)=\mathrm{e}^x f(x)$ is continuous at $x=0$?

Solution：

Let $g(x)=\mathrm{e}^x$, using the definition, we get

$$\lim_{x\to 0^-}g(x)=\lim_{x\to 0^+}g(x)=1$$

So, $g(x)$ is continuous at $x=0$.

Using the properties of continuous functions, we get $h(x)=g(x)f(x)=\mathrm{e}^x f(x)$ is continuous at $x=0$.

Example 4.7 Is the function $F(x)=\dfrac{3x^2+5x+15}{x^2+1}$ continuous everywhere?

Solution：

Let $f(x)=3x^2+5x+15$ and $g(x)=x^2+1$, then $F(x)=\dfrac{f(x)}{g(x)}$.

The functions $f(x)$ and $g(x)$ are both continuous everywhere,

$$g(x)=x^2+1>0$$

Therefore, using the properties of continuous functions, we get $F(x)=\dfrac{3x^2+5x+15}{x^2+1}$ continuous everywhere.

4.1.4 连续函数的性质 II Properties of Continuous Functions II

If $f(x)$ is continuous at x_0 and $g(x)$ is continuous at $f(x_0)$, then the composite $g\circ f$ is continuous at x_0.

假如函数 $f(x)$ 在 $x=x_0$ 处连续，$g(x)$ 在 $x=f(x_0)$ 处连续，那么复合函数（Composite Function）$g \circ f$ 在 $x=x_0$ 处连续．连续在复合函数上的性质，给"复杂"函数连续性的判断提供了方法．

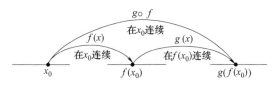

连续函数的复合函数是连续函数

Example 4.8 Show that the functions of $f(x)=\left|\dfrac{x\cos x}{x^2+1}\right|$ is continuous everywhere on its domain.

Solution：

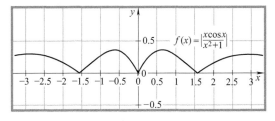

Figure 4.1.4

Because the cosine function is everywhere-continuous, the numerator term $x\cos x$ is the product of continuous functions, and denominator term x^2+1 is an everywhere-positive polynomial.

The given function is the composite of a quotient of continuous functions with the continuous absolute value.

Look at the graph of $f(x)$, see Figure 4.1.4.

4.2　间断点的分类 Kinds of Discontinuities

许多我们熟悉的物理过程都是连续进行的，我们使用连续函数描述它们．同时，在计算机科学、统计学和数学建模中大量出现间断函数，我们需要对间断点分类．

4.2.1　间断点 Discontinuity

If point $x=x_0$ at least one of the conditions of continuity is not fulfilled for the function $f(x)$, that is：

1) The function $f(x)$ is not defined at $x=x_0$；

2) $\lim\limits_{x \to x_0} f(x)$ does not exist；

3) $\lim\limits_{x \to x_0} f(x)$ exists but $\lim\limits_{x \to x_0} f(x) \neq f(x_0)$.

Then, we say $x=x_0$ is a discontinuity point of the function $f(x)$.

函数的连续性问题在实践中和理论上是有重大意义的问题之一．为了深入了解函数的连续性（Continuity），我们需要间断点．

Example 4.9 Consider the function：$f(x)=\begin{cases} x+1 & (x \leqslant 1) \\ x^2 & (x>1) \end{cases}$

Solution：

The graph of $f(x)$ is shown at the following, see Figure 4.2.1.

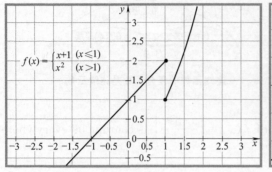
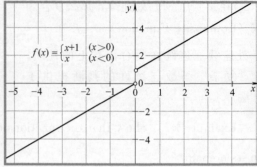

Figure 4.2.1　　　　　　　　　　　　　　Figure 4.2.2

Since $\lim\limits_{x \to 1} f(x)$ does not exist, the function is discontinuous at $x=1$. The curve is continuous everywhere except at point $x=1$.

4.2.2　跳跃间断点 Jump Discontinuity

The one-sided limits exist but have different values. In other words, $\lim\limits_{x \to x_0^-} f(x) \neq \lim\limits_{x \to x_0^+} f(x)$.

$f(x)$ 在 $x=x_0$ 处左右极限都存在，但左右极限不相等，这个间断点是跳跃间断点 (Jump Discontinuity)。

Example 4.10　Because $\lim\limits_{x \to 0^-} f(x) \neq \lim\limits_{x \to 0^+} f(x)$, the function $f(x) = \begin{cases} x+1 & (x>0) \\ x & (x<0) \end{cases}$ has a jump discontinuity at $x=0$, see Figure 4.2.2.

4.2.3　可去间断点 Removable Discontinuity

$\lim\limits_{x \to x_0} f(x)$ exists, but $\lim\limits_{x \to x_0} f(x) \neq f(x_0)$ or $f(x)$ is not defined at $x=x_0$. It can be canceled or "removable".

$f(x)$ 在 $x=x_0$ 处有间断点，左右极限都存在且相等，这个间断点是可去间断点 (Removable Discontinuity)。

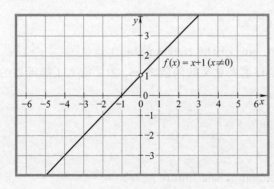

Figure 4.2.3

Example 4.11　Since $\lim\limits_{x \to 0} f(x) = 1$, but $f(x)$ is not defined at $x=0$. Then, the function $f(x) = x+1 \, (x \neq 0)$ has a removable discontinuity at $x=0$, see Figure 4.2.3.

4.2.4　无穷间断点 Infinite Discontinuity

The function $f(x)$ becomes positively or negatively infinite as $x \to x_0^-$ or

$x \to x_0^+$, then, the function $f(x)$ have an infinite discontinuity.

间断点通常分成两类：因为间断点的左右极限都存在，跳跃间断点和可去间断点属于第一类间断点；无穷间断点（Infinite Discontinuity）属于第二类间断点．

Example 4.12 Since $\lim\limits_{x \to 0^-} f(x) = -\infty$ and $\lim\limits_{x \to 0^+} f(x) = +\infty$, then $f(x) = \dfrac{1}{x}$ have an infinite discontinuity at $x = 0$, see Figure 4.2.4(a).

Example 4.13 Consider the function：$f(x) = \begin{cases} (x+2)^2 - 1, & x < 0 \\ 1, & x = 0 \\ -x, & 0 < x < 2 \\ 0, & x = 2 \\ x - 4, & 2 < x < 6 \end{cases}$

Solution：
The graph of $f(x)$ is shown at Figure 4.2.4(b).

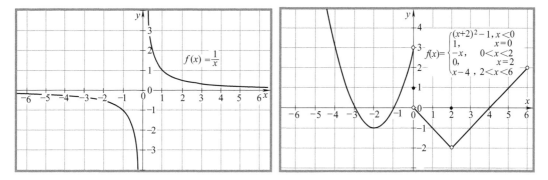

Figure 4.2.4(a)　　　　　　　Figure 4.2.4(b)

We observe that $f(x)$ is not continuous at $x = 0$, $x = 2$, and $x = 6$.

1) At $x = 0$, $f(x)$ has a domain; in fact, $f(0) = 1$.

However, since $\lim\limits_{x \to 0^-} f(x) = 3$ and $\lim\limits_{x \to 0^+} f(x) = 0$, $\lim\limits_{x \to 0} f(x)$ does not exist. Where the left-hand limit and right-hand limit exist, but are different, the function has a jump discontinuity.

2) At $x = 2$, $f(2) = 0$, and $\lim\limits_{x \to 2} f(x) = -2$, but $\lim\limits_{x \to 2} f(x) \neq f(2)$.

This discontinuity is a removable discontinuity.

3) At $x = 6$, $f(x)$ is not defined.

4.3 连续函数定理 The Continuous Functions Theorem

　　AP 微积分对比高等数学，弱化了定理证明，着重基础概念和基础理论的应用．上述定理在 AP 微积分考试中，不大可能以"证明定理"的方式出现，多以应用定理得出结论的方式呈现．

4.3.1 零点定理 The Zero- Point Theorem

If $f(x)$ is continuous on the closed interval $[a, b]$, and $f(a)f(b)<0$, then there is at least one number ξ, between a and b such that $f(\xi)=0$.

根据零点定理,若函数 $f(x)$ 在闭区间 $[a, b]$ 上连续,并且 $f(a)$ 和 $f(b)$ 异号,那么,我们总能在开区间 (a, b) 上找到方程 $f(x)=0$ 的根.

4.3.2 极值定理 The Extreme Value Theorem

If $f(x)$ is continuous on the closed interval $[a, b]$, then $f(x)$ attains a minimum value and a maximum value somewhere in that interval.

根据极值定理,如果函数 $f(x)$ 在闭区间 $[a, b]$ 上连续,那么它一定存在至少一个最大值和最小值.

4.3.3 介值定理 The Intermediate Value Theorem

If $f(x)$ is continuous on the closed interval $[a, b]$, and M is a number such that $f(a) \leqslant M \leqslant f(b)$, then there is at least one number ξ, between a and b such that $f(\xi)=M$.

介值定理可以认为是零点定理的推广. 假如函数 $f(x)$ 在闭区间 $[a, b]$ 上连续,$f(a) \leqslant M \leqslant f(b)$,那么,我们总能在开区间 (a, b) 上找到方程 $f(x)=M$ 的根.

Example 4.14 Show that there is a root of the equation $x^2+x-\dfrac{1}{x}=0$ between $\dfrac{1}{2}$ and 1.

Solution:

Let $f(x)=x^2+x-\dfrac{1}{x}$. The function $f(x)$ is continuous on the closed interval $\left[\dfrac{1}{2}, 1\right]$.

Therefore, we take $a=\dfrac{1}{2}$ and $b=1$ in The Zero-Point Theorem. We have

$$f\left(\dfrac{1}{2}\right)=\left(\dfrac{1}{2}\right)^2+\dfrac{1}{2}-2=-\dfrac{5}{4}<0$$

And

$$f(1)=1+1-1=1>0$$

Thus

$$f\left(\dfrac{1}{2}\right) \times f(1)<0$$

So, the Zero-Point Theorem says there is a number ξ between $\dfrac{1}{2}$ and 1 such that $f(\xi)=0$.

In other words, the equation $x^2 + x - \dfrac{1}{x} = 0$ has at least one root ξ in the interval $\left(\dfrac{1}{2}, 1\right)$. The graph of $f(x)$ is shown at Figure 4.3.

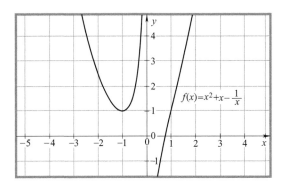

Figure 4.3

4.4 习题 Practice Exercises

(1) If the function $f(x) = \begin{cases} \dfrac{2}{x-1}, & x \neq 1 \\ 5, & x = 1 \end{cases}$ is continuous at $x = 0$?

(2) Is the function $f(x) = \begin{cases} x^2, & x \leqslant 0 \\ x-2, & x > 0 \end{cases}$ continuous at $x = 0$? Sketch the graph of the function $f(x)$.

(3) Identify the points of discontinuity of each of the following:

1) $f(x) = \dfrac{x+3}{x^2-9}$; 2) $f(x) = \begin{cases} 2x^2, & x < 3 \\ \ln x, & x \geqslant 3 \end{cases}$; 3) $f(x) = \begin{cases} \dfrac{x^3}{x^4}, & x < 0 \\ 3, & x = 0 \\ \sin 3x, & x > 0 \end{cases}$

(4) For what value of the constant a, b is the function
$$f(x) = \begin{cases} x+2, & x \leqslant 0 \\ x^2 + a, & 0 < x < 1 \\ bx, & 1 \leqslant x \end{cases}$$
is continuous on domain?

(5) The function $f(x) = \begin{cases} x+3, & -3 \leqslant x < 0 \\ \dfrac{1}{2}x^2 - 2, & 0 \leqslant x \leqslant 2 \\ x - 1, & 2 < x < 4 \\ 0, & x = 4 \\ -x + 7, & 4 < x \leqslant 6 \end{cases}$ shown in Figure 4.4.

At what point are the removable discontinuity and jump discontinuity for the function $f(x)$?

(6) Is the function $f(x) = \tan x$ continuous on the interval $\left[-\dfrac{\pi}{2}, \dfrac{\pi}{2}\right]$?

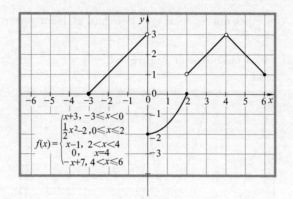

Figure 4.4

(7) Where is the function $f(x)=\dfrac{1}{x-1}$ continuous?

(8) Show that there is a root of the equation $x^5-3x=1$ on $[1,2]$.

第 5 章

导数和微分
Derivative and Differential

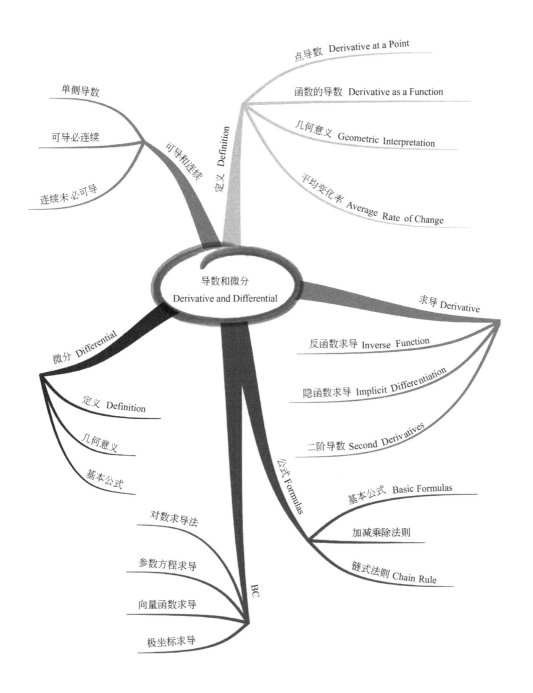

5.1 导数的定义 Definition of the Derivative

导数（Derivative）的应用范围极其广泛，同时也是 AP 微积分的重点内容，考查力度大，分值占比高，从导数的定义开始，理解并掌握导数，必须予以重视．

5.1.1 切线的斜率 Slope of the tangent line

The tangent line to the curve $y=f(x)$ at the point $P(x_0, f(x_0))$ is the line through P with slope

$$k = \lim_{\Delta x \to 0} \frac{f(x_0+\Delta x)-f(x_0)}{\Delta x}$$

Provided that this limit exists.

我们知道，直线的斜率（Slope of a Line）由两点决定，如 Figure 5.1.1(a) 所示．

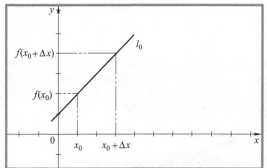

Figure 5.1.1(a)　　　　Figure 5.1.1(b)

以 k 表示直线 l_0 的斜率，则：

$$k = \frac{f(x_0+\Delta x)-f(x_0)}{\Delta x}$$

我们用 $f(x)$ 表示一条不平行于 y 轴的曲线（curve），如 Figure 5.1.1(b) 所示，$f(x)$ 在 x_0 点处的割线 l_0，k_0 表示曲线 $f(x)$ 在点 x_0 处割线的斜率（Slope of the secant line of a curve），有：

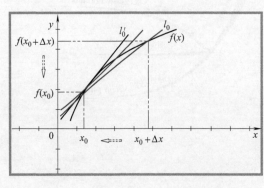

Figure 5.1.1(c)

$$k_0 = \frac{f(x_0+\Delta x)-f(x_0)}{\Delta x}$$

观察 Figure 5.1.1(b)，当 $x_0+\Delta x$ 不断靠近 x_0，$f(x_0+\Delta x)$ 也将不断向靠近 $f(x_0)$．也就是说，曲线 $f(x)$ 在 $x=x_0$ 处的割线 l_0' 将绕 x_0 点进行旋转，在极限条件下，当 $\Delta x \to 0$，得到曲线 $f(x)$ 在 $x=x_0$ 处的切线 l_0'．如 Figure 5.1.1(c) 所示．

上述过程揭示了曲线在某点的切线

系过该点割线的极限情况，依据第 3 章讲授的极限（limit）思想，结合 $f(x)$ 在 $x=x_0$ 处割线 l_0 的斜率，可得 $f(x)$ 在 $x=x_0$ 处的切线 l'_0 的斜率（Slope of the tangent line）：

$$k = \lim_{\Delta x \to 0} \frac{f(x_0 + \Delta x) - f(x_0)}{\Delta x}$$

Example 5.1　Find an equation of the tangent line to the parabola $y = x^2$ at the point $P(2,4)$.

Solution：

Here we have $x_0 = 2$ and $f(x) = x^2$, so the slope is

$$k = \lim_{\Delta x \to 0} \frac{f(2+\Delta x) - f(2)}{\Delta x} = \lim_{\Delta x \to 0} \frac{(2+\Delta x)^2 - 2^2}{\Delta x}$$

$$= \lim_{\Delta x \to 0} \frac{4 + 4\Delta x + (\Delta x)^2 - 4}{\Delta x}$$

$$= \lim_{\Delta x \to 0} (4 + \Delta x) = 4$$

Using the point-slope from of the equation of a line, we find that an equation of the tangent line at $(2,4)$ is

$$y - 4 = 4(x - 2) \text{ or } y = 4x - 4$$

5.1.2　点导数 Derivative at a Point

The derivative of a function $f(x)$ at a number x_0, denoted by $f'(x_0)$, is

$$f'(x_0) = \lim_{\Delta x \to 0} \frac{f(x_0 + \Delta x) - f(x_0)}{\Delta x}$$

If this limit exists, we say that $f(x)$ at $x = x_0$ is differentiable. Finding a derivative is called differentiation.

函数 $y = f(x)$ 在 $x = x_0$ 点处可导（Differentiable），意味着 $\lim\limits_{\Delta x \to 0} \dfrac{f(x_0 + \Delta x) - f(x_0)}{\Delta x}$ 存在．

函数 $y = f(x)$ 在 $x = x_0$ 的点导数（Derivative at a Point），除了记为 $f'(x_0)$，还可记为：$y'|_{x=x_0}$，$\dfrac{\mathrm{d}y}{\mathrm{d}x}\bigg|_{x=x_0}$ 或者 $\dfrac{\mathrm{d}}{\mathrm{d}x}f(x)|_{x=x_0}$．

Example 5.2　Using the definition, find the derivative of $f(x) = x^n$ at $x = x_0$, n is positive integer numbers.

Solution：

Use the definition of the derivative

$$f'(x_0) = \lim_{\Delta x \to 0} \frac{f(x_0 + \Delta x) - f(x_0)}{\Delta x} = \lim_{\Delta x \to 0} \frac{(x_0 + \Delta x)^n - x_0^n}{\Delta x}$$

Since Binomial theorem，

$$(x_0 + \Delta x)^n = x_0^n + n x_0^{n-1}(\Delta x) + \frac{n(n-1)}{2} x_0^{n-2}(\Delta x)^2 + \cdots +$$

$$\frac{n!}{r!(n-r)!} x_0^{n-r}(\Delta x)^r + \cdots + (\Delta x)^n$$

And simplify:

$$\frac{(x_0+\Delta x)^n - x_0^n}{\Delta x}$$

$$= \frac{x_0^n + nx_0^{n-1}(\Delta x) + \frac{n(n-1)}{2}x_0^{n-2}(\Delta x)^2 + \cdots + \frac{n!}{r!(n-r)!}x_0^{n-r}(\Delta x)^r + \cdots + (\Delta x)^n - x_0^n}{\Delta x}$$

$$= nx_0^{n-1} + \frac{n(n-1)}{2}x_0^{n-2}(\Delta x) + \cdots + \frac{n!}{r!(n-r)!}x_0^{n-r}(\Delta x)^{r-1} + \cdots + (\Delta x)^{n-1}$$

Then,

$$f'(x_0) = \lim_{\Delta x \to 0}\frac{(x_0+\Delta x)^n - x_0^n}{\Delta x}$$

$$= \lim_{\Delta x \to 0}\left[nx_0^{n-1} + \frac{n(n-1)}{2}x_0^{n-2}(\Delta x) + \cdots + \frac{n!}{r!(n-r)!}x_0^{n-r}(\Delta x)^{r-1} + \cdots + (\Delta x)^{n-1}\right]$$

$$= nx_0^{n-1}$$

Note：本题结论是导数的基本公式之一．

Example 5.3 Find the derivative of the function $f(x) = a^x$ at $x = x_0$.

Solution：

Use the definition of the derivative, we get

$$f'(x_0) = \lim_{h \to 0}\frac{f(x_0+h) - f(x_0)}{h} = \lim_{h \to 0}\frac{a^{x_0+h} - a^{x_0}}{h}$$

$$= \lim_{h \to 0}\frac{a^{x_0}(a^h - 1)}{h}$$

Let $k = a^h - 1$, $h = \log_a(k+1)$, then $k \to 0$ as $h \to 0$.

Then,

$$f'(x_0) = \lim_{h \to 0}\frac{a^{x_0}(a^h - 1)}{h} = \lim_{k \to 0}\frac{a^{x_0}k}{\log_a(k+1)}$$

$$= a^{x_0}\lim_{k \to 0}\frac{k}{\log_a(k+1)} = a^{x_0}\lim_{k \to 0}\frac{1}{\frac{1}{k}\log_a(k+1)}$$

$$= a^{x_0}\lim_{k \to 0}\frac{1}{\log_a(k+1)^{\frac{1}{k}}}$$

Since

$$\lim_{k \to 0}\frac{1}{\log_a(k+1)^{\frac{1}{k}}} = \frac{1}{\log_a\left[\lim_{k \to 0}(k+1)^{\frac{1}{k}}\right]}$$

And, $\lim_{k \to 0}(k+1)^{\frac{1}{k}} = e$,

Then,

$$f'(x_0) = a^{x_0}\frac{1}{\log_a e} = a^{x_0}\ln a$$

Note：本题结论是导数的基本公式之一，当 $a = e$ 时，有 $\ln a = \ln e = 1$，此时，$(e^{x_0})' = e^{x_0}$.

Example 5.4 Find the derivative of the function $f(x)=\log_a x$ at $x=x_0$.

Solution：

Use the definition of the derivative, we get

$$f'(x_0)=\lim_{h\to 0}\frac{f(x_0+h)-f(x_0)}{h}=\lim_{h\to 0}\frac{\log_a(x_0+h)-\log_a x_0}{h}$$

Since

$$\frac{\log_a(x_0+h)-\log_a x_0}{h}=\frac{\log_a\left(\frac{x_0+h}{x_0}\right)}{h}=\frac{1}{h}\log_a\left(\frac{x_0+h}{x_0}\right)$$

$$=\log_a\left(1+\frac{h}{x_0}\right)^{\frac{1}{h}}=\log_a\left(1+\frac{h}{x_0}\right)^{\frac{x_0}{h}\times\frac{1}{x_0}}$$

$$=\frac{1}{x_0}\log_a\left(1+\frac{h}{x_0}\right)^{\frac{x_0}{h}}$$

Then，

$$f'(x_0)=\lim_{h\to 0}\frac{1}{x_0}\log_a\left(1+\frac{h}{x_0}\right)^{\frac{x_0}{h}}=\frac{1}{x_0}\lim_{\frac{x_0}{h}\to 0}\log_a\left(1+\frac{h}{x_0}\right)^{\frac{x_0}{h}}$$

$$=\frac{1}{x_0}\log_a e=\frac{1}{x_0\ln a}$$

Note：本题结论是导数的基本公式之一，当 $a=e$ 时，有 $\log_a e=\ln e=1$，此时得到 $(\ln x_0)'=\frac{1}{x_0}$.

Example 5.5 Find the derivative of the function $f(x)=\cos x$ at $x=x_0$.

Solution：

Use the definition of the derivative, we get

$$f'(x_0)=\lim_{h\to 0}\frac{f(x_0+h)-f(x_0)}{h}=\lim_{h\to 0}\frac{\cos(x_0+h)-\cos x_0}{h}$$

$$=\lim_{h\to 0}\frac{\cos x_0\cos h-\sin x_0\sin h-\cos x_0}{h}$$

$$=\lim_{h\to 0}\frac{\cos x_0(\cos h-1)-\sin x_0\sin h}{h}$$

$$=\lim_{h\to 0}\frac{\cos x_0(\cos h-1)}{h}-\lim_{h\to 0}\frac{\sin x_0\sin h}{h}$$

$$=\cos x_0\lim_{h\to 0}\frac{(\cos h-1)}{h}-\lim_{h\to 0}\frac{\sin x_0\sin h}{h}$$

$$=0-\sin x_0=-\sin x_0$$

Note：本题结论是导数的基本公式之一，类似地，可求得 $f(x)=\sin x$ 在 $x=x_0$ 处的导数为 $f'(x_0)=\cos x_0$.

【方法总结】从导数的定义出发，求 $f(x)$ 在 $x=x_0$ 的导数（Differentiation），一般分三步走：

1）写出 $f(x_0)$ 和 $f(x_0+\Delta x)$ 的表达式；

2）展开并简化 $\dfrac{f(x_0+\Delta x)-f(x_0)}{\Delta x}$；

3）利用简化后的商式，通过计算极限 $f'(x_0)=\lim\limits_{\Delta x\to 0}\dfrac{f(x_0+\Delta x)-f(x_0)}{\Delta x}$，求得 $f'(x_0)$．

5.1.3 函数的导数 Derivative as a Function

If the function $f(x)$ is differentiable at every point of the interval (a,b), then the function is called differentiable on the interval (a,b).

$$f'(x)=\lim_{\Delta x\to 0}\frac{f(x+\Delta x)-f(x)}{\Delta x}$$

函数 $f(x)$ 在 (a,b) 内可导，则 $f(x)$ 在 (a,b) 内的每一个 x 都有一个导数值 $f'(x)$ 与之对应，称为 $f(x)$ 在 (a,b) 内对 x 的导函数．Table 5.1.3 梳理了 AP 微积分常用导函数的表示符号．

Table 5.1.3 AP 微积分常用导函数的表示符号

Function	First derivative	Second derivative	n-th derivative
y	y' or $\dfrac{dy}{dx}$	y'' or $\dfrac{d^2y}{dx^2}$	$\dfrac{d^ny}{dx^n}$
$f(x)$	$f'(x)$ or $\dfrac{df(x)}{x}$	$f''(x)$ or $\dfrac{d^2f(x)}{dx^2}$	$f^{(n)}(x)$

Example 5.6 Find the derivative of the function $f(x)=\sin x$.

Solution：

Use the definition of the derivative, we get

$$\begin{aligned}
f'(x)&=\lim_{\Delta x\to 0}\frac{f(x+\Delta x)-f(x)}{\Delta x}\\
&=\lim_{\Delta x\to 0}\frac{\sin(x+\Delta x)-\sin x}{\Delta x}\\
&=\lim_{\Delta x\to 0}\frac{\sin x\cos\Delta x+\cos x\sin\Delta x-\sin x}{\Delta x}\\
&=\lim_{\Delta x\to 0}\frac{\sin x(\cos\Delta x-1)+\cos x\sin\Delta x}{\Delta x}\\
&=\lim_{\Delta x\to 0}\frac{\sin x(\cos\Delta x-1)}{\Delta x}+\lim_{\Delta x\to 0}\frac{\cos x\sin\Delta x}{\Delta x}\\
&=\sin x\lim_{\Delta x\to 0}\frac{(\cos\Delta x-1)}{\Delta x}+\lim_{\Delta x\to 0}\frac{\cos x\sin\Delta x}{\Delta x}\\
&=0+\cos x
\end{aligned}$$

Then, $(\sin x)'=\cos x$.

Note：比较而言，Example 5.5 求 $f(x)=\cos x$ 在 $x=x_0$ 处的点导数值，而本题求 $f(x)=\sin x$ 的导函数．

Example 5.7 Find the derivative of the function $f(x)=\sqrt{x}$.

Solution:

Use the definition of the derivative, we get

$$f'(x)=\lim_{\Delta x\to 0}\frac{f(x+\Delta x)-f(x)}{\Delta x}=\lim_{\Delta x\to 0}\frac{\sqrt{x+\Delta x}-\sqrt{x}}{\Delta x}$$

$$=\lim_{\Delta x\to 0}\frac{\Delta x}{\Delta x(\sqrt{x+\Delta x}+\sqrt{x})}=\lim_{\Delta x\to 0}\frac{1}{\sqrt{x+\Delta x}+\sqrt{x}}=\frac{1}{2\sqrt{x}}$$

Then, $(\sqrt{x})'=\dfrac{1}{2\sqrt{x}}$.

Note: 类似的方法,可拿到 $(x^2)'=2x$, $\left(\dfrac{1}{x}\right)'=-\dfrac{1}{x^2}$.

5.1.4 导数的几何意义 The Geometric Interpretation of Derivative

$f'(x_0)$ expresses the slope of the tangent line of $f(x)$ at $x=x_0$, that is $f'(x_0)=\tan\alpha$ and $f'(x_0)$ does not exist if $\alpha=\dfrac{\pi}{2}$.

如 Figure 5.1.4 所示,导函数 $y=f'(x_0)$ 在 $x=x_0$ 处的值即该点的切线 l'_0 的斜率 k.

Example 5.8 Find the slope of tangent line to the curve $y=\dfrac{1}{x}$ at the point $(1,1)$.

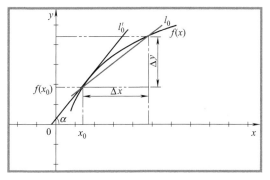

Figure 5.1.4

Solution:

The slope of tangent line to the curve is

$$y'=\lim_{\Delta x\to 0}\frac{f(x+\Delta x)-f(x)}{\Delta x}=\lim_{\Delta x\to 0}\frac{\dfrac{1}{1+\Delta x}-\dfrac{1}{x}}{\Delta x}$$

$$=\lim_{\Delta x\to 0}\frac{-1}{x(x+\Delta x)}=-\frac{1}{x^2}$$

Plug $(1,1)$ into y':

$$y'|_{x=1}=-\frac{1}{1^2}=-1$$

Note: 曲线 $y=f(x)$ 在点 $x=x_0$ 处的切线 l'_0 的斜率 k 即该点的导数值 $f'(x_0)$.

5.1.5 函数的平均变化率 Average Rate of Change of Function

We let

$$\Delta y=f(x_0+\Delta x)-f(x_0)$$

Then,

$$\frac{\Delta y}{\Delta x} = \frac{f(x_0 + \Delta x) - f(x_0)}{\Delta x}$$

The fraction $\frac{\Delta y}{\Delta x}$ is called the difference quotient for $f(x)$ at x_0 and represents the average rate of change of $f(x)$ from x_0 to $x_0 + \Delta x$.

$$f'(x_0) = \lim_{\Delta x \to 0} \frac{f(x_0 + \Delta x) - f(x_0)}{\Delta x} = \lim_{\Delta x \to 0} \frac{\Delta y}{\Delta x}$$

$\lim_{\Delta x \to 0} \frac{\Delta y}{\Delta x}$ expresses the instantaneous rate of change of a function.

差商（Difference Quotient）表示函数的平均变化率（Average Rate of Change）；导数表示函数某点处的变化率（Instantaneous Rate of Change）.

Example 5.9 Find the average rate of change of the function $f(x) = x\cos x$ on the closed interval $[0, \pi]$.

Solution：

The average rate of change is

$$\frac{\Delta y}{\Delta x} = \frac{f(\pi) - f(0)}{\pi - 0} = \frac{\pi\cos\pi - 0}{\pi - 0} = -1$$

Note：在 AP 微积分范围内，求函数的平均变化率（Average Rate of Change），就是求函数的差商（Difference Quotient）.

Example 5.10 If a particle moves along a line according to the law $s(t) = t^2 + 2t$ be the number of kilometer in t minutes. Find the average rate of change during the first 5 minutes and the rate at the end of 5 minutes.

Solution：

The average rate of change during the first 5 minutes is

$$\frac{s(5) - s(0)}{5} = \frac{(5^2 + 2 \times 5) - 0}{5} = 7 \text{km/min}$$

The instantaneous rate of change at $t = 5$ is $s'(5)$. Since

$$s'(5) = \lim_{\Delta x \to 0} \frac{s(5 + \Delta x) - s(5)}{\Delta x}$$
$$= \lim_{\Delta x \to 0} \frac{[(5 + \Delta x)^2 + 2(5 + \Delta x)] - (5^2 + 2 \times 5)}{\Delta x}$$
$$= 12$$

Thus, the rate at the end of 5 minutes is 12km/min.

Note：求质点的瞬时变化率（Instantaneous Rate of Change）等同于求该点的导数（Derivative）.

5.2 可导性和连续性 Derivability and Continuity

从单侧导数（One-Sided Derivative）看，可导比连续更严格，可导（Derivability）一定连续，连续（Continuity）不一定可导.

5.2.1 单侧导数 One-Sided Derivative

The left-hand derivative of a function $f(x)$ at a number x_0, denoted by $f'_-(x_0)$, is

$$f'_-(x_0) = \lim_{\Delta x \to 0^-} \frac{f(x_0 + \Delta x) - f(x_0)}{\Delta x}$$

And, the right-hand derivative of a function $f(x)$ at a number x_0, denoted by $f'_+(x_0)$, is

$$f'_+(x_0) = \lim_{\Delta x \to 0^+} \frac{f(x_0 + \Delta x) - f(x_0)}{\Delta x}$$

$f(x)$ is differentiable at x_0 if and only if $f'_-(x_0) = f'_+(x_0)$.

显然，当且仅当函数在一点的左、右导数都存在且相等时，函数在该点才是可导的。因此说，函数 $f(x)$ 在 $[a, b]$ 上可导，指 $f(x)$ 在 (a, b) 内处处可导，且存在 $f'_+(a)$ 及 $f'_-(b)$。

Example 5.11 Determine whether $f(x) = |x - 1|$ is differentiable at $x = 1$.

Solution: Simplify the function

$$f(x) = \begin{cases} x - 1, & x > 1 \\ 0, & x = 1 \\ 1 - x, & x < 1 \end{cases}$$

Use the definition of the one-sided derivative

$$\begin{aligned} f'_+(1) &= \lim_{\Delta x \to 0^+} \frac{f(1 + \Delta x) - f(1)}{\Delta x} \\ &= \lim_{\Delta x \to 0^+} \frac{(1 + \Delta x - 1) - (1 - 1)}{\Delta x} \\ &= 1 \end{aligned}$$

And,

$$\begin{aligned} f'_-(1) &= \lim_{\Delta x \to 0^-} \frac{f(1 + \Delta x) - f(1)}{\Delta x} \\ &= \lim_{\Delta x \to 0^+} \frac{(1 - 1 - \Delta x) - (1 - 1)}{\Delta x} \\ &= -1 \end{aligned}$$

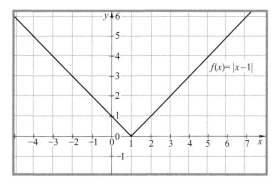

Figure 5.2.1

Since

$$f'_+(1) \neq f'_-(1)$$

The function $f(x) = |x - 1|$ is not differentiable at $x = 1$, see Figure 5.2.1.

Note: 见 Figure 5.2.1, $f(x) = |x - 1|$ 在 $x = 1$ 连续但并不可导。

5.2.2 可导一定连续 Differentiability Implies Continuity

If $f(x)$ is differentiable at x_0, then $f(x)$ is continuous at x_0.

假设函数 $y = f(x)$ 在点 x_0 处可导，即有：

$$\lim_{\Delta x \to 0} \frac{\Delta y}{\Delta x} = f'(x)$$

因为，$\Delta y = \dfrac{\Delta y}{\Delta x} \times \Delta x$，所以

$$\lim_{\Delta x \to 0} \Delta y = \lim_{\Delta x \to 0} \dfrac{\Delta y}{\Delta x} \times \Delta x = \lim_{\Delta x \to 0} \dfrac{\Delta y}{\Delta x} \times \lim_{\Delta x \to 0} \Delta x = f'(x) \times 0$$

根据单侧导数的定义，因为 $f'_-(x_0) = f'_+(x_0)$，所以 $y = f(x)$ 在点 x_0 处连续．

5.2.3 连续不一定可导 Continuous maybe Derivative

Differentiability implies continuity. The converse of this theorem is false.

下列 3 种情况，函数在定义域上连续，但是存在不可导点．

1) A corner：$f(x) = |x-1|$ is continuous at $x=1$ while $f(x) = |x-1|$ is not differentiable at $x=1$. 见 Example 5.11

2) A cusp：where the slope of $x=1$ approaches $+\infty$ from one side and $-\infty$ from the other. 见 Figure 5.2.3(a).

3) A vertical tangent：$f(x) = \sqrt[3]{x-1} - 1$ is not differentiable at $x=1$ since there is a vertical tangent line at $x=1$. 见 Figure 5.2.3(b).

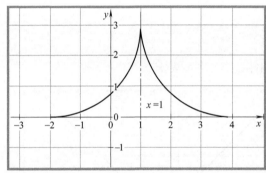

Figure 5.2.3(a)　　　　　　　　Figure 5.2.3(b)

Example 5.12 If $f(x) = \begin{cases} \dfrac{x^2-9}{x-3}, & x \neq 3 \\ 3, & x = 3 \end{cases}$, Which of the following statements about $f(x)$ are true?

Ⅰ．$f(x)$ has a limit at $x=3$；

Ⅱ．$f(x)$ is continuous at $x=3$；

Ⅲ．$f(x)$ is differentiable at $x=3$．

(A) Ⅰ only；(B) Ⅱ only；(C) Ⅲ only；(D) Ⅰ and Ⅱ only；(E) Ⅰ，Ⅱ and Ⅲ．

Solution：

From the graph, we get

$$\lim_{x \to 3^-} f(x) = \lim_{x \to 3^+} f(x) \Rightarrow \lim_{x \to 3} f(x) = 6$$

But

$$\lim_{x \to 3} f(x) \neq f(3)$$

So, $f(x)$ has a limit at $x=3$ and it is not continuous at $x=3$. Then it must be not differentiable at $x=3$, see Figure 5.2.3 (c). The correct answer is A.

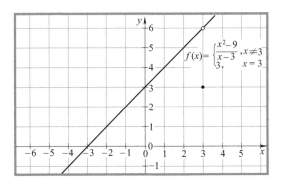

Figure 5.2.3(c)

5.3 导数的基本公式和法则 Basic Differentiation Formulas and Rules

计算函数导数是 AP 微积分中最基本的内容之一，导数（Derivative）的基本公式和加减乘除法则是完成计算函数导数的基础．

5.3.1 导数的基本公式 Basic Differentiation Formulas

从导数的定义出发，前面部分例题给出了导数的基本公式的推导过程，推导过程即证明过程，熟练掌握 Table 5.3.1 中的公式特别关键．

Table 5.3.1 导数（Derivative）的基本公式

Constants	$(C)'=0$, C is constants	
Power	$(x^n)'=nx^{n-1}$	
Exponential	$(a^x)'=a^x \ln a$	$(e^x)'=e^x$
Logarithmic	$(\log_a x)'=\dfrac{1}{x\ln a}$	$(\ln x)'=\dfrac{1}{x}$
Trigonometric	$(\sin x)'=\cos x$	$(\cos x)'=-\sin x$
	$(\tan x)'=\sec^2 x$	$(\cot x)'=-\csc^2 x$
	$(\sec x)'=\sec x \tan x$	$(\csc x)'=-\csc x \cot x$
Inverse Trigonometric	$(\arcsin x)'=\dfrac{1}{\sqrt{1-x^2}}$	$(\arccos x)'=-\dfrac{1}{\sqrt{1-x^2}}$
	$(\arctan x)'=\dfrac{1}{1+x^2}$	$(\text{arccot}\, x)'=-\dfrac{1}{1+x^2}$
	$(\text{arcsec}\, x)'=\dfrac{1}{|x|\sqrt{x^2-1}}$	$(\text{arccsc}\, x)'=-\dfrac{1}{|x|\sqrt{x^2-1}}$

Example 5.13 If $y=x^7$, then $y'=7x^6$.

Example 5.14 If $f(x)=x^\pi$, then $\dfrac{\mathrm{d}y}{\mathrm{d}x}=?$

Solution：

Since $n = \pi$, then $\dfrac{dy}{dx} = \pi x^{\pi-1}$

Example 5.15 If $y = 3x^{100}$, then $y' = ?$

Solution:
$$y' = (3x^{100})' = 3(x^{100})' = 300x^{99}$$

Example 5.16 Find the derivative of $y = \log_2 x$.

Solution:
$$y' = (\log_2 x)' = \dfrac{1}{x \ln 2}$$

Example 5.17 Find the derivative of $f(x) = 5^x$.

Solution:
$$f'(x) = (5^x)' = 5^x \ln 5$$

5.3.2 加减乘除法则 The Sum, Difference, Product and Quotient Rules

1) Constant Multiple Rule: $[ku(x)]' = k[u(x)]'$, k is constants;
2) Sum Rule: $[u(x) + v(x)]' = u'(x) + v'(x)$;
3) Difference Rule: $[u(x) - v(x)]' = u'(x) - v'(x)$;
4) Product Rule: $[u(x)v(x)]' = u'(x)v(x) + u(x)v'(x)$;
5) Quotient Rule: $\left[\dfrac{u(x)}{v(x)}\right]' = \dfrac{u'(x)v(x) - u(x)v'(x)}{v^2(x)}$, $v(x) \neq 0$.

函数求导的除法法则的一个记忆技巧是：$\dfrac{"LoDeHi - HiDeLo"}{(Lo)^2}$。并且，运用函数求导的乘法法则可以推导得到除法法则。假设 $F(x) = \dfrac{u(x)}{v(x)}$，$v(x) \neq 0$，将 $F(x)$ 写成：

$$F(x) = u(x) \dfrac{1}{v(x)}$$

运用乘法法则（Product Rule），有

$$F'(x) = u'(x) \times \dfrac{1}{v(x)} + u(x)\left(\dfrac{1}{v(x)}\right)'$$

$$= \dfrac{u'(x)v(x)}{v^2(x)} + \dfrac{u(x) \times (-1) \times v'(x)}{v^2(x)}$$

$$= \dfrac{u'(x)v(x) - u(x)v'(x)}{v^2(x)}$$

Example 5.18 If $y = 5x^5 + 7x^7 + 9$, then $\dfrac{dy}{dx} = ?$

Solution: By the Sum Rule, we have

$$\dfrac{dy}{dx} = (5x^5 + 7x^7 + 9)' = (5x^5)' + (7x^7)' + (9)'$$

$$= 5 \times 5x^4 + 7 \times 7x^6 + 0$$

$$= 25x^4 + 49x^6$$

Example 5.19 Find the derivative of $f(x) = (3x^4 + 1)(\sqrt{x} + 1)$.

Solution: Using the Product Rule, we have
$$f'(x) = (3x^4+1)'(\sqrt{x}+1) + (3x^4+1)(\sqrt{x}+1)'$$
$$= 12x^3(\sqrt{x}+1) + (3x^4+1) \times \frac{1}{2\sqrt{x}}$$

Example 5.20 Find $F'(x)$ if $F(x) = \dfrac{x^2+x-2}{x^2+3}$.

Solution: By the Quotient Rule, we have
$$F'(x) = \frac{(x^2+x-2)'(x^2+3) - (x^2+x-2)(x^2+3)'}{(x^2+3)^2}$$
$$= \frac{(2x+1)(x^2+3) - (x^2+x-2) \times 2x}{(x^2+3)^2}$$
$$= \frac{-x^2+10x+3}{(x^2+3)^2}$$

Note: 将函数看成 $F(x) = (x^2+x-2)(x^2+3)^{-1}$，运用乘法法则（Product Rule），得到的结果是一样的，请同学们自行完成.

Example 5.21 Find the derivative of $f(x) = \tan x$.

Solution: Because
$$\tan x = \frac{\sin x}{\cos x}$$

Use the Quotient Rule,
$$f'(x) = \left(\frac{\sin x}{\cos x}\right)' = \frac{\cos x \cos x - \sin x(-\sin x)}{(\cos x)^2}$$
$$= \frac{\cos^2 x + \sin^2 x}{(\cos x)^2} = \frac{1}{(\cos x)^2} = \sec^2 x$$

Note: 类似地，可以求得 $(\cot x)' = -\csc^2 x$，请同学们自行完成.

Example 5.22 Find the derivative of $f(x) = \sec x$.

Solution: Because
$$\sec x = \frac{1}{\cos x}$$

Use the Quotient Rule,
$$f'(x) = \left(\frac{1}{\cos x}\right)' = \frac{0 \times \cos x - 1 \times (-\sin x)}{(\cos x)^2}$$
$$= \frac{\sin x}{(\cos x)^2} = \frac{1}{\cos x} \times \frac{\sin x}{\cos x} = \sec x \tan x$$

Note: 类似地，可以求得 $(\csc x)' = -\csc x \cot x$，请同学们自行完成.

5.4 链式法则和反函数求导 The Chain Rule & Derivative of an Inverse Function

复合函数（Composite Functions）由两个或两个以上的基本初等函数经过复合运算得

到（见本书 2.4.3 节），其求导运算遵循链式法则（The Chain Rule）. 反函数求导法则让我们不必先求出反函数（Inverse Function）就可直接得到其导数.

5.4.1 链式法则 The Chain Rule

If y as the composite function $f[g(x)]$, where $y=f(u)$ and $u=g(x)$ are differentiable functions. Then

$$\frac{dy}{dx}=\frac{dy}{du}\times\frac{du}{dx} \text{ or } y'=f'(u)g'(x)$$

使用链式法则，先把复合函数（Composite Functions）进行换元，拆分为两个或两个以上的基本初等函数，再对各个基本初等函数求导，然后依次相乘，得到结果.

Example 5.23 Find $\dfrac{dy}{dx}$ if $y=u^3$ and $u(x)=2x$

Solution：

$$\frac{dy}{dx}=\frac{dy}{du}\times\frac{du}{dx}=3u^2\times 2=3(2x)^2\times 2=24x^2$$

Example 5.24 Find $\dfrac{dy}{dx}$ if $y=\sqrt{u}$ and $u(x)=\dfrac{1}{x+1}$

Solution：

$$\frac{dy}{dx}=\frac{dy}{du}\times\frac{du}{dx}=\frac{1}{2\sqrt{u}}\times\frac{-1}{(x+1)^2}=\frac{1}{2\sqrt{\dfrac{1}{x+1}}}\times\frac{-1}{(x+1)^2}=-\frac{1}{2(x+1)\sqrt{x+1}}$$

Example 5.25 If $f(x)=(3x^2+x+1)^5$, find $f'(x)$.

Solution：Let $f(x)=u^5$, $u(x)=3x^2+x+1$

$$\begin{aligned}f'(x)&=(u^5)'(3x^2+x+1)'\\&=5u^4(6x+1)\\&=5(3x^2+x+1)^4(6x+1)\end{aligned}$$

Example 5.26 If $y=\sqrt{(x^2+4x)(x^3-9x^2)}$, then $\dfrac{dy}{dx}=?$

Solution：

Using the Chain Rule and the Product Rule，

Let $y=\sqrt{u}$, $u=(x^2+4x)(x^3-9x^2)$

$$\begin{aligned}\frac{dy}{dx}&=(\sqrt{u})'u'=\frac{1}{2\sqrt{u}}[(x^2+4x)'(x^3-9x^2)+(x^2+4x)(x^3-9x^2)']\\&=\frac{1}{2\sqrt{u}}[(2x+4)(x^3-9x^2)+(x^2+4x)(3x^2-18x)]\\&=\frac{5x^4-20x^3-108x^2}{2\sqrt{(x^2+4x)(x^3-9x^2)}}\end{aligned}$$

Example 5.27 Find the derivative of $f(x)=\left(\dfrac{5x+3}{x^3+3x}\right)^7$.

Solution：

By the Quotient Rule and the Chain Rule, Let $f(x)=u^7$, $u(x)=\dfrac{5x+3}{x^3+3x}$

$$f'(x)=(u^7)'\left(\dfrac{5x+3}{x^3+3x}\right)'=7u^6\left[\dfrac{5(x^3+3x)-(5x+3)(3x^2+3)}{(x^3+3x)^2}\right]$$

$$=7\left(\dfrac{5x+3}{x^3+3x}\right)^6\left[\dfrac{5(x^3+3x)-(5x+3)(3x^2+3)}{(x^3+3x)^2}\right]$$

$$=-7\times\dfrac{(5x+3)^6(10x^3+9x+9)}{(x^3+3x)^8}$$

Example 5.28 Find the derivative of $y=\cos(x^3+x)$

Solution:

Let $y=\cos u$, $u=x^3+x$

$$y'=(\cos u)'u'=-\sin u(3x^2+1)=-(3x^2+1)\sin(x^3+x)$$

Example 5.29 Find the derivative of $y=\tan(x^3+\cos 2x)$

Solution:

Let $y=\tan u$, $u=x^3+\cos v$, $v=2x$

$$y'=(\tan u)'u'=\sec^2 u(3x^2-\sin v\times v')$$

$$=\sec^2 u(3x^2-2\sin v)$$

$$=\sec^2(x^3+\cos 2x)(3x^2-2\sin 2x)$$

Example 5.30 Find the derivative of $y=2xe^{\tan x}$.

Solution:

Let $y=e^u$, $u=\tan x$,

$$y'=(2x)'e^{\tan x}+2x(e^{\tan x})'$$

$$=(2x)'e^{\tan x}+2x(e^u)'(\tan x)'$$

$$=2e^{\tan x}+2x\,e^u\sec^2 x$$

$$=2e^{\tan x}(1+x\sec^2 x)$$

Example 5.31 Find the derivative of $y=\pi^{\tan x}$

Solution:

Let $y=\pi^u$, $u=\tan x$,

$$y'=(\pi^u)'(\tan x)'=\pi^u\ln\pi\sec^2 x=\pi^{\tan x}\ln\pi\sec^2 x$$

Example 5.32 Find the derivative of $y=\arccos x+x\sqrt{1-x^2}$

Solution:

$$\dfrac{dy}{dx}=-\dfrac{1}{\sqrt{1-x^2}}+\dfrac{x(-2x)}{2\sqrt{1-x^2}}+\sqrt{1-x^2}=\dfrac{-1-x^2+1-x^2}{\sqrt{1-x^2}}=\dfrac{-2x^2}{\sqrt{1-x^2}}$$

【方法总结】应用链式法则（Chain Rule）的诀窍，就是对"外面"的函数 $f(u)$ 求导并单独在"里面"的函数 $g(x)$ 处取值，然后乘上"里面"函数的导数.

5.4.2 反函数求导 The Derivative of an Inverse Function

Suppose we have a function $x=f(y)$ that is defined and differentiable at $y=a$ where $x=c$. Suppose we also know that the $f^{-1}(x)$ exists at $x=c$. Thus, $f(a)=c$ and

$f^{-1}(c) = a$. Then, because $\dfrac{dy}{dx} = \dfrac{1}{\dfrac{dx}{dy}}$,

$$\frac{d}{dx}f^{-1}(x)\Big|_{x=c} = \frac{1}{\left[\dfrac{d}{dy}f(y)\right]_{y=a}}$$

【方法总结】计算原函数 $y = f(x)$ 的反函数（Inverse Function） $x = f^{-1}(y)$ 的导数一般有两种计算方法：

1) 由 $y = f(x)$ 写出它的反函数 $x = f^{-1}(y)$ 的表达式，然后计算 $f^{-1}(y)$ 的导数；

2) 利用公式 $\dfrac{d}{dx}f^{-1}(x)\Big|_{x=c} = \dfrac{1}{\left[\dfrac{d}{dy}f(y)\right]_{y=a}}$ 计算，只要算出 $\left[\dfrac{d}{dy}f(y)\right]_{y=a}$，就可以拿到 $\dfrac{d}{dx}f^{-1}(x)\Big|_{x=c}$.

Example 5.33 Find a derivative of the inverse of $y = x^2 + 4$ when $y = 29$.

Solution：

The inverse of the function $y = x^2 + 4$ is the function
$$f^{-1}(x) = \sqrt{x-4}, \quad x \geq 4$$

The derivative of $f^{-1}(x)$ is
$$\frac{d}{dx}f^{-1}(x)\Big|_{x=29} = \frac{1}{2\sqrt{x-4}}\Big|_{x=29} = \frac{1}{10}$$

Note：运用计算方法 1) 求解：注意：原函数 $y = 29$，即其反函数 $x = 29$.

Example 5.34 Find a derivative of the inverse of $y = x^3 + 4$ when $y = 129$.

Solution：

we take the derivative of $f(x)$：
$$f'(x) = 3x^2$$

Find the value of x when $y = 129$：
$$129 = x^3 + 4, \quad x = 5$$

Using the derivative of an inverse function formula：
$$\frac{d}{dx}f^{-1}(y)\Big|_{y=129} = \frac{1}{(3x^2)\big|_{x=5}} = \frac{1}{75}$$

Note：运用计算方法 2) 求解．注意：由 $y = 129$，求出原函数中 $x = 5$.

Example 5.35 Find the derivative of the function $f(x) = \arcsin x$.

Solution：

The inverse of the function $y = \arcsin x$, $x \in [-1, 1]$ is
$$x = \sin y, \quad y \in \left[-\frac{\pi}{2}, \frac{\pi}{2}\right]$$

the derivative of $x = \sin y$ is
$$x'(y) = \frac{dx}{dy} = \cos y$$

According to the rule for the derivative of inverse functions, we know

$$\frac{\mathrm{d}y}{\mathrm{d}x} = \frac{1}{\frac{\mathrm{d}x}{\mathrm{d}y}} = \frac{1}{\cos y}$$

Since $\sin^2 x + \cos^2 x = 1$, $\cos y = \sqrt{1-\sin^2 y} = \sqrt{1-x^2}$

Then,
$$\frac{\mathrm{d}y}{\mathrm{d}x} = \frac{1}{\cos y} = \frac{1}{\sqrt{1-x^2}}, \ |x|<1$$

Note：类似地，可以求得 $f(x) = \arccos x$ 的导数为 $f'(x) = -\dfrac{1}{\sqrt{1-x^2}}$，$|x|<1$.

Example 5.36 Find the derivative of the function $f(x) = \arctan x$.

Solution：

The inverse of the function $y = \arctan x$ is
$$x = \tan y$$

the derivative of $x = \tan y$ is：
$$x'(y) = \frac{\mathrm{d}x}{\mathrm{d}y} = \sec^2 y$$

Since $1 + \tan^2 x = \sec^2 x$, $\sec^2 y = 1 + \tan^2 y = 1 + x^2$

According to the rule for the derivative of inverse functions, we know
$$\frac{\mathrm{d}y}{\mathrm{d}x} = \frac{1}{\frac{\mathrm{d}x}{\mathrm{d}y}} = \frac{1}{\sec^2 y} = \frac{1}{1+x^2}$$

Therefore,
$$(\arctan x)' = \frac{1}{1+x^2}$$

Note：类似地，可以求得 $(\text{arccot}\, x)' = -\dfrac{1}{1+x^2}$.

Example 5.37 Find the derivative of the function $f(x) = \text{arcsec}\, x$.

Solution：

we know $y = \text{arcsec}\, x$, $x \in (-\infty, -1] \cup [1, +\infty)$ and $x = \sec y$, $y \in \left[0, \dfrac{\pi}{2}\right) \cup \left(\dfrac{\pi}{2}, \pi\right]$

$$x'(y) = \frac{\mathrm{d}x}{\mathrm{d}y} = \sec y \tan y$$

Since $1 + \tan^2 x = \sec^2 x$, $\tan y = \sqrt{\sec^2 y - 1} = \sqrt{x^2 - 1}$

According to the rule for the derivative of inverse functions, we know
$$\frac{\mathrm{d}y}{\mathrm{d}x} = \frac{1}{\frac{\mathrm{d}x}{\mathrm{d}y}} = \frac{1}{\sec y \tan y} = \frac{1}{|x|\sqrt{x^2-1}}$$

Therefore,
$$(\text{arcsec}\, x)' = \frac{1}{|x|\sqrt{x^2-1}}, \ |x|>1$$

Note：类似地，可以求得 $(\operatorname{arccsc} x)' = -\dfrac{1}{|x|\sqrt{x^2-1}}$，$|x|>1$.

5.5 隐函数求导和二阶导数 Implicit Differentiation & Second Derivatives

隐函数求导（Implicit Differentiation）和二阶导数（Second Derivatives）是 AP 微积分的常考题型．微积分 BC 的同学还应该掌握对数求导法．

5.5.1 隐函数 Implicit Functions

Sometimes functions are given not in the form $y=f(x)$ but in a more complicated form in which it is difficult or impossible to express y explicitly in terms of x. Such functions are called implicit functions.

如果函数 $y=y(x)$ 是由方程 $F(x,y)=0$ 确定，函数 $y=y(x)$ 是隐函数（Implicit Functions）．例如，方程 $x^2+xy+y^2=25$ 所确定的 y 和 x 之间的函数关系叫隐函数．

隐函数不一定能表示成为显函数的形式．当 $F(x,y)=0$ 可以解出 $y=y(x)$ 时，$y=y(x)$ 的导数就容易计算；多数情况下，从 $F(x,y)=0$ 不能解出 $y=y(x)$，借助隐函数求导（Implicit Differentiation）可以得到相应的导数．

5.5.2 隐函数求导 Implicit Differentiation

When you cannot isolate y in terms of x, It is time to take the derivative implicitly. It calls implicit differentiation.

【方法总结】隐函数求导（Implicit Differentiation）的方法：将方程两端同时对 x 求导，把 y 看成复合函数，利用链式法则（Chain Rule）求导，最后解出 $\dfrac{\mathrm{d}y}{\mathrm{d}x}$．

Example 5.38 Find $\dfrac{\mathrm{d}y}{\mathrm{d}x}$, if $y^5+3y^3=x^5+5x^2$.

Solution：

Using implicit differentiation, we get：

$$\frac{\mathrm{d}}{\mathrm{d}x}(y^5+3y^3)=\frac{\mathrm{d}}{\mathrm{d}x}(x^5+5x^2)$$

$$5y^4\frac{\mathrm{d}y}{\mathrm{d}x}+9y^2\frac{\mathrm{d}y}{\mathrm{d}x}=5x^4\frac{\mathrm{d}x}{\mathrm{d}x}+10x\frac{\mathrm{d}x}{\mathrm{d}x}$$

Remember that $\dfrac{\mathrm{d}x}{\mathrm{d}x}=1$ and factor out $\dfrac{\mathrm{d}y}{\mathrm{d}x}$：

$$\frac{\mathrm{d}y}{\mathrm{d}x}(5y^4+9y^2)=5x^4+10x$$

Divide both sides by $5y^4+9y^2$：

$$\frac{\mathrm{d}y}{\mathrm{d}x}=\frac{5x^4+10x}{5y^4+9y^2}$$

Note：这里，对于 x 的多项式求导，仍旧可以看成关于 x 的复合函数并符合链式法则（Chain Rule），计算中带有 $\dfrac{dx}{dx}$，同时，$\dfrac{dx}{dx}=1$，结论是一致的．

Example 5.39 Find $\dfrac{dy}{dx}$, if $x^2+xy+y^2=25$.

Solution：

Using implicit differentiation, we get：
$$2x+\left(y+x\dfrac{dy}{dx}\right)+2y\dfrac{dy}{dx}=0$$

Put all the terms containing $\dfrac{dy}{dx}$ on the left and all the other terms on the right：
$$x\dfrac{dy}{dx}+2y\dfrac{dy}{dx}=-2x-y$$

Factor out $\dfrac{dy}{dx}$：
$$\dfrac{dy}{dx}(x+2y)=-2x-y$$

Then, isolate $\dfrac{dy}{dx}$：
$$\dfrac{dy}{dx}=-\dfrac{2x+y}{x+2y}$$

Note：运用隐函数求导，必须注意到 y 是 x 的函数，得到关于 $\dfrac{dy}{dx}$ 的方程，解这个方程得到 $\dfrac{dy}{dx}$．

Example 5.40 Find $\dfrac{dy}{dx}$, if $x+\cos(x+y)=0$.

Solution：

Using implicit differentiation, we get：
$$1-\sin(x+y)\cdot\left(1+\dfrac{dy}{dx}\right)=0$$

Then,
$$\sin(x+y)\cdot\left(1+\dfrac{dy}{dx}\right)=1$$

And isolate $\dfrac{dy}{dx}$,
$$\dfrac{dy}{dx}=\dfrac{1}{\sin(x+y)}-1=\csc(x+y)-1$$

Example 5.41 Find the derivative of $2x^2+2\sqrt{y}+y=16$ at $(2,4)$.

Solution：

Using implicit differentiation, we get：
$$4x+2\times\dfrac{1}{2\sqrt{y}}\times\dfrac{dy}{dx}+\dfrac{dy}{dx}=0$$

Plug (2,4) into the above equation:

$$4\times 2+2\times \frac{1}{2\sqrt{4}}\times \frac{dy}{dx}+\frac{dy}{dx}=0$$

Simplify:

$$\frac{dy}{dx}=-\frac{16}{3}$$

Note：这里，以（2,4）代入导数方程 $4x+2\times \frac{1}{2\sqrt{y}}\times \frac{dy}{dx}+\frac{dy}{dx}=0$ 替代分离 $\frac{dy}{dx}$，解出在（2,4）处的导数．

Example 5.42 Find the derivative of $y=\frac{3x-9x^2}{4x^3-x^2}$ at $(1,-2)$.

Solution：cross-multiply,

$$(4x^3-x^2)y=3x-9x^2$$

Take the derivative,

$$(12x^2-2x)y+(4x^3-x^2)\frac{dy}{dx}=3-18x$$

Plug $(1,-2)$ into the above equation：

$$(12\times 1-2\times 1)(-2)+(4\times 1-1)\frac{dy}{dx}=3-18\times 1$$

Simplify：

$$\frac{dy}{dx}=\frac{5}{3}$$

Note：此处不需要运用导数除法法则（Quotient Rule），避开相对繁杂的运算，运用隐函数求导（Implicit Differentiation）直接解得答案．这是一个小技巧．

5.5.3 二阶导数 Second Derivatives

If $f(x)$ is a differentiable function, then its derivative $f'(x)$ is also a function, so $f'(x)$ may have a derivative of its own, denoted by $[f'(x)]'=f''(x)$. This new function $f''(x)$ is called the second derivative of $f(x)$. We write the second derivative of $y=f(x)$ as $\frac{d}{dx}\left(\frac{dy}{dx}\right)=\frac{d^2y}{dx^2}$.

二阶导数在 AP 微积分的考查范围．二阶及二阶以上的导数统称为高阶导数（High Order Derivatives）．求高阶导数就是多次地、接连地求导数，前面学过的求导方法可以来计算高阶导数．

Example 5.43 If $f(x)=x^3+x$, find and interpret $f''(x)$.

Solution：

The first derivative is

$$f'(x)=3x^2+1$$

So, the second derivative is

$$f''(x)=6x$$

We can interpret $f''(x)$ as the slope of the curve $f'(x) = 3x^2 + 1$ at the point $(x, f'(x))$. In other words, it is the rate of change of the slope of original curve $f(x) = x^3 + x$.

Example 5.44 If $y = x^\alpha$, $\alpha > 0$, α is real number, find n-th derivative $f^{(n)}(x)$.

Solution:

The first derivative:
$$f'(x) = \alpha x^{\alpha-1}$$

The second derivative:
$$f''(x) = \alpha(\alpha-1)x^{\alpha-2}$$

The third derivative:
$$f^{(3)}(x) = \alpha(\alpha-1)(\alpha-2)x^{\alpha-3}$$
... ...

Similarly:
$$f^{(n)}(x) = \alpha(\alpha-1)(\alpha-2)\cdots(\alpha-n+1)x^{\alpha-n}$$

Specially, if $n = \alpha$, then
$$f^{(n)}(x) = (x^\alpha)^n = (x^n)^n = n(n-1)(n-2)\cdots[n-(n-1)]\times 1 = n!$$

Note：求函数的 n 阶导数是泰勒级数（Taylor Series）的基础技能，微积分 BC 的同学必须掌握．另一个方面，求二阶导数或者高阶导数（High Order Derivatives）往往运用隐函数求导的方法．

Example 5.45 Find $\dfrac{d^2 y}{dx^2}$ if $y^2 = x^2 + 2x$.

Solution:

Using implicit differentiation, we get:
$$2y \frac{dy}{dx} = 2x + 2$$

Simplify and solve for $\dfrac{dy}{dx}$:
$$\frac{dy}{dx} = \frac{2x+2}{2y} = \frac{x+1}{y}$$

Take derivative again:
$$\frac{d^2 y}{dx^2} = \frac{1 \times y - (x+1)\dfrac{dy}{dx}}{y^2}$$

Substitute for $\dfrac{dy}{dx}$ and simplify:
$$\frac{d^2 y}{dx^2} = \frac{1 \times y - (x+1)\left(\dfrac{x+1}{y}\right)}{y^2} = \frac{y^2 - (x+1)^2}{y^3}$$

Example 5.46 Find $\dfrac{d^2 y}{dx^2}$ if $y^3 + 2y = 4x^2 + 3x$.

Solution:

Take the derivative with respect to x:

$$3y^2 \frac{dy}{dx} + 2\frac{dy}{dx} = 8x + 3$$

Then solve for $\frac{dy}{dx}$:

$$\frac{dy}{dx} = \frac{8x+3}{3y^2+2}$$

The second derivative with respect to x becomes:

$$\frac{d^2y}{dx^2} = \frac{8(3y^2+2) - (8x+3)\left(6y \frac{dy}{dx}\right)}{(3y^2+2)^2}$$

Substitute for $\frac{dy}{dx}$:

$$\frac{d^2y}{dx^2} = \frac{8(3y^2+2) - (8x+3)\left(6y \frac{8x+3}{3y^2+2}\right)}{(3y^2+2)^2} = \frac{8(3y^2+2)^2 - 6y(8x+3)^2}{(3y^2+2)^3}$$

【方法总结】运用隐函数求导的方法求二阶导数的一般步骤:

1) 由方程 $F(x,y) = 0$ 的两边关于 x 求导, 得到 $\frac{dy}{dx}$ 的表达式①: $\frac{dy}{dx} = f(x,y)$;

2) 对表达式①两边关于 x 求导, 得到表达式②: $\frac{d^2y}{dx^2} = g\left(x, y, \frac{dy}{dx}\right)$;

3) 将表达式①代入式②, 得到隐函数 $y = y(x)$ 的二阶导数.

5.5.4# 对数求导法 Logarithmic Differentiation

1) 对数的运算公式 (Laws of Logarithms), 我们前面讲过 (见本书 2.3.2 节). 根据对数的运算公式, 我们可以得到:

$$\ln(AB) = \ln A + \ln B$$
$$\ln\left(\frac{A}{B}\right) = \ln A - \ln B$$
$$\ln A^B = B \ln A$$

利用上述公式, 可以简化某些函数导数的求解, 也为某些较 "难" 的求导提供了新思路.

Example 5.47 Find the derivative of $f(x) = \frac{x-1}{x+1}$.

Solution:
Take the log of the sides:

$$\ln y = \ln \frac{x-1}{x+1}$$

Using the formula $\ln\left(\frac{A}{B}\right) = \ln A - \ln B$, we get

$$\ln y = \ln(x-1) - \ln(x+1)$$

Take the derivative of both sides:

$$\frac{1}{y} \times \frac{dy}{dx} = \frac{1}{x-1} - \frac{1}{x+1}$$

Multiply both sides by y：
$$\frac{dy}{dx} = y\left(\frac{1}{x-1} - \frac{1}{x+1}\right) = \left(\frac{x-1}{x+1}\right)\left(\frac{1}{x-1} - \frac{1}{x+1}\right)$$

Note：利用对数的运算公式替代函数的除法法则（Quotient Rule），为函数求导提供了另一种思路．

Example 5.48 Find the derivative of $f(x) = (x-1)^2(x+1)^3$.

Solution：

Take the log of the sides：
$$\ln y = \ln[(x-1)^2(x+1)^3]$$

Using the formula $\ln(AB) = \ln A + \ln B$，we get
$$\ln y = 2\ln(x-1) + 3\ln(x+1)$$

Take the derivative of both sides：
$$\frac{1}{y} \times \frac{dy}{dx} = \frac{2}{x-1} + \frac{3}{x+1}$$

Multiply both sides by y：
$$\frac{dy}{dx} = y\left(\frac{2}{x-1} + \frac{3}{x+1}\right) = (x-1)^2(x+1)^3\left(\frac{2}{x-1} + \frac{3}{x+1}\right)$$

Note：利用对数的运算公式替代函数的乘法法则（Product Rule），为函数求导提供了另一种思路．

Example 5.49 Find the derivative of $y = \ln\dfrac{x-1}{x+1}$.

Solution：

Using the formula $\ln\left(\dfrac{A}{B}\right) = \ln A - \ln B$，we get
$$y = \ln(x-1) - \ln(x+1)$$

Take the derivative of both sides：
$$\frac{dy}{dx} = \frac{1}{x-1} - \frac{1}{x+1}$$

2）形如 $y = [f(x)]^{g(x)}$，$f(x) > 0$ 的函数，称为幂指函数（Power Exponential Function）．计算幂指函数导数的一般采用对数求导法．对函数 $y = [f(x)]^{g(x)}$ 两边取对数，得：
$$\ln y = g(x)\ln f(x)$$

由链式法则（Chain Rule）、四则运算法则和隐函数求导，得：
$$\frac{1}{y} \times y' = g'(x)\ln f(x) + \frac{g(x)f'(x)}{f(x)}$$

那么
$$y' = [f(x)]^{g(x)}\left[g'(x)\ln f(x) + \frac{g(x)f'(x)}{f(x)}\right]$$

Example 5.50 Find the derivative of $y = x^x$.

Solution：

Take the log of the sides：

$$\ln y = \ln x^x$$

Using the formula $\ln A^B = B\ln A$, we get

$$\ln y = x\ln x$$

Take the derivative of both sides:

$$\frac{1}{y} \times \frac{dy}{dx} = 1 \times \ln x + x\,\frac{1}{x} = \ln x + 1$$

Multiply both sides by y:

$$\frac{dy}{dx} = y(\ln x + 1) = x^x(\ln x + 1)$$

Example 5.51 Find the derivative of $y = (\tan x)^{\sin x}$.

Solution:

Take the log of the sides:

$$\ln y = \ln(\tan x)^{\sin x}$$

Using the formula $\ln A^B = B\ln A$, we get

$$\ln y = \sin x \ln \tan x$$

Take the derivative of both sides:

$$\frac{1}{y} \times \frac{dy}{dx} = \cos x \ln\tan x + \sin x\,\frac{1}{\tan x}\sec^2 x = \cos x \ln\tan x + \sec x$$

Multiply both sides by y:

$$\frac{dy}{dx} = y(\cos x \ln\tan x + \sec x) = (\tan x)^{\sin x}(\cos x \ln\tan x + \sec x)$$

5.6# 参数方程求导 Derivatives of Parametric Equations

掌握参数方程 (Parametric Equations) 并应用于微积分是微积分 BC 考纲的明确要求，本节给出参数方程求导法则及参数方程二阶求导的方法.

5.6.1# 参数方程求导 Derivatives of Parametric Equations

If $x = f(t)$ and $y = g(t)$ are differentiable functions of t, to find $\dfrac{dy}{dx}$ using parametric equations, use this rule:

$$\frac{dy}{dx} = \frac{dy/dt}{dx/dt}$$

参数方程的定义见本书章节 2.6.1. 函数 $y = y(x)$ 由参数方程 $\begin{cases} x = x(t) \\ y = y(t) \end{cases}$ 表示，其中 $x'(t)$ 和 $y'(t)$ 都是可导函数，且 $x'(t) \neq 0$，那么计算 $y = y(x)$ 的导数 $\dfrac{dy}{dx}$ 可以按上述公式计算.

Example 5.52 Find the derivative of the parametric equations $\begin{cases} x = 3\cos t \\ y = 3\sin t \end{cases}$, $0 \leqslant t \leqslant 2\pi$.

Solution:

Take the derivative of $x=x(t)$ with respect to t:
$$\frac{dx}{dt}=-3\sin t$$
Take the derivative of $y=y(t)$ with respect to t:
$$\frac{dy}{dt}=3\cos t$$
Then, the derivative of $y=y(x)$ with respect to x:
$$\frac{dy}{dx}=\frac{dy/dt}{dx/dt}=\frac{3\cos t}{-3\sin t}$$
To express the answer interms of x (or y) instead of the parameter,
$$\cos t=\frac{x}{3} \text{ and } \sin t=\frac{y}{3}$$
Then, substitute:
$$\frac{dy}{dx}=\frac{3\cos t}{-3\sin t}=\frac{3\times\frac{x}{3}}{-3\times\frac{y}{3}}=-\frac{x}{y}$$

Note：结果需要使用 x 或 y 的表达式替换 $\frac{dy}{dx}$ 中的参数 t.

Example 5.53 Find an equation of the line tangent to the curve $\begin{cases} x=4\cos t \\ y=\sin t \end{cases}$ at $t=\frac{\pi}{3}$.

Solution：
$$\frac{dy}{dt}=\cos t \text{ and } \frac{dx}{dt}=-4\sin t$$
Find the slope of the tangent line $\frac{dy}{dx}$ using the formula：
$$\frac{dy}{dx}=\frac{dy/dt}{dx/dt}=\frac{\cos t}{-4\sin t}$$
Therefore,
$$\frac{dy}{dx}=\frac{\cos\left(\frac{\pi}{3}\right)}{-4\sin\left(\frac{\pi}{3}\right)}=\frac{\frac{1}{2}}{-4\times\frac{\sqrt{3}}{2}}=-\frac{\sqrt{3}}{12}$$

At $t=\frac{\pi}{3}$, $x=2$ and $y=\frac{\sqrt{3}}{2}$. The equation of the tangent line：
$$y-\frac{\sqrt{3}}{2}=-\frac{\sqrt{3}}{12}(x-2)$$
Hence,
$$\sqrt{3}x+12y-8\sqrt{3}=0$$

5.6.2# 参数方程的二阶求导 The Second Derivatives of Parametric Equations

If $x=f(t)$ and $y=g(t)$ are differentiable functions of t, then

$$\frac{d^2y}{dx^2}=\frac{d}{dx}\left(\frac{dy}{dx}\right)=\frac{\frac{d}{dt}\left(\frac{dy}{dx}\right)}{\frac{dx}{dt}}$$

参数方程的二阶导数可以看作参数方程求导和二阶导数的结合，但本质上仍然属于二阶导数，二阶导数的相关方法同样适用（见本书 5.5.3 节）。

Example 5.54 If $\begin{cases} x=4\sin t \\ y=\cos 2t \end{cases}$, find $\frac{d^2y}{dx^2}$.

Solution:

Take the derivative with respect to x:

$$\frac{dy}{dx}=\frac{dy/dt}{dx/dt}=\frac{-2\sin 2t}{4\cos t}=\frac{-2\times 2\sin t\cos t}{4\cos t}=-\sin t$$

using the formula:

$$\frac{d^2y}{dx^2}=\frac{\frac{d}{dt}\left(\frac{dy}{dx}\right)}{\frac{dx}{dt}}=\frac{-\cos t}{4\cos t}=-\frac{1}{4}$$

Note: 理解参数方程的二阶求导公式，$\frac{d^2y}{dx^2}\neq\frac{d^2y/dt^2}{d^2x/dt^2}$，切勿犯"想当然"的错误。

Example 5.55 If a point moves on the curve $\begin{cases} x=t^2-1 \\ y=t^4-2t^3 \end{cases}$, find $\frac{d^2y}{dx^2}$ at $t=1$.

Solution:

Take the derivative with respect to x:

$$\frac{dy}{dx}=\frac{dy/dt}{dx/dt}=\frac{4t^3-6t^2}{2t}=2t^2-3t$$

using the formula:

$$\frac{d^2y}{dx^2}=\frac{\frac{d}{dt}\left(\frac{dy}{dx}\right)}{\frac{dx}{dt}}=\frac{4t-3}{2t}$$

Plug $t=1$ into the above equation:

$$\left.\frac{d^2y}{dx^2}\right|_{t=1}=\frac{4\times 1-3}{2\times 1}=\frac{1}{2}$$

5.7# 向量函数和极坐标函数求导 Derivatives of Vector Functions and Polar Functions

向量函数（Vector Functions）求导是微分在运动学的应用基础，同时，向量函数求导和极坐标函数（Polar Functions）求导也是微积分 BC 考试的易失分项，同学们注意夯实基础．

5.7.1[#] 向量函数求导 Derivatives of Vector Functions

The vector function $\vec{r(t)} = f(t)\vec{i} + g(t)\vec{j}$ has a derivative at t if $f(t)$ and $g(t)$ have derivatives at t. The derivative is the vector function

$$\vec{r'(t)} = \frac{d\vec{r}}{dt} = \frac{df}{dt}\vec{i} + \frac{dg}{dt}\vec{j} = f'(t)\vec{i} + g'(t)\vec{j}$$

因为向量函数求导是按分量逐个计算的，对可导向量函数的求导法则和对标量函数的求导法则有同样的形式，并且，向量函数求导与参数方程求导类似．向量函数（Vector Functions）请参考本书 2.6.2 节．

Example 5.56 If the vector function $\vec{r(t)} = (t\cos t)\vec{i} + (t\sin t)\vec{j}$, find the derivative of $\vec{r(t)}$.

Solution：Using the formula：
$$\vec{r'(t)} = f'(t)\vec{i} + g'(t)\vec{j} = (t\cos t)'\vec{i} + (t\sin t)'\vec{j}$$
$$= (\cos t - t\sin t)\vec{i} + (\sin t + t\cos t)\vec{j}$$

Example 5.57 If $\vec{R} = \left\langle 3\cos\frac{\pi}{3}t, 2\sin\frac{\pi}{3}t \right\rangle$ is the position vector $\langle x, y \rangle$ from the origin to moving point $P(x, y)$ at time t. When $t = 3$, find velocity vector.

Solution：

The velocity vector is the derivative of the position vector, then, using the formula：

$$\vec{R'(t)} = \frac{d\vec{R}}{dt} = \left\langle 3\cos\frac{\pi}{3}t, 2\sin\frac{\pi}{3}t \right\rangle' = \left\langle -\pi\sin\frac{\pi}{3}t, \frac{2\pi}{3}\cos\frac{\pi}{3}t \right\rangle$$

When $t = 3$, the velocity vector is

$$\vec{v(3)} = \vec{R'(3)} = \left\langle -\pi\sin\frac{\pi}{3}\times 3, \frac{2\pi}{3}\times\cos\frac{\pi}{3}\times 3 \right\rangle = \left\langle 0, -\frac{2\pi}{3} \right\rangle$$

5.7.2[#] 极坐标函数求导 Derivatives of Polar Functions

If a polar $r = f(\theta)$, we regard θ as a parameter and write its parametric equations as
$$x = r\cos\theta = f(\theta)\cos\theta \text{ and } y = r\sin\theta = f(\theta)\sin\theta$$

Then, using the method for finding Derivatives of Parametric Equation and the Product Rule, we have

$$\frac{dy}{dx} = \frac{dy/d\theta}{dx/d\theta} = \frac{\frac{dr}{d\theta}\sin\theta + r\cos\theta}{\frac{dr}{d\theta}\cos\theta - r\sin\theta}$$

【方法总结】极坐标求导的基本思路：先把极坐标系的函数表达成直角坐标系下的参数方程形式，利用参数方程求导公式拿到极坐标求导公式．我们认为向量函数求导和极坐标函数求导都是基于参数方程求导公式．

Example 5.58 If the polar functions $r = e^\theta$, find the derivative of $\frac{dy}{dx}$.

Solution:

Take the derivative $r(\theta)$ with respect to θ:

$$\frac{\mathrm{d}r}{\mathrm{d}\theta} = (e^\theta)' = e^\theta$$

Using the formula:

$$\frac{\mathrm{d}y}{\mathrm{d}x} = \frac{\frac{\mathrm{d}r}{\mathrm{d}\theta}\sin\theta + r\cos\theta}{\frac{\mathrm{d}r}{\mathrm{d}\theta}\cos\theta - r\sin\theta} = \frac{e^\theta\sin\theta + e^\theta\cos\theta}{e^\theta\cos\theta - e^\theta\sin\theta} = \frac{\sin\theta + \cos\theta}{\cos\theta - \sin\theta}$$

Example 5.59 Find the tangent line of cardioid $r = 1 + \sin\theta$ at $\theta = \frac{\pi}{3}$.

Solution:

Take the derivative $r(\theta)$ with respect to θ:

$$\frac{\mathrm{d}r}{\mathrm{d}\theta} = (1 + \sin\theta)' = \cos\theta$$

Using the formula:

$$\frac{\mathrm{d}y}{\mathrm{d}x} = \frac{\frac{\mathrm{d}r}{\mathrm{d}\theta}\sin\theta + r\cos\theta}{\frac{\mathrm{d}r}{\mathrm{d}\theta}\cos\theta - r\sin\theta} = \frac{\cos\theta\sin\theta + (1+\sin\theta)\cos\theta}{\cos\theta\cos\theta - (1+\sin\theta)\sin\theta}$$

$$= \frac{\cos\theta(1 + 2\sin\theta)}{1 - 2\sin^2\theta - \sin\theta}$$

$$= \frac{\cos\theta(1 + 2\sin\theta)}{(1 + \sin\theta)(1 - 2\sin\theta)}$$

The slope of the tangent at the point where $\theta = \frac{\pi}{3}$

$$\left.\frac{\mathrm{d}y}{\mathrm{d}x}\right|_{\theta=\frac{\pi}{3}} = \frac{\cos\theta(1 + 2\sin\theta)}{(1 + \sin\theta)(1 - 2\sin\theta)} = \frac{\cos\frac{\pi}{3}\left(1 + 2\sin\frac{\pi}{3}\right)}{\left(1 + \sin\frac{\pi}{3}\right)\left(1 - 2\sin\frac{\pi}{3}\right)} = -1$$

Express the cardioid in parametric equation. We have

$$\begin{cases} x = r\cos\theta = (1+\sin\theta)\cos\theta \\ y = r\sin\theta = (1+\sin\theta)\sin\theta \end{cases}$$

Plug $\theta = \frac{\pi}{3}$ into the parametric equation

$$\begin{cases} x = \left(1 + \sin\frac{\pi}{3}\right)\cos\frac{\pi}{3} = \frac{2+\sqrt{3}}{4} \\ y = \left(1 + \sin\frac{\pi}{3}\right)\sin\frac{\pi}{3} = \frac{3+2\sqrt{3}}{4} \end{cases}$$

The equation of the tangent line is:

$$y - \frac{3 + 2\sqrt{3}}{4} = -1 \times \left(x - \frac{2 + \sqrt{3}}{4}\right)$$

Hence,
$$4x+4y-(5+3\sqrt{3})=0$$

5.8 微分 Differential

前面我们用 $\dfrac{dy}{dx}$ 表示导数（Derivative），引进微分（Differential）概念之后，$\dfrac{dy}{dx}$ 可看作函数微分和自变量微分的商，导数即微商．同时，函数的微分就是函数的导数与自变量的微分之乘积．

5.8.1 微分的定义 Definition of the Differential

Let $y=f(x)$ be a differentiable function. The differential dx is an independent variable. The differential dy is
$$dy=f'(x)dx$$
函数的导数表示函数在点 x 处的变化率，它描述了函数在点 x 处变化的快慢程度；并且当函数在点 x 处有一个微小的改变量时，微分则描述了函数取得相应改变量的大小．

Example 5.60 Finding the differential dy if $y=\ln x$.

Solution: Find $f'(x)$:
$$f'(x)=\frac{1}{x}$$
Then,
$$dy=f'(x)dx=\frac{1}{x}dx$$

Example 5.61 If $f(x)=x^7+19x$, finding the value of dy when $x=1$ and $dx=0.2$.

Solution: Find $f'(x)$:
$$f'(x)=7x^6+19$$
Then,
$$dy=f'(x)dx=(7x^6+19)dx$$
Substituting $x=1$ and $dx=0.2$ in the expression for dy, we obtain
$$dy=(7\times1^6+19)\times0.2=5.2$$

5.8.2 微分的几何意义 The Geometric Interpretation of Differential

The differential dy represents the amount the tangent line rises or falls when x changes by an amount $dx=\Delta x$.

直角坐标系中，在函数 $y=f(x)$ 上取定一点 $M(x_0,y_0)$，过 M 点作曲线的切线，见 Figure 5.8.2．则此切线的斜率为
$$f'(x_0)=\tan\alpha$$
当自变量在点 x_0 处取得改变量 dx 时，拿到曲线上另外一点 $M'(x_0+\Delta x,y_0+\Delta f)$．那么：
$$MN=\Delta x=dx,\ M'N=\Delta f=f(x_0+dx)-f(x_0)$$

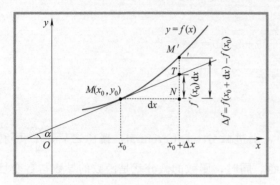

Figure 5.8.2

并且
$$NT = MN \times \tan\alpha = f'(x_0)dx = dy$$

故，函数 $y=f(x)$ 微分 dy 的几何意义：过点 $M(x_0, y_0)$ 切线纵坐标相应的改变量.

Example 5.62 If the radius e of a circle is increased by 1%, use dA to estimate the increase in the circle's area A.

Solution：

Since $A = \pi r^2$, the estimated increase is
$$dA = A'(e)dr = 2\pi(e) \times (0.01e) = 0.02\pi e^2$$

Example 5.63 The radius r of a sphere increases from $a=3$m to 3.1. Use dV to estimate the increase in the sphere's volume V. Estimate the volume of the enlarged sphere and compare your estimate to the true volume.

Solution：

Since $V = \frac{4}{3}\pi r^3$, the estimated increase is
$$dV = V'(a)dr = 4\pi a^2 dr$$
$$= 4\pi \times 3^2 \times 0.1 = 3.6\pi (\text{m}^3)$$

Thus，
$$V(3+0.1) \approx V(3) + 3.6\pi$$
$$= \frac{4}{3}\pi \times 3^3 + 3.6\pi = 39.6\pi(\text{m}^3)$$

The volume of a sphere of radius 3.1m is approximately $39.6\pi \text{m}^3$.

The true volume is
$$V(3.1) = \frac{4}{3}\pi \times 3.1^3 = 39.721\pi(\text{m}^3)$$

The error in our estimate is $1.21\pi \text{ m}^3$.

5.8.3 微分的基本公式 Basic Differential Formulas

微分（Differential）和导数（Derivative）表现形式和几何意义不一样，但两者有很强的联系，Table 5.8.3 给出微分的基本公式. 微分和导数的基本公式类似，求微分运算和求导运算类似.

Example 5.64 Find the differential dy if $y = \ln\sin 3x$.

Solution：

Let $u = \sin 3x$, then $y = \ln u$
$$dy = \frac{1}{u}du = \frac{1}{\sin 3x}d(\sin 3x)$$

Let $v = 3x$, then
$$dy = \frac{1}{\sin 3x}d(\sin 3x) = \frac{1}{\sin 3x}\cos 3x\, d(3x) = 3 \times \frac{\cos 3x}{\sin 3x}dx = 3\cot 3x\, dx$$

Table 5.8.3 微分（Differential）的基本公式

Constants	$d(C)=0$, C is constants	
Power	$d(x^n) = nx^{n-1}dx$	
Exponential	$d(a^x) = a^x \ln a\, dx$	$d(e^x) = e^x dx$
Logarithmic	$d(\log_a x) = \dfrac{dx}{x \ln a}$	$d(\ln x) = \dfrac{dx}{x}$
Trigonometric	$d(\sin x) = \cos x\, dx$	$d(\cos x) = -\sin x\, dx$
	$d(\tan x) = \sec^2 x\, dx$	$d(\cot x) = -\csc^2 x\, dx$
	$d(\sec x) = \sec x \tan x\, dx$	$d(\csc x) = -\csc x \cot x\, dx$
Inverse Trigonometric	$d(\arcsin x) = \dfrac{dx}{\sqrt{1-x^2}}$	$d(\arccos x) = -\dfrac{dx}{\sqrt{1-x^2}}$
	$d(\arctan x) = \dfrac{dx}{1+x^2}$	$d(\text{arccot}\, x) = -\dfrac{dx}{1+x^2}$
	$d(\text{arcsec}\, x) = \dfrac{dx}{|x|\sqrt{x^2-1}}$	$d(\text{arccsc}\, x) = -\dfrac{dx}{|x|\sqrt{x^2-1}}$

Note：求微分的时候切勿把 dx 丢掉了.

5.8.4 微分形式不变性 Differential Form Invariance

假设 $y = f(u)$ 和 $u = \varphi(x)$ 都可导（Differentiable），那么复合函数 $y = f[\varphi(x)]$ 的微分（Differential）为：

$$dy = f'(u)du = f'(u)\varphi'(x)dx$$

无论 u 是自变量还是另一个变量的可微函数，微分形式 $dy = f'(u)du$ 保持不变，这也称为微分形式不变性（Differential Form Invariance）.

Example 5.65 Find the differential dy if $y = \sqrt{x^3 + x + 2}$.

Solution：

$$dy = \frac{1}{2\sqrt{x^3+x+2}} d(\sqrt{x^3+x+2}) = \frac{3x^2+1}{2\sqrt{x^3+x+2}} dx$$

与微分形式不变性类似，微分的加减乘除法则（Sum, Difference, Product and Quotient Rules）与导数的加减乘除法则类似. 假设 $u = u(x)$ 和 $v = v(x)$ 都可导，根据函数微分的表达式有：

$$d(uv) = (uv)'dx = (u'v + uv')dx = u'v\,dx + uv'\,dx = v\,du + u\,dv$$

得到微分的乘法法则（Product Rule）：

$$d(uv) = v\,du + u\,dv$$

类似地，可得到微分的除法法则（Quotient Rule）：

$$d\left(\frac{u}{v}\right) = \frac{v\,du - u\,dv}{v^2}$$

其他微分运算法则都可用类似方法得到.

Example 5.66 Find the differential dy if $y = e^{1-2x}\cos x$.

Solution：

$$dy = \cos x\, d(e^{1-2x}) + e^{1-2x} d(\cos x)$$
$$= \cos x\, e^{1-2x} d(1-2x) + e^{1-2x}(-\sin x)dx$$

$$= \cos x \, e^{1-2x}(-2)dx - \sin x \, e^{1-2x} dx$$
$$= -e^{1-2x}(2\cos x + \sin x)dx$$

5.9 习题 Practice Exercises

(1) Find the derivative of the following functions.

1) $f(x) = x^7 + x^\pi$;

2) $f(x) = \tan x + \cot x$;

3) $f(x) = \log_2 x + 5^x$;

4) $f(x) = \cos x + \pi^x$;

5) $f(x) = x^3 + x^2 + x + 1$.

(2) If $f(x) = \begin{cases} \dfrac{3x^2 - 12}{x+2}, & x \neq -2 \\ x+2, & x = -2 \end{cases}$, Which of the following statements about $f(x)$ are true?

I. $\lim\limits_{x \to -2} f(x)$ exists;

II. $f(x)$ is continuous at $x = -2$;

III. $f(x)$ is differentiable at $x = -2$.

(A) I only; (B) II only; (C) III only; (D) I and II only; (E) I, II and III.

(3) Find the derivative of the following functions.

1) $y = x^3 \sin x$;

2) $y = \dfrac{1-x}{3x+1}$;

3) $y = \dfrac{\sqrt{x}}{\sin x}$;

4) $y = \dfrac{e^x - e^{-x}}{e^x + e^{-x}}$;

5) $y = \sqrt{x} - \dfrac{1}{\sqrt{x}}$.

(4) Find the derivative of the following functions.

1) $y = \cos\left(\dfrac{1}{x}\right)$;

2) $y = \ln\sqrt{x^3 + 1}$;

3) $y = x \ln^3 x$;

4) $y = 3^{x^2}$;

5) $y = \arcsin x - \sqrt{1 - x^2}$.

(5) Find a derivative of the inverse of $y = x^2 + x$ when $y = 12$ $(x > 0)$.

(6) Find $\dfrac{dy}{dx}$, if $y^3 + 3xy - x^2 = 0$.

(7) Find $\dfrac{d^2 y}{dx^2}$ if $y^2 = x^2 + 1$.

(8)# Use logarithmic differentiation to find the derivative of $y = \dfrac{(x^2 - 3x)^2 (5x^3 + 2x + 1)^3}{(x^2 + x)^2}$.

(9)# Find the equation of the tangent line to $\begin{cases} x = t^2 + 4 \\ y = 8t \end{cases}$, at $t = 6$.

(10)$^{\#}$ If $\vec{R}=\left\langle 6\cos\dfrac{\pi}{6}t, 3\sin\dfrac{\pi}{3}t \right\rangle$ is the position vector $\langle x,y \rangle$ from the origin to moving point $P(x,y)$ at time t. When $t=3$, find velocity vector.

(11)$^{\#}$ If the polar functions $r=2^{\theta}$, find the derivative of $\dfrac{\mathrm{d}y}{\mathrm{d}x}$.

(12) If $f(x)=x^2\sin x$, finding the value of $\mathrm{d}y$ when $x=\dfrac{\pi}{2}$ and $\mathrm{d}x=0.1$.

(13) Find the differential $\mathrm{d}y$ of the following functions.

1) $y=\ln x^2$;

2) $y=\mathrm{e}^{2x}\cos x$;

3) $y=\arctan\dfrac{x}{3}$;

4) $y=\ln(\sec x+\tan x)$;

5) $y=\sqrt{x^3+2}$.

(14) If the radius 10 of a circle is increased by 1%, use $\mathrm{d}A$ to estimate the increase in the circle's area A.

(15)$^{\#}$ Find the tangent line of cardioid $r=1+\cos\theta$ at $\theta=\dfrac{\pi}{6}$.

第 6 章

微分的应用
Applications of Differential Calculus

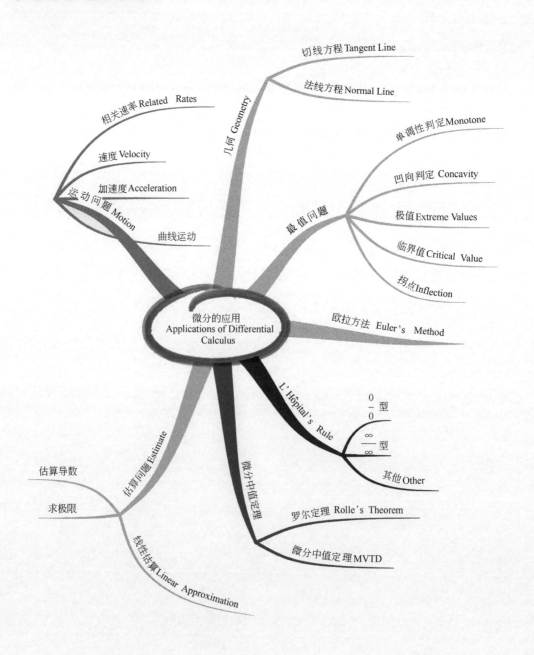

6.1 切线方程和法线方程 Equations of Tangent and Normal

AP 微积分中,切线方程问题大致分两类:1) 计算曲线 $y=f(x)$ 在某个点 $M(x_0, y_0)$ 处的切线方程;2) 计算曲线 $y=f(x)$ 上某个点的切线方程垂直或平行某条直线(如坐标轴),求这个点 $M(x_0, y_0)$. 法线方程和切线方程有类似之处.

6.1.1 切线方程 Equations of Tangent Line

The equation of the tangent to the curve $y=f(x)$ at point $M(x_0, y_0)$ is
$$y-y_0=f'(x_0)(x-x_0)$$
函数 $f(x)$ 在 $x=x_0$ 处的导函数值 $f'(x_0)$ 即曲线在该点的切线 l'_0 的斜率 k,见本书 5.1.4 节.

Example 6.1 Find the tangent line of parabola $y=3x^2+2x+1$ at point $M(1,6)$.

Solution:

Find the derivative of the function
$$y'=6x+2$$
Plug $x=1$ into y':
$$f'(1)=y'|_{x=1}=6\times 1+2=8$$
The equation of the tangent line is:
$$y-6=8(x-1)$$
Therefore,
$$8x-y-2=0$$

Note:计算曲线 $y=f(x)$ 在某个点 $M(x_0, y_0)$ 处的切线方程.

Example 6.2 Find the equation of the tangent to $y^3=3x^2y+2x+27$ at point where $x=0$.

Solution:

Using implicit differentiation, we get:
$$3y^2\frac{dy}{dx}=3\left(2xy+x^2\frac{dy}{dx}\right)+2$$
Simplify:
$$\frac{dy}{dx}=\frac{6xy+2}{3(y^2-x^2)}$$
When $x=0$, $y=3$ and Plug $(0, 3)$ into the above equation:
$$\frac{dy}{dx}|_{x=0}=\frac{6\times 0\times 3+2}{3(3^2-0^2)}=\frac{2}{27}$$
The equation of the tangent line is:
$$y-3=\frac{2}{27}(x-0) \text{ or } 2x-27y+81=0$$

Note:隐函数求导结合切线方程的考法是 AP 微积分一种常见的考法. 隐函数求导相

关内容见本书 5.5.2 节.

Example 6.3# Find The equation of the tangent to the curve with parametric equations $x=2t+1$, $y=3-t^3$ at the point where $t=2$.

Solution：

Take the derivative of $x=x(t)$ with respect to t：

$$\frac{dx}{dt}=2$$

Take the derivative of $y=y(t)$ with respect to t：

$$\frac{dy}{dt}=-3t^2$$

Then, the derivative of $y=y(x)$ with respect to x：

$$\frac{dy}{dx}=\frac{dy/dt}{dx/dt}=\frac{-3t^2}{2}$$

Since $\begin{cases} x=5 \\ y=-5 \end{cases}$ when $t=2$ and the slope at this point is：

$$\frac{dy}{dx}\bigg|_{t=2}=\frac{-3\times 2^2}{2}=-6$$

The equation of the tangent line：

$$y+5=-6(x-5) \text{ or } 6x+y-25=0$$

Note：类似的例题参考本书 Example 5.53，参数方程求导参见本书 5.6.1 节.

Example 6.4 Find the coordinate of any point on the curve of $y^2-x^2=6xy+10$ for which the tangent is parallel to the x-axis.

Solution：

Using implicit differentiation, we get：

$$2y\frac{dy}{dx}-2x=6\left(y+x\frac{dy}{dx}\right)$$

Simplify：

$$\frac{dy}{dx}=\frac{3y+x}{y-3x}$$

Since the tangent is parallel to the x-axis when $\frac{dy}{dx}=0$.

Then,

$$x=-3y$$

substitute $x=-3y$ in the equation of the curve, we get

$$y^2-(-3y)^2=6(-3y)y+10$$

Thus $y=\pm 1$ and $x=\pm 3$.

The point, then, are $(-3,1)$ and $(3,-1)$.

Note：计算曲线 $y=f(x)$ 上某个点的切线方程垂直或平行某条直线（如坐标轴），求这个点 $M(x_0,y_0)$.

6.1.2 法线方程 Equations of Normal Line

The equation of the normal to the curve $y=f(x)$ at point $M(x_0,y_0)$ is

$$y - y_0 = -\frac{1}{f'(x_0)}(x - x_0)$$

法线方程垂直于切线方程．因为相互垂直两直线的斜率乘积等于-1，所以法线方程的斜率为$-\frac{1}{f'(x_0)}$，除此之外，法线方程和切线方程有类似之处．

Example 6.5 Find the normal line of parabola $y = x^2 + 2x + 1$ at point $M(1, 4)$.

Solution：

Find the derivative of the function
$$y' = 2x + 2$$

Plug $x = 1$ into y'：
$$f'(1) = y'|_{x=1} = 2 \times 1 + 2 = 4$$

The equation of the normal line is：
$$y - 4 = -\frac{1}{4}(x - 1)$$

Therefore，
$$x - 4y + 17 = 0$$

Example 6.6$^\#$ Find the equation of the normal to $F(t) = (\sin t, \cos^2 t)$ at the point where $t = \frac{\pi}{6}$.

Solution：

Take the derivative of $x = x(t)$ with respect to t：
$$\frac{dx}{dt} = \cos t$$

Take the derivative of $y = y(t)$ with respect to t：
$$\frac{dy}{dt} = 2\cos t(-\sin t) = -2\sin t \cos t$$

Then，the derivative of $y = y(x)$ with respect to x：
$$\frac{dy}{dx} = \frac{dy/dt}{dx/dt} = \frac{-2\sin t \cos t}{\cos t} = -2\sin t$$

Since $\begin{cases} x = \frac{1}{2} \\ y = \frac{3}{4} \end{cases}$ when $t = \frac{\pi}{6}$ and the slope at this point is：

$$-\frac{1}{\frac{dy}{dx}\Big|_{t=\frac{\pi}{6}}} = -\frac{1}{-2 \times \frac{1}{2}} = 1$$

The equation of the normal line：
$$y - \frac{3}{4} = x - \frac{1}{2} \text{ or } 4x - 4y + 1 = 0$$

6.2 最值问题 The Problems of Maxima and Minima

微分（Differential）是解决最值问题的利器．最值问题涉及函数的单调性

(Monotone) 和凹向（Concavity）的判定，极值点（Local Extreme Values）、临界值（Critical Value）和拐点（Inflection），以及极大值（Local Maximum）/极小值（Local Minimum）等知识点，是微分应用的经典题型．

6.2.1 增函数和减函数 Increasing and Decreasing Functions

Suppose $f(x)$ is continuous on the closed interval $[a,b]$ and differentiable on the open interval (a,b)：

1) If $f'(x)>0$, then $f(x)$ increasing on $[a,b]$ and over which the curve rises.

2) If $f'(x)<0$, then $f(x)$ decreasing on $[a,b]$ and over which the curve falls.

单调性是函数重要的性质之一（见本书 2.2.1 节）．一阶导数 $f'(x)$ 的重要应用之一就是判定函数单调性．多数情况下，得到函数单调性（Monotone），才能求解函数的极值（最值）．

Example 6.7 Let $f(x)=x^3+3x^2$. Find the intervals on which $f(x)$ is increasing or decreasing.

Solution：

The domain of function $f(x)$ is R and differentiable on the domain.

Find the derivative of the function
$$f'(x)=3x^2+6x$$

Which is zero when $x=-2$ or $x=0$. We analyze the signs of $f'(x)$ in three intervals：

x	$(-\infty,-2)$	$(-2,0)$	$(0,+\infty)$
$f'(x)$	+	−	+
$f(x)$	increasing	decreasing	increasing

Sketch the graph of $f(x)$, see Figure 6.2.1(a).

Thus,

$f(x)$ is increasing on $(-\infty,-2)\cup(0,+\infty)$ and decreasing on $(-2,0)$.

Note：本题，函数 $f(x)$ 在定义域上可导（Differentiability）．假如函数定义域上存在不可导点，先"抠除"不可导点，再讨论各个子区间的单调性．

Example 6.8 Let $f(x)=\begin{cases}(x+3)^2-4, x\leqslant-1\\ 3x+3, x>-1\end{cases}$. Find the intervals on which $f(x)$ is increasing or decreasing.

Solution：

The domain of function $f(x)$ is $(-\infty,+\infty)$ and continuous on the domain.

Find the derivative of the function
$$f'(x)=\begin{cases}2(x+3), x<-1\\ 3, x>-1\end{cases}$$

Since, $f'_+(-1)\neq f'_-(-1)$.

Then, the function $f(x)$ is not differentiable at $x=-1$.

Which is zero when $x=-3$. We analyze the signs of $f'(x)$ in three intervals:

x	$(-\infty,-3)$	$(-3,-1)$	$(-1,+\infty)$
$f'(x)$	$-$	$+$	$+$
$f(x)$	decreasing	increasing	increasing

Sketch the graph of $f(x)$, see Figure 6.2.1（b）.

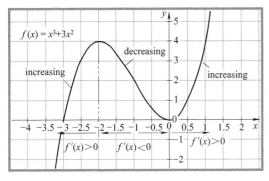

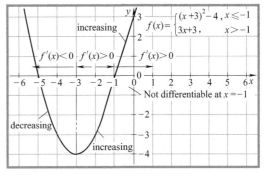

Figure 6.2.1(a)　　　　　　　　　　Figure 6.2.1(b)

Thus,

$f(x)$ is increasing on $(-3,-1) \cup (-1,+\infty)$ and decreasing on $(-\infty,-3)$.

【方法总结】确定 $f(x)$ 的单调区间通常有 3 个步骤：

1) 确定 $f(x)$ 的定义域 D；

2) 确定不可导点，求导函数 $f'(x)$，并令 $f'(x)=0$，用所得到的解和不可导点将定义域 D 分割成子区间；

3) 使得 $f'(x)>0$ 的区间为单调递增区间，使得 $f'(x)<0$ 的区间为单调递减区间. 当区间较多时，通过列表，表达更加清晰.

6.2.2　函数凹向的判定 Concavity Test

If the graph of $f(x)$ lies above all its tangents on an interval I, then it is called concave upward on I.

If the graph of $f(x)$ lies below all its tangents on an interval I, then it is called concave downward on I.

上凹的和下凹的统称为曲线的凹向. Figure 6.2.2(a) 中，曲线 $f(x)$ 上弧 $\overset{\frown}{AC}$ 位于其上任一点的切线的上方，该曲线弧在区间 I 内是上凹的；Figure 6.2.2(b) 中，曲线 $f(x)$ 上弧 $\overset{\frown}{A'C'}$ 位于其上任一点的切线的下方，该曲线弧在区间 I' 内是下凹的.

The Second Derivative Test for Concavity：

Let $y=f(x)$ be twice-differentiable on an interval I：

1) If $f''(x)>0$ on I, the graph of $f(x)$ over I is concave up.

2) If $f''(x)<0$ on I, the graph of $f(x)$ over I is concave down.

判断函数的凹向是函数二阶求导重要的应用，同时，函数的凹向取决于其二阶导函数值的正负号.

第 6 章　微分的应用 Applications of Differential Calculus

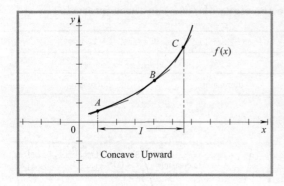

Figure 6.2.2(a)

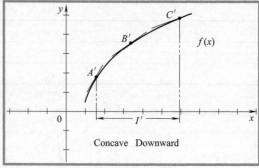

Figure 6.2.2(b)

Example 6.9 Determine the concavity of $f(x) = \dfrac{2x^2}{(1-x)^2}$.

Solution:

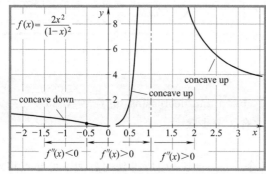

Figure 6.2.2(c)

The domain of function $f(x)$ is $\{x \mid x \neq 1\}$.

Then, the function $f(x)$ is not twice-differentiable at $x=1$.

The first derivative is
$$f'(x) = \frac{4x}{(1-x)^3}$$

The second derivative is
$$f''(x) = \frac{8x+4}{(1-x)^4}$$

Let $f''(x) = 0$, then, $x = -\dfrac{1}{2}$

We analyze the signs of $f''(x)$ in three intervals:

x	$\left(-\infty, -\dfrac{1}{2}\right)$	$\left(-\dfrac{1}{2}, 1\right)$	$(1, +\infty)$
$f''(x)$	$-$	$+$	$+$
$f(x)$	concave down	concave up	concave up

Sketch the graph of $f(x)$, see Figure 6.2.2(c).

Thus,

$f(x)$ is concave down on $\left(-\infty, -\dfrac{1}{2}\right)$ and concave up on $\left(-\dfrac{1}{2}, 1\right) \cup (1, +\infty)$.

Note：确定曲线的凹向（Concavity）与确定函数单调区间步骤类似，同学们可以参照解答．

6.2.3 极小值和极大值 Local Maximum and Local Minimum

Let x_0 be a number in the domain D of a function $f(x)$. The $f(x_0)$ is

1) local maximum value of $f(x)$ if $f(x) \leqslant f(x_0)$ when x is near x_0;
2) local minimum value of $f(x)$ if $f(x) \geqslant f(x_0)$ when x is near x_0;

极小值和极大值统称为极值（Local Extreme Values）. Figure 6.2.3(a) 中，曲线 $f(x)$ 在 A 点取得极小值；Figure 6.2.3(b) 中，曲线 $f(x)$ 在 B 点取得极大值. 函数可能在 $f'(x)=0$、不可导点或者端点处取得极值.

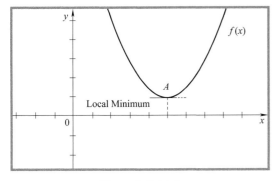

Figure 6.2.3(a)

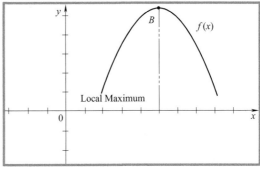

Figure 6.2.3(b)

Example 6.10 Find the local extreme values of $y = x^3 - x$ on the interval $[-4, 4]$.

Solution:

Take the derivative and set it equal to zero:
$$\frac{dy}{dx} = 3x^2 - 1 = 0$$

Solve for x:
$$x_1 = \frac{\sqrt{3}}{3} \text{ and } x_2 = -\frac{\sqrt{3}}{3}$$

Take the second derivative of the function:
$$\frac{d^2 y}{dx^2} = 6x$$

Since
$$\left.\frac{d^2 y}{dx^2}\right|_{x=\frac{\sqrt{3}}{3}} = 6 \times \frac{\sqrt{3}}{3} > 0$$

then, $f(x)$ has a local minimum at $x = \frac{\sqrt{3}}{3}$:
$$f\left(\frac{\sqrt{3}}{3}\right) = \left(\frac{\sqrt{3}}{3}\right)^3 - \left(\frac{\sqrt{3}}{3}\right) = -\frac{2\sqrt{3}}{9}$$

Since
$$\left.\frac{d^2 y}{dx^2}\right|_{x=-\frac{\sqrt{3}}{3}} = 6 \times \left(-\frac{\sqrt{3}}{3}\right) < 0$$

then, $f(x)$ has a local maximum at $x = -\frac{\sqrt{3}}{3}$:
$$f\left(\frac{\sqrt{3}}{3}\right) = \left(-\frac{\sqrt{3}}{3}\right)^3 - \left(-\frac{\sqrt{3}}{3}\right) = \frac{2\sqrt{3}}{9}$$

Note：二阶导检验法确定函数极值点（Local Extreme Values）.

If $f(x)$ has a local maximum or minimum value at an interior point x_0 of its domain, and if $f'(x)$ is defined at x_0, then

$$f'(x)=0$$

如果导数存在，函数极值点的一阶导数必然等于零．注意前提是导数存在，否则极值点的一阶导数不一定为零，因为这个极值点处导数可能不存在或是 $f(x)$ 定义域的端点．

Example 6.11　Find the local minimum and maximum values of $y=2x^3-3x+1$ on the interval $[-4,4]$.

Solution：

Take the derivative and set it equal to zero：

$$\frac{dy}{dx}=6x^2-3=0$$

Solve for x：

$$x_1=\frac{\sqrt{2}}{2} \text{ and } x_2=-\frac{\sqrt{2}}{2}$$

We analyze the signs of $f'(x)$ in three intervals：

x	$[-4,-\frac{\sqrt{2}}{2})$	$-\frac{\sqrt{2}}{2}$	$(-\frac{\sqrt{2}}{2},\frac{\sqrt{2}}{2})$	$\frac{\sqrt{2}}{2}$	$(\frac{\sqrt{2}}{2},4]$
$f'(x)$	+	0	−	0	+
$f(x)$	↗	$1+\sqrt{2}$	↘	$1-\sqrt{2}$	↗

Then, the local minimum values of $f(x)$ is $1-\sqrt{2}$ and the local maximum values of $f(x)$ is $1+\sqrt{2}$.

【方法总结】假如 $f'(x_0)=0$ 且 $x=x_0$ 为 $f(x)$ 的极值点，有两个方法判断 $f(x_0)$ 为极大值还是极小值：

1) 二阶导检验法：Suppose $f''(x_0)$ is continuous near x_0 and $f'(x_0)=0$.

If $f''(x_0)>0$, then $f(x_0)$ has a local minimum at x_0.

If $f''(x_0)<0$, then $f(x_0)$ has a local maximum at x_0.

2) 考查 $x=x_0$ 左右邻域内 $f'(x)$ 的正负：Suppose $f'(x_0)=0$.

If $\begin{cases} f'(x)<0, x<x_0 \\ f'(x)>0, x>x_0 \end{cases}$, it is a local minimum.

If $\begin{cases} f'(x)>0, x<x_0 \\ f'(x)<0, x>x_0 \end{cases}$, it is a local maximum.

6.2.4　临界值和拐点 Critical Value and Inflection Point

1) Any x_0 in the domain of $f(x)$ such that either $f'(x_0)=0$ or $f'(x_0)$ is undefined is called a critical value.

2) A point of inflection is a point where the curve changes its concavity from upward to downward or from downward to upward.

临界点（Critical Point）和拐点（Inflection）是函数两类重要的特殊点．临界点和拐点都代表着函数某方面的性质．临界点是某邻域内函数值停止变化的点，而拐点是某邻域内函数图像凹向突变的点．

Figure 6.2.4(a) 中，点 B、C、D 和 E 是函数 $f(x)$ 的临界点．其中，点 C 和 D 是 $f(x)$ 的不可导点；在点 B 和 E 处，$f'(x)=0$. 另外，点 A 和 F 是 $f(x)$ 的端点．Figure 6.2.4(b) 中，点 A' 和 B' 是函数 $f(x)$ 的拐点．

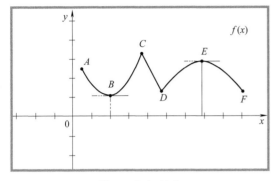

Figure 6.2.4(a)

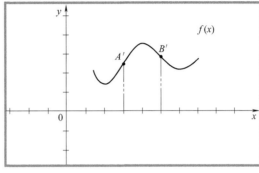

Figure 6.2.4(b)

Example 6.12　Find the critical point of $f(x)=\sqrt[5]{x^3}(4-x)$.

Solution：

Using the Product Rule, we get

$$f'(x)=\sqrt[5]{x^3}\times(-1)+(4-x)\left(\frac{3}{5\sqrt[5]{x^2}}\right)$$

$$=\frac{-5x+3(4-x)}{5\sqrt[5]{x^2}}$$

$$=\frac{12-8x}{5\sqrt[5]{x^2}}$$

Therefore $f'(x)=0$ if $12-8x=0$

That is, $x=\dfrac{3}{2}$, and $f'(x)$ does not exist when $x=0$.

Thus, the critical numbers are $\dfrac{3}{2}$ and 0.

值得注意的是：尽管函数的极值只可能在临界点和端点处取得，Figure 6.2.4(a) 中，点 B、C、D 和 E 都是函数 $f(x)$ 的临界点和极值点．但并非每个临界点或端点都表示在该点一定取得极值．Figure 6.2.4(c) 中，$\dfrac{dy}{dx}=3x^2$ 在 $x=0$ 处等于 0，但是 $f(x)=x^3$ 在 $x=0$ 处的值不是极值．Figure 6.2.4(d) 中，$f(x)=\sqrt[3]{x}$ 在 $x=0$ 处导数不存在且不是极值．

Example 6.13　Find the curve $f(x)=x^4-4x^3$ points of inflection.

Solution：

The domain of function $f(x)$ is R.

Find the first derivative of the function

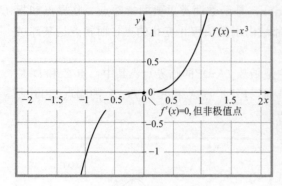
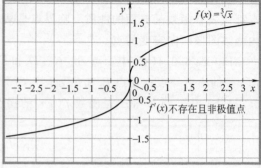

Figure 6.2.4(c) Figure 6.2.4(d)

$$f'(x) = 4x^3 - 12x^2 = 4x^2(x-3)$$

Find the second derivative of the function

$$f''(x) = 12x^2 - 24x = 12x(x-2)$$

We set $f''(x) = 0$, then, $x = 0$ or $x = 2$.

We analyze the signs of $f''(x)$ in three intervals:

x	$(-\infty, 0)$	0	$(0, 2)$	2	$(2, +\infty)$
$f''(x)$	+	0	−	0	+
$f(x)$	concave up	0	concave down	−16	concave up

Sine the curve changes from concave upward to concave downward, the point $(0, 0)$ is an inflection point. Also $(2, -16)$ is an inflection point since the curve changes from concave downward to concave upward there.

6.2.5 最大值和最小值 Absolute Maximum and Minimum

Let x_0 be a number in the domain D of a function $f(x)$. The $f(x_0)$ is

1) absolute maximum value of on D if $f(x_0) \geqslant f(x)$ for all x in D;
2) absolute minimum value of on D if $f(x_0) \leqslant f(x)$ for all x in D.

函数在某区间上的最大值（Absolute Maximum）是所有极大值和区间端点函数值中最大的那个值；最小值（Absolute Minimum）是所有极小值与区间端点函数值中最小的那个值.

见 Figure 6.2.5(a)，曲线 $f(x)$ 上点 A 和 F 是端点，点 B、C、D 和 E 是临界点，其中，点 C 是 $f(x)$ 的最大值点，点 F 是最小值点.

Example 6.14 Find the absolute minimum and maximum values of $y = x^3 - 3x + 1$ on the interval $\left[-\dfrac{3}{2}, \dfrac{3}{2}\right]$.

Solution:

Take the derivative and set it equal to zero:

$$\frac{dy}{dx} = 3x^2 - 3 = 0$$

Solve for x:

$$x_1 = 1 \text{ and } x_2 = -1$$

Take the second derivative of the function:
$$\frac{d^2y}{dx^2} = 6x$$

Since $\left.\frac{d^2y}{dx^2}\right|_{x=1} = 6 \times 1 > 0$, $f(x)$ has a local minimum at $x=1$:
$$f(1) = 1^3 - 3 \times 1 + 1 = -1$$

Since $\left.\frac{d^2y}{dx^2}\right|_{x=-1} = 6 \times (-1) < 0$, $f(x)$ has a local maximum at $x=-1$:
$$f(-1) = (-1)^3 - 3 \times (-1) + 1 = 3$$

Check the endpoint of the interval:
$$\text{At } x = -\frac{3}{2}, \ y = \frac{17}{8} \text{ and at } x = \frac{3}{2}, \ y = -\frac{1}{8}$$

Then, the absolute minimum values of $f(x)$ is -1 and the absolute maximum values of $f(x)$ is 3. See Figure 6.2.5(b).

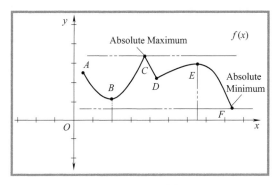

Figure 6.2.5(a)   Figure 6.2.5(b)

Note：求最大值或最小值，经常结合应用题来考查．解决这类型的题目，首先根据条件列出函数表达式，然后运用本讲相关知识解答．

Example 6.15 Find the point on the curve $y = \sqrt{x+1}$ that is a minimum distance from the point $(5,0)$.

Solution:

Using the distance formula:
$$D^2 = (x-5)^2 + (y-0)^2 = x^2 - 10x + 25 + y^2$$

Because $y = \sqrt{x+1}$
$$D^2 = x^2 - 10x + 25 + (x+1) = x^2 - 9x + 26$$

Let $f(x) = D^2$, then
$$f(x) = x^2 - 9x + 26$$

Take the derivative and set it equal to zero:
$$\frac{dy}{dx} = 2x - 9 = 0$$

Solve for x:

第6章 微分的应用 Applications of Differential Calculus

$$x = \frac{9}{2}$$

Solving for y, we get $y = \sqrt{\frac{11}{2}}$.

Because

$$\frac{d^2 y}{dx^2} = 2 > 0$$

The point $\left(\frac{9}{2}, \sqrt{\frac{11}{2}}\right)$ is the minimum distance from the point $(5,0)$.

【方法总结】求最大值和最小值的 3 个步骤：

1) 求得 $f'(x)$ 并令 $f'(x) = 0$，求得 $f(x)$ 的临界点（包含不可导点）；

2) 求得 $f''(x)$ 并判断各个临界点是极大值还是极小值；

3) 比较临界值、端点值在定义域上的大小，给出最值.

6.2.6　绘制曲线 Sketching a Curve

AP 微积分中，解决很多问题，需要绘制曲线草图. 绘制曲线大致是以下几个方面：

1) 定义域（Domain）：见 2.1.1 节.

2) 截距（Intercepts）：曲线 $f(x)$ 在 y 轴上的截距是 $f(0)$；令 $f(x) = 0$，可得曲线 $f(x)$ 在 x 轴上截距.

3) 对称（Symmetry）：主要考查函数的奇偶性以及周期性. 如果 $f(-x) = f(x)$，则曲线 $f(x)$ 关于 y 轴对称；如果 $f(-x) = -f(x)$，则曲线 $f(x)$ 关于原点 $(0,0)$ 对称；如果 $f(x+T) = f(x)$，则曲线 $f(x)$ 呈现周期变化；见本书 2.2.2 节和 2.2.3 节.

4) 渐近线（Asymptotes）：垂直渐近线和水平渐近线，见本书 3.4.1 节和 3.4.2 节.

5) 区间上的单调性（Intervals of Increase or Decrease），见本书 6.2.1 节.

6) 极大值和极小值（Local Maximum and Minimum），见本书 6.2.3 节.

7) 拐点和曲线的凹向（Points of Inflection and Concavity），见本书 6.2.2 节.

Example 6.16　Sketch the equation $y = x^3 - 9x$.

Solution：The domain is R.

Find the x-intercepts：

$$x^3 - 9x = 0$$

The curve has x-intercepts at $(-3,0)$, $(3,0)$ and $(0,0)$

Find the y-intercepts：

$$y = 0^3 - 9 \times 0 = 0$$

The curve has y-intercepts at $(0,0)$

Since $f(-x) = -f(x)$, the function $f(x)$ is odd. The curve is symmetric about the origin.

Take the derivative of the function to find the critical points：

$$\frac{dy}{dx} = 3x^2 - 9$$

Set the derivative equal to zero and solve for x：

$$3x^2 - 9 = 0$$

So
$$x_1 = -\sqrt{3} \text{ and } x_2 = \sqrt{3}$$

Plug $x_1 = -\sqrt{3}$ and $x_2 = \sqrt{3}$ into the original equation to find the y-coordinates of the critical points:

$$y_1 = (-\sqrt{3})^3 - 9 \times (-\sqrt{3}) = 6\sqrt{3}$$
$$y_2 = (\sqrt{3})^3 - 9 \times \sqrt{3} = -6\sqrt{3}$$

Thus, the critical points at $(-\sqrt{3}, 6\sqrt{3})$ and $(\sqrt{3}, -6\sqrt{3})$.

Take the second derivative to find any point of inflection:

$$\frac{d^2 y}{dx^2} = 6x$$

Let $6x = 0$, the curve has a point of inflection at $(0,0)$.

Since $\frac{d^2 y}{dx^2}\big|_{x=-\sqrt{3}} = -6\sqrt{3} < 0$ and $\frac{d^2 y}{dx^2}\big|_{x=\sqrt{3}} = 6\sqrt{3} > 0$

The curve has a local minimum at $(\sqrt{3}, -6\sqrt{3})$ and a local maximum at $(-\sqrt{3}, 6\sqrt{3})$.

We can plot the graph [see Figure 6.2.6(a)]:

Example 6.17 Sketch the curve $y = \dfrac{3x^2}{x^2 - 1}$

Solution:

The domain is $\{x \mid x^2 - 1 \neq 0\} = \{x \mid x \neq \pm 1\}$

The x-intercepts and y-intercepts are both at $(0, 0)$.

Since $f(-x) = f(x)$, the function $f(x)$ is even. The curve is symmetric about the y-axis.

$$\lim_{x \to \infty} \frac{3x^2}{x^2 - 1} = 3$$

Therefore, the line $y = 3$ is a horizontal asymptote.

Since the denominator is 0 when $x = \pm 1$, we get:

$$\lim_{x \to 1^+} \frac{3x^2}{x^2 - 1} = \infty \text{ and } \lim_{x \to -1^+} \frac{3x^2}{x^2 - 1} = -\infty$$

Therefore, the line $x = 1$ and $x = -1$ are vertical asymptotes.

Take the derivative of the function to find the critical points:

$$\frac{dy}{dx} = \frac{6x(x^2 - 1) - 3x^2 \times 2x}{(x^2 - 1)^2} = \frac{-6x}{(x^2 - 1)^2}$$

Since $f'(x) > 0$ when $x < 0 (x \neq -1)$ and $f'(x) < 0$ when $x > 0 (x \neq 1)$, $f(x)$ is increasing on $(-\infty, -1)$ and $(-1, 0)$ and decreasing on $(0, 1)$ and $(1, \infty)$.

Thus, the critical points at $(0, 0)$.

Take the second derivative:

$$\frac{d^2 y}{dx^2} = \frac{-6(x^2 - 1)^2 + 6x \times 2(x^2 - 1) \times 2x}{(x^2 - 1)^4} = \frac{18x^2 + 6}{(x^2 - 1)^3}$$

Since $18x^2+6>0$ for all x, we have
$$f''(x)>0 \iff x^2-1>0 \iff |x|>1$$
$$f''(x)<0 \iff x^2-1<0 \iff |x|<1$$

Thus, the curve is concave up on the intervals $(-\infty,-1)$ and $(1,\infty)$ and concave down on $(-1,1)$. It has no point of inflection since 1 and -1 are not in the domain of $f(x)$.

We can plot the graph [see Figure 6.2.6(b)]:

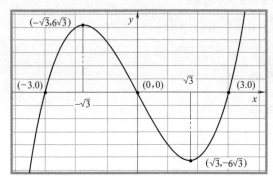
Figure 6.2.6(a)

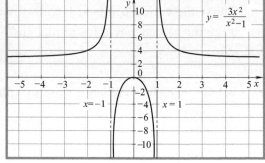
Figure 6.2.6(b)

6.3 运动问题 The Problems of Motion

相关速率（Related Rates）、位置（Position）、速度（Velocity）和加速度（Acceleration）几乎会出现在每场 AP 微积分考试上．微积分 BC 还要应用微分处理曲线运动（Motion Along a Curve）问题．

6.3.1 相关速率 Related Rates

假如 $y=f(x)$ 是关于 x 的函数，由导数的几何意义可知，y 关于 x 的瞬时速率为 $\dfrac{dy}{dx}$（Instantaneous Rate of Change）．同时，假如 $y=y(t)$ 是关于时间 t 的函数，则 y 关于 t 的瞬时速率为 $\dfrac{dy}{dt}$；假如 $x=x(t)$ 是关于时间 t 的函数，则 x 关于 t 的瞬时速率为 $\dfrac{dx}{dt}$．

把 $y=f(x)$ 看做 y 和 x 的方程，则 y 和 x 相关，上述瞬时速率 $\dfrac{dy}{dt}$ 和 $\dfrac{dx}{dt}$ 也就相关，都是关于时间 t 的瞬时速率．多个变量组成的方程有一定相关性，方程两边同时对时间求导之后，多个变量的瞬时速率也相关，这就是相关速率（Related Rates）．

Example 6.18 A ladder 13 ft long rests against a vertical wall. If the bottom of the ladder slides away from the wall at a rate of 2 ft/sec, how fast is the top of the ladder sliding down the wall when the bottom of the ladder is 5 ft from the wall?

Solution:

Let x feet be the distance from the bottom of the ladder to the wall and y feet the dis-

tance from the top of the ladder to the ground, see Figure 6.3.1(a). the relationship between x and y is given by the Pythagorean Theorem:
$$x^2 + y^2 = 13^2$$

When $x=5$, the Pythagorean Theorem gives $y=12$.

Note that x and y are both functions of t. Take the derivative both sides with respect to t:
$$2x\frac{\mathrm{d}x}{\mathrm{d}t} + 2y\frac{\mathrm{d}y}{\mathrm{d}t} = 0$$

We are given that $\frac{\mathrm{d}x}{\mathrm{d}t} = 2 \mathrm{ft/sec}$ and we are asked to find $\frac{\mathrm{d}y}{\mathrm{d}t}$ when $x=5\mathrm{ft}$, see Figure 6.3.1(b).

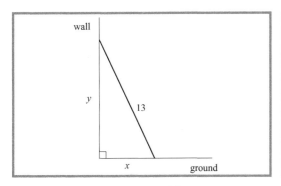

Figure 6.3.1(a)　　　　　　Figure 6.3.1(b)

Then,
$$2 \times 5 \times 2 + 2 \times 12 \times \frac{\mathrm{d}y}{\mathrm{d}t} = 0$$
$$\frac{\mathrm{d}y}{\mathrm{d}t} = -\frac{5}{6}\mathrm{ft/sec}$$

In other words, the top of the ladder is sliding down the wall at the rate of $\frac{5}{6}\mathrm{ft/sec}$.

Note: $\frac{\mathrm{d}y}{\mathrm{d}t}$ 为负数，表明此时梯子头部到与地板的距离在减少.

Example 6.19 A mixture of air and hydrogen is being pumped into a spherical balloon so that its volume increases at a rate of 150 cm³/sec. How fast is the radius of the balloon increasing when the diameter is 30cm?

Solution:

Let V be the volume of the balloon and let r be its radius.

Relate V and r by the formula for the volume of a sphere:
$$V = \frac{4}{3}\pi r^3$$

Differentiate each side of this equation with respect to t:
$$\frac{\mathrm{d}V}{\mathrm{d}t} = 4\pi r^2 \frac{\mathrm{d}r}{\mathrm{d}t}$$

Since $2r=30$cm, $r=15$cm.

Put $r=15$ and $\dfrac{dV}{dt}=150$ in this equation, we obtain

$$150=4\pi\times 15^2\times \dfrac{dr}{dt}$$

$$\dfrac{dr}{dt}=\dfrac{1}{6\pi}$$

The radius of the balloon is increasing at the rate of $\dfrac{1}{6\pi}\approx 0.053$cm/sec.

Note：AP 微积分考试中，没有特殊说明，最后结果一般保留小数点后 3 位；题中给出的量带了单位，最后结果也必须带单位．

Example 6.20 A water tank is in the shape of an inverted cone. The height of the tank is 10 meters and the diameter of the base is 6 meters. Water is pumped into the tank at the rate of $2\pi \text{m}^3/\text{min}$. How fast is the water level rising when the water is 4 meters deep?

Solution：

Let V, r and h be the volume of the water, the radius of the surface, and the height of the water at time t, where t is measured in minutes. See Figure 6.3.1(c).

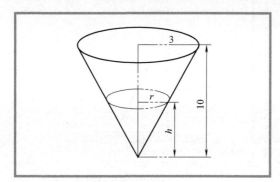

Figure 6.3.1(c)

The quantities V and h are related by equation

$$V=\dfrac{1}{3}\pi r^2 h$$

Using the similar triangles in Figure 6.3.1(c) to write：

$$\dfrac{h}{r}=\dfrac{10}{3} \text{ and } r=\dfrac{3}{10}h$$

Then,

$$V=\dfrac{1}{3}\pi\left(\dfrac{3}{10}h\right)^2 h=\dfrac{3\pi}{100}\times h^3$$

Differentiate each side of this equation with respect to t：

$$\dfrac{dV}{dt}=\dfrac{3\pi}{100}\times (3h^2)\dfrac{dh}{dt}=\dfrac{9\pi}{100}h^2\dfrac{dh}{dt}$$

Substituting $h=4$m and $\dfrac{dV}{dt}=2\pi \text{m}^3/\text{min}$, we have

$$2\pi=\dfrac{9\pi}{100}\times 4^2\times \dfrac{dh}{dt}$$

Then,

$$\dfrac{dh}{dt}=\dfrac{25}{18}$$

The water level is rising at a rate of $\frac{25}{18} \approx 1.389 \text{m/min}$.

【方法总结】求相关速率（Related Rates）问题的步骤：
1) 确定所有相关变量，列出方程；
2) 根据隐函数求导法则，对相关方程关于时间 t 求导；
3) 代入已知条件或数值，解得相关速率.

6.3.2 速度和加速度 Velocity and Acceleration

Recall that if the object has position function $s=f(t)$, then the velocity function is $v(t)=s'(t)=\frac{ds}{dt}$. Likewise, the acceleration function is $a(t)=v'(t)=\frac{dv}{dt}$ or $a(t)=s''(t)=\frac{d^2s}{dt^2}$. The speed of the object is $|v(t)|$, the magnitude of $v(t)$.

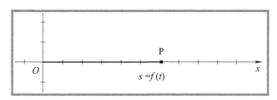

Figure 6.3.2(a)

位置（Position）、速度（Velocity）和加速度（Acceleration）涉及微分和积分两部分的应用，我们先讲微分应用部分的，积分的应用见本书 9.4.1 节. 假如规定直线运动的物体向右运动为正方向（与 x 轴的方向相同），见 Figure 6.3.2(a).

1) 假如 $v(t)>0$，物体 P 向右运动，$s(t)$ 在增大；假如 $v(t)<0$，物体 P 向左运动，$s(t)$ 在减小.
2) 假如 $a(t)>0$，物体 P 的速度 $v(t)$ 在增大；假如 $a(t)<0$，$v(t)$ 在减小.
3) 假如 $v(t)$ 和 $a(t)$ 同号，物体的速率（Speed）在增大；假如 $v(t)$ 和 $a(t)$ 异号，物体的速率在减小.
4) 当物体 P 的速度 $v(t)$ 一步一步减小为零并且加速度 $a(t)$ 不为零，则物体 P 将短暂停止之后，改变运动方向.

Example 6.21 If the position of a particle at a time t is given by the equation $f(t)=2t^3-18t^2+36t+5$, find the velocity and the acceleration of the particle at time $t=3$.

Solution：
Take the derivative of $f(t)$：
$$v(t)=f'(t)=6t^2-36t+36$$
Plug in $t=3$ to find the velocity at that time：
$$v(3)=6\times 3^2-36\times 3+36=-18$$
Take the derivative of $v(t)$：
$$a(t)=v'(t)=12t-36$$
Plug in $t=3$ to find the acceleration at that time：
$$a(3)=12\times 3-36=0$$
The velocity is -18 and the acceleration is 0.

Example 6.22 If the position of a particle at a time is given by the equation $f(t)=t^3-9t^2+24t-10$, for $t\geqslant 0$.

1) Find all t for which the distance is decreasing.

2) Find all t for which the velocity is increasing.

3) Find all t for which the speed of the particle is increasing.

4) Find the total distance traveled between $t=0$ and $t=5$.

Solution:

Take the derivative of $f(t)$:
$$v(t)=f'(t)=3t^2-18t+24=3(t-2)(t-4)$$

Take the derivative of $v(t)$:
$$a(t)=v'(t)=6t-18=6(t-3)$$

Velocity $v(t)=0$ at $t=2$ and $t=4$, and:

t	$[0, 2)$	2	$(2, 4)$	4	$(4, +\infty)$
$v(t)$	+	0	−	0	+

Acceleration $a(t)=0$ at $t=3$, and:

t	$[0, 3)$	3	$(3, +\infty)$
$a(t)$	−	0	+

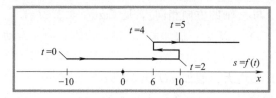

Figure 6.3.2(b)

These signs of $v(t)$ and $a(t)$ immediately yield the answers, as follows:

1) The distance $f(t)$ decreases when t on $(2,4)$.

2) The velocity $v(t)$ increases when t on $(3, +\infty)$.

3) The speed $|v|$ is increasing when $v(t)$ and $a(t)$ are both positive, that is, for t on $(4, +\infty)$, and when $v(t)$ and $a(t)$ are both negative, that is, for t on $(2, 3)$.

4) P's motion can be indicated as show in Figure 6.3.2(b). P moves to the right if t on $[0, 2)$, reverses its direction at $t=2$, moves to the left when t on $(2, 4)$, reverses again at $t=4$, and continues to the right for all $t>4$. The position of P at certain time t are show in the following table:

t	0	2	4	5
$f(t)$	−10	10	6	10

Thus, P travels a total of 28 units between times $t=0$ and $t=5$.

Example 6.23 If the position of a particle is given by the equation $f(t)=t^3-\dfrac{21}{2}t^2+30t+7$, for $t\geqslant 0$, find the point at which the particle changes direction.

Solution:

Take the derivative of $f(t)$:
$$v(t)=f'(t)=3t^2-21t+30=3(t-2)(t-5)$$

Set it equal to zero and solve for t:
$$t=2 \text{ or } t=5$$

Since $f''(t)=6t-21$, it equals 0 at $t=\dfrac{7}{2}$.

Therefore, the particle is changing direction at $t=2$ and $t=5$.

6.3.3# 曲线运动 Motion Along a Curve

If the object moves along a curve defined parametrical by position vector $\vec{r}(t)=\langle x(t), y(t)\rangle$, then velocity vector $\vec{v}(t)=\dfrac{\mathrm{d}\vec{r}}{\mathrm{d}t}=\left\langle\dfrac{\mathrm{d}x}{\mathrm{d}t},\dfrac{\mathrm{d}y}{\mathrm{d}t}\right\rangle$, $v_x=\dfrac{\mathrm{d}x}{\mathrm{d}t}$, $v_y=\dfrac{\mathrm{d}y}{\mathrm{d}t}$. Likewise, the acceleration vector $\vec{a}(t)=\dfrac{\mathrm{d}\vec{v}}{\mathrm{d}t}=\left\langle\dfrac{\mathrm{d}v_x}{\mathrm{d}t},\dfrac{\mathrm{d}v_y}{\mathrm{d}t}\right\rangle$ or $\vec{a}(t)=\dfrac{\mathrm{d}^2\vec{r}}{\mathrm{d}t^2}=\left\langle\dfrac{\mathrm{d}^2x}{\mathrm{d}t^2},\dfrac{\mathrm{d}^2y}{\mathrm{d}t^2}\right\rangle$. The speed of the object is the magnitude of $v(t)$, $|\vec{v}(t)|=\sqrt{\left(\dfrac{\mathrm{d}x}{\mathrm{d}t}\right)^2+\left(\dfrac{\mathrm{d}y}{\mathrm{d}t}\right)^2}=\sqrt{v_x^2+v_y^2}$.

物体 P 的位置向量为 $\langle x(t), y(t)\rangle$，见 Figure 6.3.3. AP 微积分中，曲线运动通常围绕物体 P 的速度或加速度向量展开，速度向量为 $\vec{v}(t)$，可分解为水平方向的向量 v_x 和垂直方向的向量 v_y，实质是向量函数求导（见本书 5.7.1 节）的具体应用.

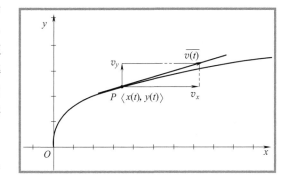

Figure 6.3.3

Example 6.24# $\vec{R}=\left\langle 3\cos\dfrac{\pi}{6}t, 2\sin\dfrac{\pi}{6}t\right\rangle$ is the position vector $\langle x, y\rangle$ from the origin to a moving point $P(x, y)$ at time t.

1) Find the velocity and acceleration at the time $t=6$.
2) Find the speed of the particle when $t=6$.
3) Find the magnitude of the acceleration when $t=6$.
4) Find the slope of the curve along which the particle moves at the point where $t=2$.

Solution：

Since $\vec{R}=\left\langle 3\cos\dfrac{\pi}{6}t, 2\sin\dfrac{\pi}{6}t\right\rangle$, $x(t)=3\cos\dfrac{\pi}{6}t$ and $y(t)=2\sin\dfrac{\pi}{6}t$

Take the derivative of $x=x(t)$ with respect to t：
$$\dfrac{\mathrm{d}x}{\mathrm{d}t}=-\dfrac{\pi}{2}\sin\dfrac{\pi}{6}t \text{ and } \dfrac{\mathrm{d}^2x}{\mathrm{d}t^2}=-\dfrac{\pi^2}{12}\cos\dfrac{\pi}{6}t$$

Take the derivative of $y=y(t)$ with respect to t：
$$\dfrac{\mathrm{d}y}{\mathrm{d}t}=\dfrac{\pi}{3}\cos\dfrac{\pi}{6}t \text{ and } \dfrac{\mathrm{d}^2y}{\mathrm{d}t^2}=-\dfrac{\pi^2}{18}\sin\dfrac{\pi}{6}t$$

1) The velocity $\vec{v(t)} = \langle \dfrac{dx}{dt}, \dfrac{dy}{dt} \rangle = \langle -\dfrac{\pi}{2}\sin\dfrac{\pi}{6}t, \dfrac{\pi}{3}\cos\dfrac{\pi}{6}t \rangle$,

$$\vec{v(6)} = \langle -\dfrac{\pi}{2}\sin\left(\dfrac{\pi}{6}\times 6\right), \dfrac{\pi}{3}\cos\left(\dfrac{\pi}{6}\times 6\right) \rangle = \langle 0, -\dfrac{\pi}{3} \rangle$$

The acceleration $\vec{a(t)} = \langle \dfrac{d^2x}{dt^2}, \dfrac{d^2y}{dt^2} \rangle = \langle -\dfrac{\pi^2}{12}\cos\dfrac{\pi}{6}t, -\dfrac{\pi^2}{18}\sin\dfrac{\pi}{6}t \rangle$,

$$\vec{a(6)} = \langle -\dfrac{\pi^2}{12}\cos\left(\dfrac{\pi}{6}\times 6\right), -\dfrac{\pi^2}{18}\sin\left(\dfrac{\pi}{6}\times 6\right) \rangle = \langle \dfrac{\pi^2}{12}, 0 \rangle$$

2) The speed of the particle when $t=6$

$$|\vec{v(6)}| = \sqrt{(0)^2 + \left(-\dfrac{\pi}{3}\right)^2} = \dfrac{\pi}{3}$$

3) The magnitude of the acceleration when $t=6$:

$$|\vec{a(6)}| = \sqrt{\left(\dfrac{\pi^2}{12}\right)^2 + (0)^2} = \dfrac{\pi^2}{12}$$

4) The slope of the curve is the slope of $\vec{v(t)}$,

$$\dfrac{dy}{dx} = \dfrac{\dfrac{dy}{dt}}{\dfrac{dx}{dt}} = \dfrac{\dfrac{\pi}{3}\cos\dfrac{\pi}{6}t}{-\dfrac{\pi}{2}\sin\dfrac{\pi}{6}t} = -\dfrac{2}{3}\cot\dfrac{\pi}{6}t$$

$$\left.\dfrac{dy}{dx}\right|_{t=2} = -\dfrac{2}{3}\cot\left(\dfrac{\pi}{6}\times 2\right) = -\dfrac{2\sqrt{3}}{9}$$

The slope of the curve along which the particle moves at the point where $t=2$ is $-\dfrac{2\sqrt{3}}{9}$.

6.4 微分中值定理 The Mean Value Theorem for Derivatives

微分中值定理把函数在区间上的平均变化率（Average Rate of Change）和该区间内一点处的瞬时变化率（Instantaneous Rate of Change）联系起来．

6.4.1 罗尔定理 Rolle's Theorem

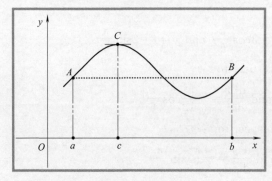

Figure 6.4.1(a)

Suppose that $y = f(x)$ is continuous at every point of the closed interval $[a, b]$ and differentiable at every point of its interior (a, b). If

$$f(a) = f(b)$$

Then there is at least one number c in (a, b) at which $f'(x) = 0$.

根据罗尔定理，如果连续光滑曲线 $y=f(x)$ 在点 A、B 处的纵坐标相等，那么，在弧$\overset{\frown}{AB}$上至少有一个点$C[c,f(c)]$，使得曲线在C点的切线平行于x轴，见 Figure 6.4.1(a)。

Example 6.25 Find the values of c that satisfy MVTD for $f(x)=\cos x$ on the interval $\left[-\dfrac{\pi}{2}, \dfrac{\pi}{2}\right]$.

【注】MVTD：The Mean Value Theorem for Derivatives

Solution：

The function $f(x)$ is continuous on $\left[-\dfrac{\pi}{2}, \dfrac{\pi}{2}\right]$ and is differentiable on $\left(-\dfrac{\pi}{2}, \dfrac{\pi}{2}\right)$.
$$f\left(-\dfrac{\pi}{2}\right)=\cos\left(-\dfrac{\pi}{2}\right)=0 \text{ and } f\left(\dfrac{\pi}{2}\right)=\cos\left(\dfrac{\pi}{2}\right)=0$$

According to the Rolle's Theorem, we obtain：
$$f'(c)=0$$

Find $f'(x)$：
$$f'(x)=-\sin x$$

Thus，
$$f'(c)=-\sin c=0, \text{ and } c=0$$

【方法总结】罗尔定理包含三个条件：1) 在闭区间 $[a,b]$ 上连续；2) 在开区间 (a,b) 内可导；3) $f(a)=f(b)$. 缺少其中任何一个条件，定理的结论不一定成立，见

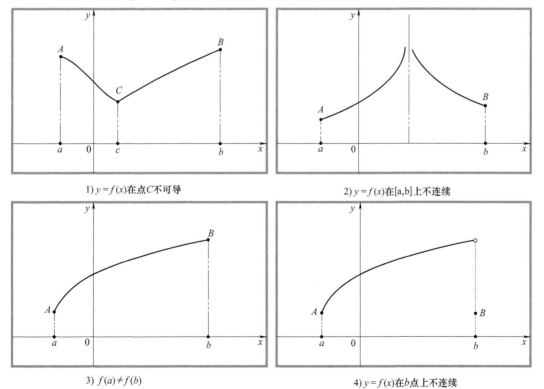

Figure 6.4.1(b)

Figure 6.4.1(b).

Example 6.26　Find the values of c that satisfy MVTD for $f(x)=1+\dfrac{1}{x}$ on the interval $[-3,3]$.

Solution：

Because $f(x)$ is not continuous at $x=0$, which is in the interval.

Then, there is no solution for c.

6.4.2　微分中值定理 The Mean Value Theorem for Derivatives

Suppose $y=f(x)$ is continuous on a closed interval $[a,b]$ and differentiable on the interval's interior (a,b). Then there is at least one point c in (a,b) at which

$$\frac{f(b)-f(a)}{b-a}=f'(c)$$

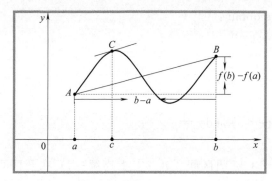

Figure 6.4.2

假设连续曲线 $y=f(x)$ 的弧 $\overset{\frown}{AB}$ 上处处具有不垂直于 x 轴的切线，见 Figure 6.4.2，可知 $\dfrac{f(b)-f(a)}{b-a}$ 是割线 AB 的斜率，而 $f'(c)$ 是曲线 $y=f(x)$ 上横坐标 $x=c$ 点的切线斜率．这样，根据微分中值定理，弧 $\overset{\frown}{AB}$ 上至少能找到一点 $C(c,f(c))$，使得曲线在 C 点的切线平行于割线 AB．

Example 6.27　Find the values of c that satisfy MVTD for $f(x)=x^3-x^2$ on the interval $[-3,3]$.

Solution：

The function $f(x)$ is continuous on $[-3,3]$ and is differentiable on $(-3,3)$.

$$\frac{f(3)-f(-3)}{3-(-3)}=\frac{18-(-36)}{6}=9$$

According to The Mean Value Theorem for Derivatives, we obtain：

$$f'(c)=9$$

Find $f'(x)$：

$$f'(x)=3x^2-2x$$

Thus,

$$f'(c)=3c^2-2c=9$$

$$c=\frac{1\pm 2\sqrt{7}}{6}$$

6.5　洛必达法则 L'Hôpital's Rule

到目前为止，经验告诉我们，求不定式的极限（Indeterminate），代数上的难易程度差别非常大. 但有了 L'Hôpital 法则，计算不定式的极限可以取得明显的成功.

6.5.1 不定式 $\frac{0}{0}$ 型 Indeterminate Form $\frac{0}{0}$

If $\lim\limits_{x \to x_0} f(x) = \lim\limits_{x \to x_0} f(x) = 0$, $f'(x_0)$ and $g'(x_0)$ exists, and that $g'(x_0) \neq 0$. Then,

$$\lim_{x \to x_0} \frac{f(x)}{g(x)} = \lim_{x \to x_0} \frac{f'(x)}{g'(x)}$$

如果连续函数 $f(x)$ 和 $g(x)$ 二者在 $x=x_0$ 都是零，那么 $\lim\limits_{x \to x_0} \frac{f(x)}{g(x)}$ 不能通过代入 $x=x_0$ 而求得. 代入产生 $\frac{0}{0}$，这是一个无意义的表达式，称为不定式（Indeterminate）. 类似的不定式有：$\frac{\infty}{\infty}$，$0 \times \infty$，$\infty - \infty$，1^{∞}，0^0 和 ∞^0.

Example 6.28 Find $\lim\limits_{x \to 0} \frac{(1+x)^7 - 1}{x}$.

Solution：

$[(1+x)^7 - 1] \to 0$ and $x \to 0$ as $x \to 0$

$$\lim_{x \to 0} \frac{(1+x)^7 - 1}{x} = \lim_{x \to 0} \frac{[(1+x)^7 - 1]'}{x'} = \lim_{x \to 0} \frac{7(1+x)^6}{1} = 7$$

Example 6.29 Find $\lim\limits_{x \to 0} \frac{1-\cos x}{2x^2}$.

Solution：

$(1-\cos x) \to 0$ and $x^3 \to 0$ as $x \to 0$

$$\lim_{x \to 0} \frac{1-\cos x}{2x^2} = \lim_{x \to 0} \frac{(1-\cos x)'}{(2x^2)'} = \lim_{x \to 0} \frac{\sin x}{4x}$$
$$= \frac{1}{4} \lim_{x \to 0} \frac{\sin x}{x} = \frac{1}{4}$$

Example 6.30 Find $\lim\limits_{x \to 0} \frac{x - \sin x}{x - x\cos x}$.

Solution：

$(x - \sin x) \to 0$ and $(x - x\cos x) \to 0$ as $x \to 0$

$$\lim_{x \to 0} \frac{x - \sin x}{x - x\cos x} = \lim_{x \to 0} \frac{(x - \sin x)'}{(x - x\cos x)'}$$
$$= \lim_{x \to 0} \frac{1 - \cos x}{1 - \cos x + x\sin x}$$

$(1 - \cos x) \to 0$ and $(1 - \cos x + x\sin x) \to 0$ as $x \to 0$

$$\lim_{x \to 0} \frac{1 - \cos x}{1 - \cos x + x\sin x} = \lim_{x \to 0} \frac{(1 - \cos x)'}{(1 - \cos x + x\sin x)'}$$
$$= \lim_{x \to 0} \frac{\sin x}{2\sin x + x\cos x}$$

$\sin x \to 0$ and $(2\sin x + x\cos x) \to 0$ as $x \to 0$

$$\lim_{x\to 0}\frac{\sin x}{2\sin x+x\cos x}=\lim_{x\to 0}\frac{(\sin x)'}{(2\sin x+x\cos x)'}$$
$$=\lim_{x\to 0}\frac{\cos x}{3\cos x-x\sin x}$$
$$=\frac{1}{3}$$

注：连续 3 次运用 L'Hôpital 法则．

6.5.2 不定式 $\frac{\infty}{\infty}$ 型 Indeterminate Form $\frac{\infty}{\infty}$

If $\lim\limits_{x\to x_0}f(x)=\lim\limits_{x\to x_0}f(x)=\infty$，$f'(x_0)$ and $g'(x_0)$ exists，and that $g'(x_0)\neq 0$．Then，

$$\lim_{x\to x_0}\frac{f(x)}{g(x)}=\lim_{x\to x_0}\frac{f'(x)}{g'(x)}$$

类似于不定式 $\frac{0}{0}$ 型求极限，不定式 $\frac{\infty}{\infty}$ 型求极限可直接运用 L'Hôpital 法则．这个方法的关键是：在符合 L'Hôpital 法则前提下，将 $\frac{f(x)}{g(x)}$ 的计算问题转化为 $\frac{f'(x)}{g'(x)}$ 的计算．

Example 6.31 Find $\lim\limits_{x\to\infty}\frac{x-3x^2}{2x^2-5x}$．

Solution：

$(x-3x^2)\to\infty$ and $(2x^2-5x)\to\infty$ as $x\to\infty$

$$\lim_{x\to\infty}\frac{x-3x^2}{2x^2-5x}=\lim_{x\to\infty}\frac{(x-3x^2)'}{(2x^2-5x)'}$$
$$=\lim_{x\to\infty}\frac{1-6x}{4x-5}=\lim_{x\to\infty}\frac{(1-6x)'}{(4x-5)'}$$
$$=\lim_{x\to\infty}\frac{-6}{4}=-\frac{3}{2}$$

Example 6.32 Find $\lim\limits_{x\to\frac{\pi}{2}}\frac{\tan x}{\tan 5x}$．

Solution：$\tan x\to\infty$ and $\tan 5x\to\infty$ as $x\to\frac{\pi}{2}$

$$\lim_{x\to\frac{\pi}{2}}\frac{\tan x}{\tan 5x}=\lim_{x\to\frac{\pi}{2}}\frac{(\tan x)'}{(\tan 5x)'}$$
$$=\lim_{x\to\frac{\pi}{2}}\frac{\sec^2 x}{5\sec^2 5x}=\frac{1}{5}\lim_{x\to\frac{\pi}{2}}\frac{\cos^2 5x}{\cos^2 x}$$
$$=\frac{1}{5}\lim_{x\to\frac{\pi}{2}}\frac{(\cos^2 5x)'}{(\cos^2 x)'}=\frac{1}{5}\lim_{x\to\frac{\pi}{2}}\frac{2\cos 5x(-5\sin 5x)}{2\cos x(-\sin x)}$$
$$=\lim_{x\to\frac{\pi}{2}}\frac{\sin 10x}{\sin 2x}=\lim_{x\to\frac{\pi}{2}}\frac{(\sin 10x)'}{(\sin 2x)'}$$

$$= \lim_{x \to \frac{\pi}{2}} \frac{10\cos 10x}{2\cos 2x} = 5$$

6.5.3# 其他不定式 Other Indeterminate Form

除了不定式 $\frac{0}{0}$ 和 $\frac{\infty}{\infty}$ 型外，不定式还有 5 种：1) $0 \times \infty$ 型；2) $\infty - \infty$ 型；3) 1^∞ 型；4) 0^0 型；5) ∞^0 型．上述 5 种类型的不定式，都可以通过转化为 $\frac{0}{0}$ 或 $\frac{\infty}{\infty}$ 型不定式进行计算．下面举例讲解：

Example 6.33#　Find $\lim\limits_{x \to \infty} x \sin \frac{1}{x}$.

Solution：

$$\lim_{x \to \infty} x \sin \frac{1}{x} = \lim_{x \to \infty} \frac{\sin \frac{1}{x}}{\frac{1}{x}}$$

Let $k = \frac{1}{x}$, then $k \to 0$ and $\sin k \to 0$ as $x \to \infty$

$$\lim_{x \to \infty} \frac{\sin \frac{1}{x}}{\frac{1}{x}} = \lim_{k \to 0} \frac{\sin k}{k} = \lim_{k \to 0} \frac{\cos k}{1} = 1$$

Note：本题是 $\infty \times 0$ 型，转化为 $\frac{0}{0}$ 型求解．

Example 6.34#　Find $\lim\limits_{x \to 0^+} x \ln x$.

Solution：

$$\lim_{x \to 0^+} x \ln x = \lim_{x \to 0^+} \frac{\ln x}{\frac{1}{x}}$$

$\frac{1}{x} \to \infty$ and $\ln x \to -\infty$ as $x \to 0^+$

$$\lim_{x \to 0^+} \frac{\ln x}{\frac{1}{x}} = \lim_{x \to 0^+} \frac{\frac{1}{x}}{-\frac{1}{x^2}} = \lim_{x \to 0^+} \left(-\frac{x^2}{x}\right) = 0$$

Note：本题是 $0 \times \infty$ 型，转化为 $\frac{\infty}{\infty}$ 型求解．

Example 6.35#　Find $\lim\limits_{x \to 0} \left(\csc x - \frac{1}{x}\right)$.

Solution：

$$\lim_{x \to 0} \left(\csc x - \frac{1}{x}\right) = \lim_{x \to 0} \left(\frac{1}{\sin x} - \frac{1}{x}\right) = \lim_{x \to 0} \frac{x - \sin x}{x \sin x}$$

$(x - \sin x) \to 0$ and $x \sin x \to 0$ as $x \to 0$

$$\lim_{x\to 0}\frac{x-\sin x}{x\sin x}=\lim_{x\to 0}\frac{(x-\sin x)'}{(x\sin x)'}$$

$$=\lim_{x\to 0}\frac{1-\cos x}{\sin x+x\cos x}=\lim_{x\to 0}\frac{(1-\cos x)'}{(\sin x+x\cos x)'}$$

$$=\lim_{x\to 0}\frac{\sin x}{2\cos x-x\sin x}=\frac{0}{2}=0$$

Note：本题是 $\infty-\infty$ 型，转化为 $\dfrac{0}{0}$ 型求解．

Example 6.36$^{\#}$　Find $\lim\limits_{x\to\infty}\left(1+\dfrac{1}{x}\right)^{x}$.

Solution：

Let $y=\left(1+\dfrac{1}{x}\right)^{x}$, so that $\ln y=\ln\left(1+\dfrac{1}{x}\right)^{x}=x\ln\left(1+\dfrac{1}{x}\right)$

Then,

$$\lim_{x\to\infty}\ln y=\lim_{x\to\infty}\left[x\ln\left(1+\frac{1}{x}\right)\right]=\lim_{x\to\infty}\frac{\ln\left(1+\dfrac{1}{x}\right)}{\dfrac{1}{x}}$$

$\dfrac{1}{x}\to 0$ and $\ln\left(1+\dfrac{1}{x}\right)\to 0$ as $x\to\infty$

$$\lim_{x\to\infty}\frac{\ln\left(1+\dfrac{1}{x}\right)}{\dfrac{1}{x}}=\lim_{x\to\infty}\frac{\dfrac{x}{1+x}\times\dfrac{x-(1+x)}{x^{2}}}{-\dfrac{1}{x^{2}}}$$

$$=\lim_{x\to\infty}\frac{x}{1+x}=1$$

Since $\lim\limits_{x\to\infty}\ln y=1$, $\lim\limits_{x\to\infty}\left(1+\dfrac{1}{x}\right)^{x}=\lim\limits_{x\to\infty}y=e^{1}=e$

Note：本题是 1^{∞} 型，转化为 $\dfrac{0}{0}$ 型求解．有时通过首先取对数可以处理不定式 1^{∞} 型、0^{0} 型和 ∞^{0} 型，我们使用 L'Hôpital 法则求出对数的极限，再取指数求出原极限．

Example 6.37$^{\#}$　Find $\lim\limits_{x\to 0^{+}}x^{x}$.

Solution：

Let $y=x^{x}$, so that $\ln y=\ln x^{x}=x\ln x$, then,

$$\lim_{x\to 0^{+}}\ln y=\lim_{x\to 0^{+}}(x\ln x)=\lim_{x\to 0^{+}}\frac{\ln x}{\dfrac{1}{x}}$$

$\dfrac{1}{x}\to\infty$ and $\ln x\to-\infty$ as $x\to 0^{+}$

$$\lim_{x\to 0^{+}}\frac{\ln x}{\dfrac{1}{x}}=\lim_{x\to 0^{+}}\frac{\dfrac{1}{x}}{-\dfrac{1}{x^{2}}}$$

$$= \lim_{x \to 0^+}(-x) = 0$$

Since $\lim\limits_{x \to 0^+} \ln y = 0$, $\lim\limits_{x \to 0^+} x^x = \lim\limits_{x \to 0^+} y = e^0 = 1$

Note：本题是 0^0 型，转化为 $\dfrac{\infty}{\infty}$ 型求解．

Example 6.38$^\#$ Find $\lim\limits_{x \to \infty} x^{\frac{1}{x}}$.

Solution：

Let $y = x^{\frac{1}{x}}$, so that $\ln y = \ln x^{\frac{1}{x}} = \dfrac{\ln x}{x}$, then,

$$\lim_{x \to \infty} \ln y = \lim_{x \to \infty}\left(\dfrac{\ln x}{x}\right)$$

$\ln x \to \infty$ as $x \to \infty$

$$\lim_{x \to \infty}\left(\dfrac{\ln x}{x}\right) = \lim_{x \to \infty} \dfrac{\dfrac{1}{x}}{1} = 0$$

Since $\lim\limits_{x \to \infty} \ln y = 0$, $\lim\limits_{x \to \infty} x^{\frac{1}{x}} = \lim\limits_{x \to \infty} y = e^0 = 1$

Note：本题是 ∞^0 型，转化为 $\dfrac{\infty}{\infty}$ 型求解．

6.6 估算问题 The Problems of Estimate

应用微分，可以使用数值法（Numerically）、图像法（Graphically）和对称差商法（From a Symmetric Difference Quotient）估算一个导数，构造导数估算极限值以及线性估算（Linear Approximation）．

6.6.1 估算一个导数 Estimating A Derivative

1）数值法和图像法 Numerically and Graphically

Since $f'(x_0) = \lim\limits_{\Delta x \to 0} \dfrac{f(x_0 + \Delta x) - f(x_0)}{\Delta x}$ and if that Δx is very small increment of x_0, then

$$f'(x_0) \approx \dfrac{f(x_0 + \Delta x) - f(x_0)}{\Delta x}$$

AP 微积分要求同学们会以给定的数值或函数图像估算某一点的导数值．

Example 6.39 If $f(x) = 6^x$ and $6^{1.001} \approx 6.011$, estimate $f'(1)$.

Solution：

$$f'(1) \approx \dfrac{6^{1.001} - 6^1}{1.001 - 1} = \dfrac{6.011 - 6}{0.001} = 11$$

Then,

$$f'(1) \approx 11$$

Example 6.40 A differentiable function $f(x)$ has the values shown. Estimate $f'(2.3)$.

x	2.0	2.2	2.4	2.6	2.8
$f(x)$	6	7	9	10	11

Solution:
$$f'(2.3) \approx \frac{f(2.4)-f(2.2)}{2.4-2.2} = \frac{9-7}{0.2} = 10$$

Then, $f'(2.3) \approx 10$.

Example 6.41 Use the Figure 6.6.1(a) to find $f'(5)$. The graph of $f(x)$ consists of a line segments and a semicircle.

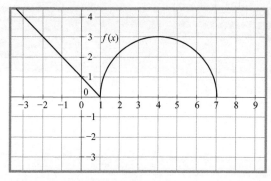

Figure 6.6.1(a)

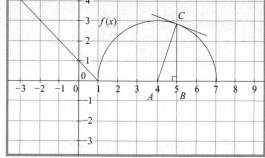

Figure 6.6.1(b)

Solution:
See Figure 6.6.1(b). Consider triangle ABC:
$$AB = 1 \text{ and radius } AC = 3$$
Thus,
$$BC = \sqrt{AC^2 - AB^2} = 2\sqrt{2}$$
The slope of AC:
$$k_{AC} = \frac{BC}{AB} = \frac{2\sqrt{2}}{1} = 2\sqrt{2}$$
The tangent line is perpendicular to the radius, then
$$k_{CF} = -\frac{1}{k_{AC}} = -\frac{\sqrt{2}}{4}$$
Thus, $f'(5) = -\frac{\sqrt{2}}{4}$.

2) 对称差商法 From a Symmetric Difference Quotient.

Estimate $f'(x_0)$ numerically using the symmetric difference quotient, which is defined as follows:[7]

$$f'(x_0) \approx \frac{f(x_0 + \Delta x) - f(x_0 - \Delta x)}{2\Delta x}$$

对称差商法是部分计算器估算函数导数的原理. 因为被估算的 $f'(x_0)$ 在 x 轴被置于 $x = x_0 + \Delta x$ 和 $x = x_0 - \Delta x$ 的中点，操作性强，准确度高.

Example 6.42 If the function $f(x) = x^5$, estimate $f'(1)$ using the symmetric difference quotient with $\Delta x = 0.001$.

Solution:
$$f'(1) \approx \frac{f(1+0.001) - f(1-0.001)}{2 \times 0.001} = \frac{1.001^5 - 0.999^5}{0.002} = 5.00001$$

Note：AP 微积分中，最后结果一般保留小数点后 3 位，中间结果保留小数点后 4 位．

6.6.2 构造导数估算极限值 Recognizing A Given Limit as A Derivative

$$\lim_{h \to 0} \frac{f(x_0 + h) - f(x_0)}{h} = f'(x_0)$$

根据导数的定义，某些趋于零的点极限，可以借助相对应的函数导数估算极限值．

Example 6.43 Find $\lim\limits_{x \to 0} \dfrac{(3+x)^5 - 3^5}{x}$.

Solution:
Let $f(x) = x^5$, then
$$f'(3) = \lim_{x \to 0} \frac{(3+x)^5 - 3^5}{x}$$

Find $f'(x)$:
$$f'(x) = 5x^4$$

Thus,
$$\lim_{x \to 0} \frac{(3+x)^5 - 3^5}{x} = f'(3) = 5 \times 3^4 = 405$$

Example 6.44 Find $\lim\limits_{x \to 0} \dfrac{\sin x}{x}$.

Solution:
Let $f(x) = \sin x$, then
$$f'(0) = \lim_{x \to 0} \frac{\sin(0+x) - \sin 0}{x} = \lim_{x \to 0} \frac{\sin x}{x}$$

Find $f'(x)$:
$$f'(x) = \cos x$$

Thus,
$$\lim_{x \to 0} \frac{\sin x}{x} = f'(0) = \cos 0 = 1$$

Example 6.45 Find $\lim\limits_{x \to 0} \dfrac{1}{x}\left(\dfrac{1}{5+x} - \dfrac{1}{5}\right)$.

Solution:
Let $f(x) = \dfrac{1}{x}$, then
$$f'(5) = \lim_{x \to 0} \frac{\dfrac{1}{5+x} - \dfrac{1}{5}}{x} = \lim_{x \to 0} \frac{1}{x}\left(\frac{1}{5+x} - \frac{1}{5}\right)$$

Find $f'(x)$:

$$f'(x) = -\frac{1}{x^2}$$

Thus,

$$\lim_{x \to 0} \frac{1}{x}\left(\frac{1}{5+x} - \frac{1}{5}\right) = f'(5) = -\frac{1}{5^2} = -\frac{1}{25}$$

6.6.3 线性估算 Linear Approximation

If $f(x)$ is differentiable at $x = x_0$, then the approximating function

$$L(x) = f(x_0) + f'(x_0)(x - x_0)$$

Is the linearization of $f(x)$ at x_0. The approximation

$$f(x) \approx L(x)$$

of $f(x)$ by $L(x)$ is the linear approximation of $f(x)$ at x_0.

函数 $y = f(x)$ 在 $x = x_0$ 处的切线为：

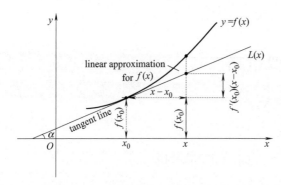

Figure 6.6.3

$$y - f(x_0) = f'(x_0)(x - x_0)$$

显然，当 x 靠近 x_0 时，切线 $L(x)$ 上的 y 值近似于函数值 $f(x)$，见 Figure 6.6.3. 故我们以

$$L(x) = f(x_0) + f'(x_0)(x - x_0)$$

线性估算（Linear Approximation）$f(x)$ 的值，这是微分几何意义的实际运用.

Example 6.46 Use differentials to approximate $\sqrt{25.01}$.

Solution：

Let $f(x) = \sqrt{x}$, $x_0 = 25$, $x = 25.01$. The derivative of the function

$$f'(x) = \frac{1}{2\sqrt{x}}$$

Then,

$$f(25.01) \approx f(25) + f'(25) \times (25.01 - 25)$$
$$= \sqrt{25} + \frac{1}{2\sqrt{25}} \times (25.01 - 25) = 5.001$$

Thus,

$$\sqrt{25.01} \approx 5.001$$

Example 6.47 Find the best linear approximation for $f(x) = \tan x$ near $x = \frac{\pi}{6}$.

Solution：

The derivative of the function

$$f'(x) = \sec^2 x$$

Then,

$$L(x) = f\left(\frac{\pi}{6}\right) + f'\left(\frac{\pi}{6}\right)\left(x - \frac{\pi}{6}\right)$$
$$= \tan\left(\frac{\pi}{6}\right) + \sec^2\left(\frac{\pi}{6}\right)\left(x - \frac{\pi}{6}\right)$$
$$= \frac{\sqrt{3}}{3} + \frac{4}{3}\left(x - \frac{\pi}{6}\right)$$

Thus, the best linear approximation for $f(x) = \tan x$ near $x = \frac{\pi}{6}$ is

$$\tan x \approx \frac{\sqrt{3}}{3} + \frac{4}{3}\left(x - \frac{\pi}{6}\right)$$

6.7# 欧拉方法 Euler's Method

Approximate values for the solution of the initial-value problem y'_n, $y(x_0) = y_0$, with step size h, at $x_n = x_{n-1} + h$, are

$$y_n = y_{n-1} + h\, y'_{n-1}$$

Repeat for $n = 1, 2, 3, \cdots$

欧拉方法可看做线性估算的进一步完善．线性估算随着 $\Delta x = x - x_0$ 增大，精确程度会快速下降，只有当 Δx 很小时才被应用．当 Δx 比较大时，应用欧拉方法，实施分步估算，提高估算精确度．

见 Figure 6.7(a)，若以 x_0 线性估算 $f(x_2)$，即
$$f(x_2) \approx L(x_0) = f(x_0) + f'(x_0)(x_2 - x_0) = BC$$
其误差（error）为：$f(x_2) - L(x_0) = AB$；

取 x_0 和 x_2 的中点 x_1，采用欧拉方法，依次估算 $f(x_1)$ 和 $f(x_2)$，见 Figure 6.7(b)，
$$f(x_1) \approx L'(x_0) = f(x_0) + f'(x_0)(x_1 - x_0) = ED = CF$$
$$f(x_2) \approx L(x_1) = L'(x_0) + f'(x_1)(x_2 - x_1) = CF + FG = CG$$

其中，$EF // l'$，l' 为 $y = f(x)$ 在 $x = x_1$ 处的切线，则其误差为：$f(x_2) - L(x_1) = AG$；显然，$AG < AB$，即采用欧拉方法分两步估算比线性估算精确度更高，并且，增加分步数可提高估算精度．

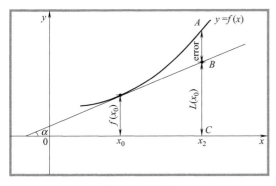

Figure 6.7(a)

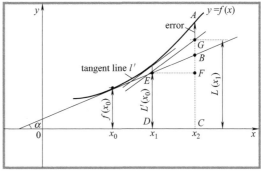

Figure 6.7(b)

Example 6.48#　Use Euler's Method, with $h = 0.4$, to estimate $y(1.2)$ if $y'(x) =$

$y-1$ and $y(0)=2$.

Solution: We get,
$$x_0=0, \text{ and } y_0=2, y'_0=y_0-1=2-1=1$$

Use Euler's Method,
$$x_1=x_0+h=0+0.4=0.4$$
$$y_1=y_0+h\,y'_0=2+0.4\times 1=2.4$$
$$y'_1=y_1-1=2.4-1=1.4$$

Next
$$x_2=x_1+h=0.4+0.4=0.8$$
$$y_2=y_1+h\,y'_1=2.4+0.4\times 1.4=2.96$$
$$y'_2=y_2-1=2.96-1=1.96$$

Next
$$x_3=x_2+h=0.8+0.4=1.2$$
$$y_3=y_2+h\,y'_2=2.96+0.4\times 1.96=3.744$$

Thus,
$$y(1.2)\approx 3.744$$

Example 6.49[#] Given that $y(0)=1$ and $y'(x)=x+y$, use Euler's Method, with $h=0.1$, to estimate $y(0.3)$.

Solution: We get,
$$x_0=0, \text{ and } y_0=1, y'_0=x_0+y_0=0+1=1$$

Use Euler's Method,
$$x_1=x_0+h=0+0.1=0.1$$
$$y_1=y_0+hy'_0=1+0.1\times 1=1.1$$
$$y'_1=x_1+y_1=0.1+1.1=1.2$$

Next
$$x_2=x_1+h=0.1+0.1=0.2$$
$$y_2=y_1+hy'_1=1.1+0.1\times 1.2=1.22$$
$$y'_2=x_2+y_2=0.2+1.22=1.42$$

Next
$$x_3=x_2+h=0.2+0.1=0.3$$
$$y_3=y_2+hy'_2=1.22+0.1\times 1.42=1.362$$

Thus,
$$y(0.3)\approx 1.362$$

6.8 习题 Practice Exercises

(1) Find the tangent line of parabola $y=2x^2+3x+1$ at point $M(1,6)$.

(2) Find the coordinate of any point on the curve of $y^2+x^2=xy+9$ for which the tangent is parallel to the x-axis.

(3)# Find the equation of the normal to $F(t)=(\sin^2 t, \cos t)$ at the point where $t=\frac{\pi}{3}$.

(4) Let $f(x)=2x^3+x^2$. Find the intervals on which $f(x)$ is increasing or decreasing.

(5) Find the local minimum and maximum values of $y=x^3-3x+1$ on the interval $[-4, 4]$.

(6) Find the absolute minimum and maximum values of $y=2x^3-x+1$ on the interval $[-1, 1]$.

(7) A spherical balloon is inflating at a rate 64π in^3/sec. How fast is the radius of the balloon increasing when the radius is 4 in?

(8) If the position of a particle at a time is given by the equation $f(t)=2t^3-15t^2+24t-5$, for $t\geqslant 0$.

1) Find all t for which the distance is decreasing.

2) Find all t for which the velocity is increasing.

3) Find all t for which the speed of the particle is increasing.

4) Find the total distance traveled between $t=0$ and $t=5$.

(9) Find the values of c that satisfy MVTD for $f(x)=\sin x$ on the interval $[-\pi, 0]$.

(10)# Find these limits using L'Hôpital's Rule.

1) $\lim\limits_{x\to 0}\dfrac{(1+2x)^6-1}{x}$;

2) $\lim\limits_{x\to\infty}\dfrac{x^2}{e^{2x}}$;

3) $\lim\limits_{x\to 0}\dfrac{x-\sin x}{x^2}$;

4) $\lim\limits_{x\to 1} x^{\frac{1}{1-x}}$;

5) $\lim\limits_{x\to 0}\dfrac{\sin 2x}{3x}$.

(11) If $f(x)=5^x$ and $5^{1.001}\approx 5.008$, estimate $f'(1)$.

(12) Use differentials to approximate $\sqrt{36.01}$.

(13)# Given that $y(0)=2$ and $y'(x)=x-y$, use Euler's Method, with $h=0.1$, to estimate $y(0.3)$.

第 7 章

不定积分
The Indefinite Integral

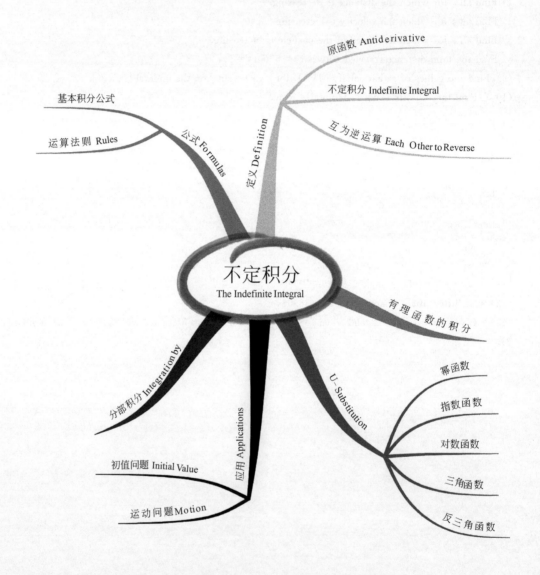

7.1 不定积分的定义 Definition of The Indefinite Integral

前面我们讨论了如何求一个函数的导函数（Derivative）的问题，本节我们将讨论它的逆过程，即寻找一个原函数（Antiderivative），使得它的导函数等于已知函数，这是积分学的基本问题之一．

7.1.1 原函数 Antiderivative

If $F'(x)=f(x)$, Then $F(x)$ is called an antiderivative of $f(x)$.

设 $f(x)$ 是区间 I 上的已知函数．如果存在函数 $F(x)$，使得任意 $x \in I$ 都有 $F'(x)=f(x)$，或者 $dF(x)=f(x)dx$，则称 $F(x)$ 是 $f(x)$ 在 I 上的原函数（Antiderivative）．

Example 7.1 Find an antiderivative for each of the following functions.

a) $f(x)=4x$；b) $g(x)=\sin x$；c) $h(x)=4x+\sin x$；d) $k(x)=\dfrac{1}{x\ln 2}$，$(x>0)$．

Solution：

a) Since $(2x^2)'=4x$, then $F(x)=2x^2$.

b) Since $(-\cos x)'=\sin x$, then $G(x)=-\cos x$.

c) Since $(2x^2-\cos x)'=4x+\sin x$, then $H(x)=2x^2-\cos x$.

d) Since $(\log_2 x)'=\dfrac{1}{x\ln 2}$, $(x>0)$, then $K(x)=\log_2 x$.

If $F(x)$ is an antiderivative of $f(x)$ on an interval I, then the most general antiderivative of $f(x)$ on I is

$$F(x)+C$$

where C is an arbitrary constant.

当 $f(x)$ 在区间 I 上连续（Continuity）时，它在 I 上的原函数（Antiderivative）必定存在．另外，求不定积分时，切莫漏掉常数 C，多个不定积分相加可能有多个不同常数相加，统一用一个常数 C 表示．

Example 7.2 Find an antiderivative of $f(x)=\cos x$ that satisfies $F\left(\dfrac{\pi}{2}\right)=4$.

Solution：

Since the derivative of $\sin x$ is $\cos x$, the general antiderivative

$$F(x)=\sin x+C$$

Gives all the antiderivatives of $f(x)$.

Substituting $x=\dfrac{\pi}{2}$ into $F(x)=\sin x+C$ gives

$$F\left(\dfrac{\pi}{2}\right)=\sin\left(\dfrac{\pi}{2}\right)+C=1+C$$

Since $F\left(\dfrac{\pi}{2}\right)=4$, solving for C gives $C=3$. Thus

$$F(x)=\sin x+3$$

is the antiderivative satisfying $F\left(\dfrac{\pi}{2}\right)=4$.

7.1.2 不定积分 Indefinite Integral

The set of all antiderivatives of $f(x)$ is the indefinite integral of $f(x)$ with respect to x, denoted by

$$\int f(x)\,\mathrm{d}x$$

The symbol $\int$ is an integral sign. The function $f(x)$ is the integrand of the integral, and x is the variable of integration. The $\mathrm{d}x$ is the differential divisor.

Suppose $\dfrac{\mathrm{d}F(x)}{\mathrm{d}x}=f(x)$, then

$$\int f(x)\,\mathrm{d}x = F(x)+C$$

设 $F(x)$ 是 $f(x)$ 的一个原函数（Antiderivative），则称 $f(x)$ 的所有原函数 $F(x)+C$（C 是任意常数）为 $f(x)$ 的不定积分（Indefinite Integral）。

Example 7.3 Evaluate $\int 2x\,\mathrm{d}x$.

Solution：

Since $(x^2)'=2x$, x^2 is an antiderivative of $2x$.

Thus,

$$\int 2x\,\mathrm{d}x = x^2+C$$

【方法总结】根据函数求导法则，$(x^2)'=2x$，$(x^2-1)'=2x$，$(x^2+1)'=2x$，…，也就是说，x^2 是 $2x$ 的一个原函数（Antiderivative），C 表示任意常数，$2x$ 关于 x 的不定积分（Indefinite Integral）是 x^2+C，即 $\int 2x\,\mathrm{d}x = x^2+C$。

7.1.3 互为逆运算 Each Other to Reverse

An antiderivative is a derivative in reverse. Finding an antiderivative is also called an indefinite integral.

Since $\int f(x)\,\mathrm{d}x = F(x)+C$ and $F'(x)=f(x)$, we get

$$\int F'(x)\,\mathrm{d}x = F(x)+C$$

That is

$$\int f'(x)\,\mathrm{d}x = f(x)+C$$

函数 $f(x)$ 求导之后得到导函数 $f'(x)$，再对导函数 $f'(x)$ 不定积分，得到原函数 $f(x)$。因此，类似于加法与减法、乘法与除法、乘法与开方，求导（Differentiation）与不定积分（Indefinite Integral）互为逆运算。

Example 7.4 Find $f(x)$ if $f'(x)=x$ and $f(1)=2$.

Solution:

The general antiderivative of
$$f'(x) = x$$
is $f(x) = \frac{1}{2}x^2 + C$, therefore,
$$f(x) = \int f'(x)\,dx = \int x\,dx = \frac{1}{2}x^2 + C$$

To determine C we use the fact that $f(1) = 2$:
$$f(1) = \frac{1}{2}(1)^2 + C = 2$$

Solving for C, we get $C = \frac{3}{2}$, so the solution is
$$f(x) = \frac{1}{2}x^2 + \frac{3}{2}$$

原理上，因为求导和不定积分互为逆运算，每个求导公式可写出其对应的积分公式，以便于不定积分运算．比如：

由 $(x^n)' = nx^{n-1}$，可得 $\int nx^{n-1}\,dx = x^n + C$

即：
$$\int x^{n-1}\,dx = \frac{x^n + C}{n}$$

为了便于记忆和使用，上式可写成（Power Rule for Indefinite Integral）：
$$\int x^n\,dx = \frac{x^{n+1}}{n+1} + C \quad (n \neq -1)$$

这是一道重要的不定积分公式．运用此公式，Example 7.3 和 Example 7.4 能更方便快速得到解答，接下来我们将给出所有基本积分公式．

至此，我们开始触及微积分核心．同学们，跟紧我的脚步，微积分殿堂大门即将为你开启……

7.2 不定积分公式 Formulas of The Indefinite Integral

不定积分（Indefinite Integral）是 AP 微积分中最基本的运算之一，基本积分（Integration）公式和不定积分运算法则是完成不定积分计算的基础，同学们务必结合记忆规律，"巧记" ＋ "死记"，熟练掌握．

7.2.1 基本积分公式 Basic Integration Formulas

因为求不定积分是求导数的逆运算，所以由导数（Derivative）的基本公式对应地可以得到基本积分（Integration）公式，见 Table 7.2.1.

Example 7.5　Find $\int x^5\,dx$.

Solution:

Table 7.2.1 基本积分 (Integration) 公式

Constants	$\int k\,dx = kx + C$, C is constants	
Power	$\int x^n\,dx = \dfrac{x^{n+1}}{n+1} + C\ (n \neq -1)$	
Exponential	$\int a^x\,dx = \dfrac{a^x}{\ln a} + C$	$\int e^x\,dx = e^x + C$
Logarithmic	$\int \dfrac{1}{x}\,dx = \ln|x| + C$	
Trigonometric	$\int \sin x\,dx = -\cos x + C$	$\int \cos x\,dx = \sin x + C$
	$\int \tan x\,dx = -\ln|\cos x| + C$	$\int \cot x\,dx = \ln|\sin x| + C$
	$\int \sec x\,dx = \ln|\sec x + \tan x| + C$	$\int \csc x\,dx = \ln|\csc x - \cot x| + C$
	$\int \sec^2 x\,dx = \tan x + C$	$\int \csc^2 x\,dx = -\cot x + C$
	$\int \sec x \tan x\,dx = \sec x + C$	$\int \csc x \cot x\,dx = -\csc x + C$
Inverse Trigonometric	$\int \dfrac{dx}{\sqrt{1-x^2}} = \arcsin x + C$	
	$\int \dfrac{dx}{1+x^2} = \arctan x + C$	
	$\int \dfrac{dx}{x\sqrt{x^2-1}} = \text{arcsec}\,x + C$	

Using the Power Rule $\int x^n\,dx = \dfrac{x^{n+1}}{n+1} + C$, we get:

$$\int x^5\,dx = \dfrac{x^6}{6} + C$$

Example 7.6 Find $\int x^{-5}\,dx$.

Solution:

$$\int x^{-5}\,dx = \dfrac{x^{-4}}{-4} + C$$

Example 7.7 Find $\int \sqrt{x}\,dx$.

Solution:

$$\int \sqrt{x}\,dx = \int x^{\frac{1}{2}}\,dx = \dfrac{x^{\frac{1}{2}+1}}{\frac{1}{2}+1} + C = \dfrac{2}{3}\sqrt{x^3} + C$$

Example 7.8 Find $\int 3x^7\,dx$.

Solution:

$$\int 3x^7\,dx = 3\int x^7\,dx = 3 \times \dfrac{x^8}{8} + C$$

7.2.2 不定积分运算法则 Rules for Finding Indefinite Integral

1)	$\int kf(x)\mathrm{d}x = k\int f(x)\mathrm{d}x$		
2)	$\int [f(x)+g(x)]\mathrm{d}x = \int f(x)\mathrm{d}x + \int g(x)\mathrm{d}x$		
	$\int [f(x)-g(x)]\mathrm{d}x = \int f(x)\mathrm{d}x - \int g(x)\mathrm{d}x$		
3)	$\left[\int f(x)\mathrm{d}x\right]' = f(x)$		$\mathrm{d}\int f(x)\mathrm{d}x = f(x)\mathrm{d}x$
4)	$\int f'(x)\mathrm{d}x = f(x)+C$		$\int \mathrm{d}f(x) = f(x)+C$

不定积分只有加法公式（Sum Rule）和减法公式（Difference Rule），没有乘法和除法公式，这与求导法则有区别.

Example 7.9 Find $\int (x^3 + \sin x + 7)\mathrm{d}x$.

Solution：

Break this into separate integrals, we get

$$\int (x^3 + \sin x + 7)\mathrm{d}x = \int x^3 \mathrm{d}x + \int \sin x \mathrm{d}x + \int 7\mathrm{d}x$$

Integrate each of these individually：

$$\int x^3 \mathrm{d}x + \int \sin x \mathrm{d}x + \int 7\mathrm{d}x = \frac{x^4}{4} - \cos x + 7x + C$$

Thus,

$$\int (x^3 + \sin x + 7)\mathrm{d}x = \frac{x^4}{4} - \cos x + 7x + C$$

Example 7.10 Find $\int (5-x^3)^2 \mathrm{d}x$.

Solution：

Expand the integrand：

$$\int (5-x^3)^2 \mathrm{d}x = \int (25 - 10x^3 + x^6)\mathrm{d}x$$

Break this up into several integrals：

$$\int (25 - 10x^3 + x^6)\mathrm{d}x = \int 25\mathrm{d}x - \int 10x^3 \mathrm{d}x + \int x^6 \mathrm{d}x$$

And integrate according to the power Rule：

$$\int 25\mathrm{d}x - \int 10x^3 \mathrm{d}x + \int x^6 \mathrm{d}x = 25x - \frac{5x^4}{2} + \frac{x^7}{7} + C$$

Thus,

$$\int (5-x^3)^2 \mathrm{d}x = 25x - \frac{5x^4}{2} + \frac{x^7}{7} + C$$

Example 7.11 Find $f(x)$ if $f'(x) = 2\sin x + \cos x$ and $f\left(\frac{\pi}{3}\right) = 4$.

Solution：

Since $\int f'(x)\mathrm{d}x = f(x) + C_1$ and

$$\int f'(x)\mathrm{d}x = \int (2\sin x + \cos x)\mathrm{d}x$$
$$= 2\int \sin x\,\mathrm{d}x + \int \cos x\,\mathrm{d}x$$
$$= -2\cos x + \sin x + C_2$$

Then,
$$f(x) = -2\cos x + \sin x + C_2 - C_1$$

Let $C = C_2 - C_1$, then
$$f(x) = -2\cos x + \sin x + C$$

To determine C we use the fact that $f\left(\dfrac{\pi}{3}\right) = 4$:

$$f\left(\dfrac{\pi}{3}\right) = -2\cos\left(\dfrac{\pi}{3}\right) + \sin\left(\dfrac{\pi}{3}\right) + C = 4$$

Solving for C, we get $C = 5 - \dfrac{\sqrt{3}}{2}$, so the solution is

$$f(x) = -2\cos x + \sin x + 5 - \dfrac{\sqrt{3}}{2}$$

Example 7.12 Find $\int e^{2x} f'(e^x)\mathrm{d}x$, if $x\ln x - x$ is an antiderivative of the function $f(x)$.

Solution:

Since $x\ln x - x$ is an antiderivative of the function $f(x)$, then

$$\int f(x)\mathrm{d}x = x\ln x - x$$

And

$$\left[\int f(x)\mathrm{d}x\right]' = f(x)$$

$$f(x) = \left[\int f(x)\mathrm{d}x\right]' = (x\ln x - x)' = \ln x + 1 - 1 = \ln x$$

Then,

$$f'(x) = (\ln x)' = \dfrac{1}{x} \text{ and } f'(e^x) = \dfrac{1}{e^x}$$

Thus,

$$\int e^{2x} f'(e^x)\mathrm{d}x = \int e^{2x} \dfrac{1}{e^x}\mathrm{d}x = \int e^x\,\mathrm{d}x = e^x + C$$

7.3 U-替换法 U-Substitution

If $u = g(x)$ is a differentiable function whose range is an interval I and $f(x)$ is continuous on I, then

$$\int f[g(x)]g'(x)\mathrm{d}x = \int f[g(x)]\mathrm{d}g(x)$$

Therefore,

$$\int f[g(x)]g'(x)\mathrm{d}x = \int f(u)\mathrm{d}u$$

利用基本积分公式和运算法则，所能计算的不定积分是非常有限的，我们需要进一步研究不定积分的求法．

在导数和微分章节，我们已经学习了针对复合函数（Composite Function）求导的链式法则（Chain Rule）．本节把复合函数求导的链式法则倒过来，利用中间变量 $u(x)$ 替换得到复合函数的积分法，称为 U-Substitution，是不定积分求法必要的扩充．

假设 $f(u)$ 具有原函数 $F(u)$，$u = g(x)$ 可导，运用 U-Substitution，所求不定积分需要具备以下特征：

$$\int f[g(x)]g'(x)\mathrm{d}x$$

被积函数（Integrand）可以看成一个复合函数 $f[g(x)]$ 与其中间变量导数 $g'(x)$ 的乘积，那么，令 $u = g(x)$，此时，若 $\int f(u)\mathrm{d}u$ 可以从基本积分公式中查到，得到

$$\int f(u)\mathrm{d}u = F(u) + C$$

再将 $u = g(x)$ 替代上式右端 u，便拿到所求的积分．

7.3.1 幂函数积分 Integrals of Power Functions

$$\text{If } f(u) = u^n, \text{ then } \int u^n \mathrm{d}u = \frac{u^{n+1}}{n+1} + C \,(n \neq -1)$$

Whenever u is a function of x, we define $\mathrm{d}u$ to be $u'(x)\ \mathrm{d}x$; when u is a function of t, we define $\mathrm{d}u$ to be $u'(t)\ \mathrm{d}t$; and so on.

幂函数种类繁多，幂函数积分在 AP 微积分考试中一般难度不大，却是每年必考之积分类型，其中有些细节，请同学们仔细把握，跟着笔者再多学一点．

Example 7.13 Evaluate $\int (3x+5)^7 \mathrm{d}x$.

Solution：

Let $u = 3x + 5$, then $\mathrm{d}u = 3\mathrm{d}x$

$$\int (3x+5)^7 \mathrm{d}x = \int \frac{1}{3}u^7 \mathrm{d}u$$

Integrate：

$$\int \frac{1}{3}u^7 \mathrm{d}u = \frac{1}{3} \times \frac{u^8}{8} + C$$

Substitute back for u：

$$\int (3x+5)^7 \mathrm{d}x = \frac{(3x+5)^8}{24} + C$$

Example 7.14 Evaluate $\int 8x(4x^2+5)^7 \mathrm{d}x$.

Solution：

Let $u=4x^2+5$, then $du=8x\,dx$

$$\int 8x(4x^2+5)^7\,dx = \int u^7\,du$$

Integrate:

$$\int u^7\,du = \frac{u^8}{8}+C$$

Substitute back for u:

$$\int 8x(4x^2+5)^7\,dx = \frac{(4x^2+5)^8}{8}+C$$

Example 7.15　Evaluate $\int 3x^2\sqrt{x^3+8}\,dx$.

Solution:

Let $u=x^3+8$, then $du=3x^2\,dx$

$$\int 3x^2\sqrt{x^3+8}\,dx = \int \sqrt{u}\,du$$

Integrate:

$$\int \sqrt{u}\,du = \frac{2\sqrt{u^3}}{3}+C$$

Substitute back for u:

$$\int 3x^2\sqrt{x^3+8}\,dx = \frac{2\sqrt{(x^3+8)^3}}{3}+C$$

7.3.2　指数函数和对数函数积分 Integrals of Exponential and Logarithmic Functions

$$\int a^u\,du = \frac{a^u}{\ln a}+C\,;\quad \int e^u\,du = e^u+C\,;\quad \int \frac{1}{u}\,du = \ln|u|+C$$

假定 $u=u(x)$，指数函数 $f(u)$ 是复合函数，其不定积分需要运用 U-Substitution，部分同学容易在此出错. 并且，函数 $f(u)=e^u$ 很有意思，求导或是求积分，都回到其本身. $\int \frac{1}{u}\,du = \ln|u|+C$ 不单运用于对数函数不定积分，在三角函数积分中也有较多的运用.

Example 7.16　Evaluate $\int 2^{3x+4}\,dx$.

Solution:

Let $u=3x+4$, then $du=3\,dx$

$$\int 2^{3x+4}\,dx = \int \frac{1}{3}\times 2^u\,du$$

Integrate:

$$\int \frac{1}{3}\times 2^u\,du = \frac{1}{3}\times \frac{2^u}{\ln 2}+C$$

Substitute back for u:

$$\int 2^{3x+4}\,dx = \frac{1}{3}\times \frac{2^{3x+4}}{\ln 2}+C$$

Example 7.17   Evaluate $\int e^{2x} dx$.

Solution:

Let $u = 2x$, then $du = 2dx$

$$\int e^{2x} dx = \int \frac{1}{2} e^u du$$

Integrate:

$$\int \frac{1}{2} e^u du = \frac{1}{2} e^u + C$$

Substitute back for u:

$$\int e^{2x} dx = \frac{1}{2} e^{2x} + C$$

Example 7.18   Evaluate $\int x e^{2x^2+1} dx$.

Solution:

Let $u = 2x^2 + 1$, then $du = 4x dx$

$$\int x e^{2x^2+1} dx = \int \frac{1}{4} e^u du$$

Integrate:

$$\int \frac{1}{4} e^u du = \frac{1}{4} e^u + C$$

Substitute back for u:

$$\int x e^{2x^2+1} dx = \frac{1}{4} e^{2x^2+1} + C$$

Example 7.19   Evaluate $\int \frac{7}{x+1} dx$.

Solution:

Let $u = x+1$, then $du = x dx$

$$\int \frac{7}{x+1} dx = 7 \int \frac{1}{u} du$$

Integrate:

$$7 \int \frac{1}{u} du = 7 \ln |u| + C$$

Substitute back for u:

$$\int \frac{7}{x+1} dx = 7 \ln |x+1| + C$$

Example 7.20   Evaluate $\int \frac{2x^2}{1+x^3} dx$.

Solution:

Let $u = 1 + x^3$, then $du = 3x^2 dx$

$$\int \frac{2x^2}{1+x^3} dx = \int \frac{2}{3} \times \frac{1}{u} du$$

Integrate:
$$\int \frac{2}{3} \times \frac{1}{u} du = \frac{2}{3} \ln|u| + C$$

Substitute back for u:
$$\int \frac{2x^2}{1+x^3} dx = \frac{2}{3} \ln|1+x^3| + C$$

7.3.3　三角函数积分 I　Integrals of Trigonometric Functions I

$$\int \sin u\, du = -\cos u + C \qquad \int \cos u\, du = \sin u + C$$

以上是三角函数基础的积分公式．就我的教学经验而言，三角函数积分是中国大陆学生的一个薄弱点．其中一个原因是它的公式比较多，下面导出（Proof）$\int \tan x\, dx$、$\int \cot x\, dx$、$\int \sec x\, dx$ 和 $\int \csc x\, dx$，帮助同学们揭示三角函数积分公式的内在联系．

Example 7.21　Evaluate $\int \tan x\, dx$.

Solution:
Let $u = \cos x$, then $du = -\sin x\, dx$
$$\int \tan x\, dx = \int \frac{\sin x}{\cos x} dx = \int \frac{-1}{u} du$$

Integrate:
$$\int \frac{-1}{u} du = -\ln|u| + C$$

Substitute back for u:
$$\int \tan x\, dx = -\ln|\cos x| + C$$

Example 7.22　Evaluate $\int \cot x\, dx$.

Solution:
Let $u = \sin x$, then $du = \cos x\, dx$
$$\int \cot x\, dx = \int \frac{\cos x}{\sin x} dx = \int \frac{1}{u} du$$

Integrate:
$$\int \frac{1}{u} du = \ln|u| + C$$

Substitute back for u:
$$\int \cot x\, dx = \ln|\sin x| + C$$

Example 7.23　Evaluate $\int \sec x\, dx$.

Solution:
Multiply the integrand $\sec x$ by

$$\frac{\sec x + \tan x}{\sec x + \tan x}$$

Then,
$$\int \sec x \, dx = \int \sec x \, \frac{\sec x + \tan x}{\sec x + \tan x} dx = \int \frac{\sec^2 x + \sec x \tan x}{\sec x + \tan x} dx$$

Let $u = \sec x + \tan x$, then
$$du = (\sec^2 x + \sec x \tan x) dx$$
$$\int \frac{\sec^2 x + \sec x \tan x}{\sec x + \tan x} dx = \int \frac{1}{u} du$$

Integrate:
$$\int \frac{1}{u} du = \ln|u| + C$$

Substitute back for u:
$$\int \sec x \, dx = \ln|\sec x + \tan x| + C$$

Example 7.24 Evaluate $\int \csc x \, dx$.

Solution: Since
$$\csc x = \frac{1}{\sin x} = \frac{1}{2\sin\frac{x}{2}\cos\frac{x}{2}} = \frac{\sec^2 \frac{x}{2}}{2\tan\frac{x}{2}}$$

Then,
$$\int \csc x \, dx = \int \frac{\sec^2 \frac{x}{2}}{2\tan\frac{x}{2}} dx$$

Let $u = \tan\frac{x}{2}$, then
$$du = \frac{1}{2} \sec^2 \frac{x}{2} dx$$
$$\int \frac{\sec^2 \frac{x}{2}}{2\tan\frac{x}{2}} dx = \int \frac{1}{u} du$$

Integrate:
$$\int \frac{1}{u} du = \ln|u| + C$$

Substitute back for u:
$$\int \csc x \, dx = \ln\left|\tan\frac{x}{2}\right| + C$$

Since
$$\tan\frac{x}{2} = \frac{\sin\frac{x}{2}}{\cos\frac{x}{2}} = \frac{2\sin^2\frac{x}{2}}{2\sin\frac{x}{2}\cos\frac{x}{2}} = \frac{1-\cos x}{\sin x} = \frac{1}{\sin x} - \frac{\cos x}{\sin x} = \csc x - \cot x$$

Then,
$$\int \csc x \, dx = \ln|\csc x - \cot x| + C$$

【方法总结】事实上，仅用 $\sin x$ 和 $\cos x$ 的不定积分公式，$\int \tan x \, dx$、$\int \cot x \, dx$、$\int \sec x \, dx$ 和 $\int \csc x \, dx$ 的推导过程给了解答三角函数不定积分一个典型思路：结合三角函数基本公式和 U-Substitution，运用

$$\int \frac{1}{u} du = \ln|u| + C$$

最后，整理过程，得到答案.

7.3.4 三角函数积分 II Integrals of Trigonometric Functions II

$$\int \sec^2 x \, dx = \tan x + C \qquad \int \csc^2 x \, dx = -\cot x + C$$
$$\int \sec x \tan x \, dx = \sec x + C \qquad \int \csc x \cot x \, dx = -\csc x + C$$

我们已经讲过：求导（Differentiation）与不定积分（Indefinite Integral）互为逆运算. 并且，导数的基本公式：

$$(\tan x)' = \sec^2 x \qquad (\cot x)' = -\csc^2 x$$
$$(\sec x)' = \sec x \tan x \qquad (\csc x)' = -\csc x \cot x$$

所以，由上述导数的基本公式可对应写出其三角函数基本积分公式. 求 $\tan x$、$\cot x$、$\sec x$ 和 $\csc x$ 的导数的原函数，是一类常见的三角函数积分.

Example 7.25 Evaluate $\int \tan x \sec^2 x \, dx$.

Solution:

Let $u = \tan x$, then $du = \sec^2 x \, dx$
$$\int \tan x \sec^2 x \, dx = \int u \, du$$

Integrate:
$$\int u \, du = \frac{u^2}{2} + C$$

Substitute back for u:
$$\int \tan x \sec^2 x \, dx = \frac{1}{2} \tan^2 x + C$$

Example 7.26 Evaluate $\int \tan^2 3x \, dx$.

Solution:

Use the trigonometric substitution $\tan^2 x = \sec^2 x - 1$:
$$\int \tan^2 3x \, dx = \int (\sec^2 3x - 1) \, dx$$

Break this up into two integrals:
$$\int (\sec^2 3x - 1) \, dx = \int \sec^2 3x \, dx - \int dx$$

Let $u=3x$, then $du=3dx$

$$\int \sec^2 3x\,dx - \int dx = \int \frac{1}{3}\sec^2 u\,du - \int dx$$

Integrate:

$$\int \frac{1}{3}\sec^2 u\,du - \int dx = \frac{1}{3}\tan u - x + C$$

Substitute back for u:

$$\int \tan^2 3x\,dx = \frac{1}{3}\tan 3x - x + C$$

Example 7.27 Evaluate $\int \tan^3 x\,dx$.

Solution:

Use the trigonometric substitution $\tan^2 x = \sec^2 x - 1$:

$$\tan^3 x = \tan^2 x \tan x = (\sec^2 x - 1)\tan x = \tan x \sec^2 x - \tan x$$

Then,

$$\int \tan^3 x\,dx = \int (\tan x \sec^2 x - \tan x)\,dx = \int \tan x \sec^2 x\,dx - \int \tan x\,dx$$

Let $u = \tan x$, then $du = \sec^2 x\,dx$

$$\int \tan x \sec^2 x\,dx - \int \tan x\,dx = \int u\,du - \int \tan x\,dx$$

Integrate:

$$\int u\,du - \int \tan x\,dx = \frac{u^2}{2} - (-\ln|\cos x|) + C = \frac{u^2}{2} + \ln|\cos x| + C$$

Substitute back for u:

$$\int \tan^3 x\,dx = \frac{\tan^2 x}{2} + \ln|\cos x| + C$$

7.3.5 三角函数积分Ⅲ Integrals of Trigonometric Functions Ⅲ

三角函数不定积分中有一类积分，其被积函数（Integrand）形如$(\sin x)^p(\cos x)^q$。这类积分在 AP 微积分考试中经常出现，部分同学却感到有些困难，接下来分两类情况讨论：

1) p、q 至少有一个为奇数时，参考 Example 7.28；

2) p、q 全为偶数时，参考 Example 7.29。

Example 7.28 Evaluate $\int \sin^2 x \cos x\,dx$.

Solution:

Let $u = \sin x$, then $du = \cos x\,dx$

$$\int \sin^2 x \cos x\,dx = \int u^2\,du$$

Integrate:

$$\int u^2\,du = \frac{u^3}{3} + C$$

第 7 章 不定积分 The Indefinite Integral

Substitute back for u:
$$\int \sin^2 x \cos x \, dx = \frac{\sin^3 x}{3} + C$$

Example 7.29 Evaluate $\int \sin^4 x \, dx$.

Solution:

Since $\cos 2x = 1 - 2\sin^2 x$, then
$$\sin^4 x = (\sin^2 x)^2 = \left(\frac{1-\cos 2x}{2}\right)^2 = \frac{1}{4}(1 - 2\cos 2x + \cos^2 2x)$$

And since $\cos 2x = 2\cos^2 x - 1$, then
$$\cos^2 2x = \frac{1+\cos 4x}{2}$$

So,
$$\sin^4 x = \frac{1}{4}\left(1 - 2\cos 2x + \frac{1+\cos 4x}{2}\right) = \frac{1}{8}(3 - 4\cos 2x + \cos 4x)$$

Then,
$$\int \sin^4 x \, dx = \int \frac{1}{8}(3 - 4\cos 2x + \cos 4x) \, dx$$

Break this up into several integrals:
$$\int \frac{1}{8}(3 - 4\cos 2x + \cos 4x) \, dx = \frac{1}{8}\left(\int 3 \, dx - \int 4\cos 2x \, dx + \int \cos 4x \, dx\right)$$

Integrate:
$$\int \sin^4 x \, dx = \frac{3x}{8} - \frac{1}{4}\sin 2x + \frac{1}{32}\sin 4x + C$$

Example 7.30 Evaluate $\int \sin x \, e^{\cos x} \, dx$.

Solution:

Let $u = \cos x$, then $du = -\sin x \, dx$
$$\int \sin x \, e^{\cos x} \, dx = \int -e^u \, du$$

Integrate:
$$\int -e^u \, du = -e^u + C$$

Substitute back for u:
$$\int \sin x \, e^{\cos x} \, dx = -e^{\cos x} + C$$

【方法总结】忽略正负号影响，$\sin x$ 和 $\cos x$ 求导（Derivative）或不定积分（Indefinite Integral）互换，即：

$$\sin x \xrightarrow[\text{Indefinite Integral}]{\text{Derivative}} \cos x$$

7.3.6 反三角函数积分 Integrals of Inverse Trigonometric Functions

$$\int \frac{du}{\sqrt{1-u^2}} = \arcsin u + C; \quad \int \frac{du}{1+u^2} = \arctan u + C; \quad \int \frac{du}{u\sqrt{u^2-1}} = \text{arcsec}\, u + C$$

反三角函数积分的被积函数（Integrand）有相对明确的特点，AP 微积分考试中，被积函数系含有无理式的分式，可优先考虑反三角函数积分．

Example 7.31 Evaluate $\int \dfrac{2x}{\sqrt{1-x^4}} dx$.

Solution：

Let $u = x^2$, then $du = 2x\,dx$

$$\int \frac{2x}{\sqrt{1-x^4}} dx = \int \frac{1}{\sqrt{1-u^2}} du$$

Integrate：

$$\int \frac{1}{\sqrt{1-u^2}} du = \arcsin u + C$$

Substitute back for u：

$$\int \frac{2x}{\sqrt{1-x^4}} dx = \arcsin(x^2) + C$$

Example 7.32 Evaluate $\int \dfrac{x}{\sqrt{9-x^4}} dx$.

Solution：

$$\int \frac{x}{\sqrt{9-x^4}} dx = \int \frac{x}{3\sqrt{1-\left(\frac{x^2}{3}\right)^2}} dx$$

Let $u = \dfrac{x^2}{3}$, then $du = \dfrac{2x}{3} dx$

$$\int \frac{x}{3\sqrt{1-\left(\frac{x^2}{3}\right)^2}} dx = \frac{1}{2} \int \frac{1}{\sqrt{1-u^2}} du$$

Integrate：

$$\frac{1}{2} \int \frac{1}{\sqrt{1-u^2}} du = \frac{1}{2} \arcsin u + C$$

Substitute back for u：

$$\int \frac{x}{\sqrt{9-x^4}} dx = \frac{1}{2} \arcsin\left(\frac{x^2}{3}\right) + C$$

Example 7.33 Evaluate $\int \dfrac{1}{x^2+6x+10} dx$.

Solution：

$$\int \frac{1}{x^2+6x+10} dx = \int \frac{1}{1+(x+3)^2} dx$$

Let $u = x+3$, then $du = dx$

$$\int \frac{1}{1+(x+3)^2} dx = \int \frac{1}{1+u^2} du$$

Integrate：

$$\int \frac{1}{1+u^2} du = \arctan u + C$$

Substitute back for u:

$$\int \frac{1}{x^2+6x+10} dx = \arctan(x+3) + C$$

Example 7.34 Evaluate $\int \frac{1}{x+3} \times \frac{1}{\sqrt{x^2+6x+8}} dx$.

Solution:

$$\int \frac{1}{x+3} \times \frac{1}{\sqrt{x^2+6x+8}} dx = \int \frac{1}{(u+3)\sqrt{(x+3)^2-1}} dx$$

Let $u = x+3$, then $du = dx$

$$\int \frac{1}{(u+3)\sqrt{(x+3)^2-1}} dx = \int \frac{1}{u\sqrt{u^2-1}} du$$

Integrate:

$$\int \frac{1}{u\sqrt{u^2-1}} du = \text{arcsec}\, u + C$$

Substitute back for u:

$$\int \frac{1}{x+3} \times \frac{1}{\sqrt{x^2+6x+8}} dx = \text{arcsec}(x+3) + C$$

7.4# 分部积分法 Integration by Parts

If $u = u(x)$ and $v = v(x)$ are differentiable function of x, the Product Rule:

$$\frac{d}{dx}(uv) = u\frac{dv}{dx} + v\frac{du}{dx}$$

Multiply both sides by dx:

$$d(uv) = u\, dv + v\, du$$

that is:

$$u\, dv = d(uv) - v\, du$$

Integrating both sides:

$$\int u\, dv = uv - \int v\, du \quad \text{or} \quad \int u(x) dv(x) = uv - \int v\, du$$

运用分部积分法，如果 u 和 dv 选取不当就求不出结果，恰当选取 u 和 dv 是一个关键．选取 u 和 dv 时考虑以下两点：①v 要容易求出；②$\int v\, du$ 要比 $\int u\, dv$ 容易积分．

7.4.1# 运用分部积分法 I Using Integration by Parts I

被积函数（Integrand）是幂函数和正弦函数、幂函数和余弦函数或幂函数和指数函数的乘积，可以考虑用分部积分法（Using Integration by Parts），并多数假设幂函数（Power Function）为 u．

Example 7.35 Evaluate $\int x\sin x\, dx$.

Solution:

Let $u=x$ and $dv=\sin x\,dx$, then $du=dx$ and $v=-\cos x$

Thus,
$$\int x\sin x\,dx = x(-\cos x)-\int(-\cos x)\,dx$$
$$= -x\cos x+\int\cos x\,dx$$
$$= -x\cos x+\sin x+C$$

Note：注意到，假设 $u=\sin x$, $dv=x\,dx$，那么，$du=\cos x\,dx$, $v=\dfrac{x^2}{2}$，于是

$$\int x\sin x\,dx = \dfrac{x^2}{2}\sin x-\int\dfrac{x^2}{2}\cos x\,dx$$

右端的积分比原积分更不容易求出，分部积分法失去意义.

Example 7.36　Evaluate $\int x e^x\,dx$.

Solution:

Let $u=x$ and $dv=e^x\,dx$, then $du=dx$ and $v=e^x$

Thus,
$$\int x e^x\,dx = x e^x-\int e^x\,dx$$
$$= x e^x-e^x+C=e^x(x-1)+C$$

7.4.2[#]　运用分部积分法 Ⅱ Using Integration by Parts Ⅱ

被积函数（Integrand）是幂函数和对数函数、幂函数和反三角函数的乘积，可以考虑用分部积分法（Using Integration by Parts），并多数假设对数函数（Logarithmic Function）或反三角函数（Inverse Trigonometric Function）为 u.

Example 7.37　Evaluate $\int x^2\ln x\,dx$.

Solution:

Let $u=\ln x$ and $dv=x^2\,dx$, then $du=\dfrac{1}{x}dx$ and $v=\dfrac{x^3}{3}$

Thus,
$$\int x^2\ln x\,dx = \ln x\,\dfrac{x^3}{3}-\int\dfrac{x^3}{3}\times\dfrac{1}{x}dx = \ln x\,\dfrac{x^3}{3}-\dfrac{x^3}{9}+C$$

Example 7.38　Evaluate $\int \ln x\,dx$.

Solution:

Let $u=\ln x$ and $dv=dx$, then $du=\dfrac{1}{x}dx$ and $v=x$

Thus,
$$\int \ln x\,dx = x\ln x-\int x\,\dfrac{1}{x}dx = x\ln x-x+C$$

Example 7.39　Evaluate $\int \arccos x\,dx$.

第 7 章　不定积分 The Indefinite Integral

Solution:

Let $u = \arccos x$ and $dv = dx$, then $du = -\dfrac{1}{\sqrt{1-x^2}} dx$ and $v = x$

Thus,
$$\int \arccos x \, dx = x \arccos x - \int x \left(-\dfrac{1}{\sqrt{1-x^2}}\right) dx$$

Let $u = 1 - x^2$, then $du = -2x \, dx$
$$\int x \left(-\dfrac{1}{\sqrt{1-x^2}}\right) dx = \int \dfrac{1}{2} \times \dfrac{1}{\sqrt{u}} du$$
$$= \sqrt{u} + C = \sqrt{1-x^2} + C$$

Hence,
$$\int \cos^{-1} x \, dx = x \arccos x - \int x \left(-\dfrac{1}{\sqrt{1-x^2}}\right) dx$$
$$= x \arccos x - \sqrt{1-x^2} + C$$

Example 7.40 Evaluate $\int x \arctan x \, dx$.

Solution:

Let $u = \arctan x$ and $dv = x \, dx$, then $du = \dfrac{1}{1+x^2} dx$ and $v = \dfrac{x^2}{2}$

Thus,
$$\int x \arctan x \, dx = \dfrac{x^2}{2} \arctan x - \dfrac{1}{2} \int \dfrac{x^2}{1+x^2} dx$$

Since
$$\int \dfrac{x^2}{1+x^2} dx = \int \dfrac{x^2+1-1}{1+x^2} dx = \int 1 - \dfrac{1}{1+x^2} dx$$
$$= \int dx - \int \dfrac{1}{1+x^2} dx = x - \arctan x + C$$

Then,
$$\int x \arctan x \, dx = \dfrac{x^2}{2} \arctan x - \dfrac{1}{2}(x - \arctan x) + C$$

7.4.3# 重复使用 Repeated Use of Integration by Parts

有时，我们需要不止一次地使用分部积分法．典型地，比如：假设幂函数（Power Function）为 u，使用分部积分法，幂函数的指数降低一次，再使用分部积分法，幂函数的指数再降低一次．

Example 7.41 Evaluate $\int x^2 e^x \, dx$.

Solution:

Let $u = x^2$ and $dv = e^x \, dx$, then $du = 2x \, dx$ and $v = e^x$

Thus,

$$\int x^2 e^x \, dx = x^2 e^x - 2 \int x e^x \, dx$$

Integrate by parts again with $u = x$ and $dv = e^x \, dx$, then $du = dx$ and $v = e^x$

$$\int x e^x \, dx = x e^x - \int e^x \, dx = x e^x - e^x + C = e^x (x - 1) + C$$

Hence,

$$\int x^2 e^x \, dx = x^2 e^x - 2 \int x e^x \, dx = x^2 e^x - 2 e^x (x - 1) + C$$

7.4.4[#]　解出未知积分 Solving for the Unknown Integral

被积函数（Integrand）是三角函数（Trigonometric Function）和对数函数（Logarithmic Function）的乘积，考虑用分部积分法，任何一方都可以作为 u，至少两次进行分部积分，形成方程，再解出未知积分．

Example 7.42　Evaluate $\int e^x \sin x \, dx$.

Solution：

Let $u = e^x$ and $dv = \sin x \, dx$, then $du = e^x \, dx$ and $v = -\cos x$

Thus,

$$\int e^x \sin x \, dx = -e^x \cos x + \int \cos x \, e^x \, dx$$

Again, we let $u = e^x$ and $dv = \cos x \, dx$, then $du = e^x \, dx$ and $v = \sin x$

Thus,

$$\int \cos x \, e^x \, dx = e^x \sin x - \int e^x \sin x \, dx$$

Hence,

$$\int e^x \sin x \, dx = -e^x \cos x + e^x \sin x - \int e^x \sin x \, dx$$

Adding the integral to both sides and adding the constant of integration gives,

$$2 \int e^x \sin x \, dx = -e^x \cos x + e^x \sin x + C_1$$

Dividing by 2 and renaming the constant of integration gives

$$\int e^x \sin x \, dx = \frac{1}{2} e^x (\sin x - \cos x) + C$$

Example 7.43　Evaluate $\int \cos x \, e^{-x} \, dx$.

Solution：

Let $u = \cos x$ and $dv = e^{-x} \, dx$, then $du = -\sin x \, dx$ and $v = -e^{-x}$. Thus,

$$\int \cos x \, e^{-x} \, dx = \cos x (-e^{-x}) - \int \sin x \, e^{-x} \, dx$$

Again, we let $u = \sin x$ and $dv = e^{-x} \, dx$, then $du = \cos x \, dx$ and $v = -e^{-x}$. Thus,

$$\int \sin x \, e^{-x} \, dx = \sin x (-e^{-x}) + \int \cos x \, e^{-x} \, dx$$

Hence,

$$\int \cos x \, e^{-x} \, dx = \cos x (-e^{-x}) - \sin x (-e^{-x}) - \int \cos x \, e^{-x} \, dx$$

Adding the integral to both sides and adding the constant of integration gives,

$$2\int \cos x\, e^{-x}\, dx = \cos x(-e^{-x}) - \sin x(-e^{-x}) + C_1$$

Dividing by 2 and renaming the constant of integration gives

$$\int \cos x\, e^{-x}\, dx = \frac{1}{2}\cos x(-e^{-x}) - \frac{1}{2}\sin x(-e^{-x}) + C$$

7.4.5# 列表积分法 Using Tabular Integration

形如 $\int f(x)g(x)\,dx$ 的积分，其中 $f(x)$ 可以重复求导直至出现零，而 $g(x)$ 可以容易地重复积分，考虑分部积分法，并且列表计算，简化计算方式，这是列表积分法.

Example 7.44　Evaluate $\int x^2 e^x\, dx$.

Solution：

Let $f(x)=x^2$ and $g(x)=e^x$，we list：

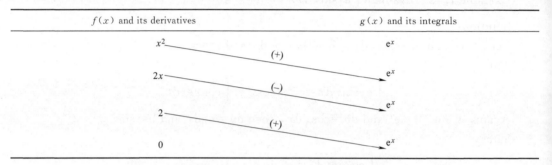

Combine the products of the functions connected by the arrows according to the operation signs the arrows to obtain

$$\int x^2 e^x\, dx = x^2 e^x - 2x e^x + 2e^x + C$$

Note：本题解法和 Example 7.41 解法比较，更为简洁.

Example 7.45　Evaluate $\int x^4 \cos x\, dx$.

Solution：

Let $f(x)=x^4$ and $g(x)=\cos x$，we list：

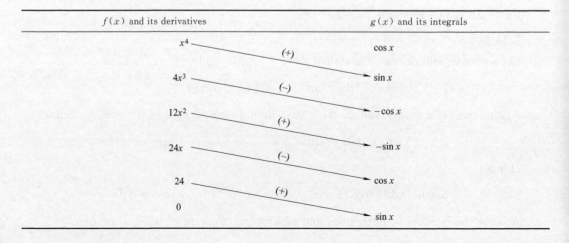

Combine the products of the functions connected by the arrows according to the operation signs the arrows to obtain

$$\int x^4 \cos x \, \mathrm{d}x = x^4 \sin x - 4x^3(-\cos x) + 12x^2(-\sin x) - 24x \cos x + 24\sin x + C$$
$$= x^4 \sin x + 4x^3 \cos x - 12x^2 \sin x - 24x \cos x + 24\sin x + C$$

7.5[#] 有理函数的积分 Integration of Rational Functions

假设 $P(x)$、$Q(x)$ 分别是 m、n 多项式，当被积函数（Integrand）是有理函数时，即：

$$\int \frac{P(x)}{Q(x)} \mathrm{d}x$$

称有理函数不定积分（The Rational Function of Indefinite Integral）.

有理函数（Rational Functions）是常用的函数类型（见本书 2.3.3 节）. 有理函数不定积分，是常见的不定积分. 其常用方法是：分式拆分求积分（Integration by Partial Fractions）.

7.5.1[#] 线性因子不同 Distinct Linear Factors

$$\int \frac{P(x)}{(x-x_1)(x-x_2)} \mathrm{d}x = \int \frac{A}{x-x_1} \mathrm{d}x + \int \frac{B}{x-x_2} \mathrm{d}x$$

把被积函数按不同线性因子分式拆分，根据待定系数法（The Method of Undetermined Coefficients）解得 A、B，再求不定积分.

Example 7.46 Evaluate $\int \frac{9x-13}{(x-1)(x-2)} \mathrm{d}x$.

Solution：

The partial fraction decomposition has the from

$$\frac{9x-13}{(x-1)(x-2)} = \frac{A}{x-1} + \frac{B}{x-2}$$

To find the values of the undetermined coefficients A、B, multiply through by $(x-1)(x-2)$, we clear fractions and get,

$$9x - 13 = A(x-2) + B(x-1)$$

Simplify and group the terms：

$$9x - 13 = (A+B)x - 2A - B$$

The polynomials on both sides of the above equation are identical, so we equate coefficients of like powers of x obtaining,

$$\begin{cases} A + B = 9 \\ -2A - B = -13 \end{cases}$$

Solve the equations, we get

$$A = 4, B = 5$$

That is,

$$\frac{9x-13}{(x-1)(x-2)} = \frac{4}{x-1} + \frac{5}{x-2}$$

Hence,

$$\int \frac{9x-13}{(x-1)(x-2)} dx = \int \frac{4}{x-1} dx + \int \frac{5}{x-2} dx$$
$$= 4\ln|x-1| + 5\ln|x-2| + C$$

7.5.2# 线性因子相同 Repeated Linear Factors

$$\int \frac{P(x)}{(x-a)^2} dx = \int \frac{A}{x-a} dx + \int \frac{B}{(x-a)^2} dx$$

Example 7.47 Evaluate $\int \frac{x+6}{(x+2)^2} dx$.

Solution：

The partial fraction decomposition has the from

$$\frac{x+6}{(x+2)^2} = \frac{A}{x+2} + \frac{B}{(x+2)^2}$$

To find the values of the undetermined coefficients A、B, multiply through by $(x+2)^2$, we clear fractions and get,

$$x+6 = Ax + 2A + B$$

The polynomials on both sides of the above equation are identical, so we equate coefficients of like powers of x obtaining,

$$\begin{cases} A = 1 \\ 2A + B = 6 \end{cases}$$

Solve the equations, we get

$$A = 1, B = 4$$

That is,

$$\frac{x+6}{(x+2)^2} = \frac{1}{x+2} + \frac{4}{(x+2)^2}$$

Hence,

$$\int \frac{x+6}{(x+2)^2} dx = \int \left[\frac{1}{x+2} + \frac{4}{(x+2)^2} \right] dx$$
$$= \int \frac{1}{x+2} dx + \int \frac{4}{(x+2)^2} dx$$
$$= \ln|x+2| - \frac{4}{x+2} + C$$

7.5.3# 存在二次因子 A Single Quadratic Factor

$$\int \frac{P(x)}{(x-a)(x^2+px+q)} dx = \int \frac{A}{x-a} dx + \int \frac{Bx+C}{x^2+px+q} dx$$

Example 7.48 Evaluate $\int \frac{x^2-x+1}{x(x+1)} dx$.

Solution：

The partial fraction decomposition has the from
$$\frac{x^2-x+1}{x(x+1)} = \frac{A}{x} + \frac{Bx+C}{x+1}$$
To find the values of the undetermined coefficients A、B、C，multiply through by $x(x+1)$，we clear fractions and get，
$$x^2-x+1 = A(x+1) + (Bx+C)x$$
Simplify and group the terms：
$$x^2-x+1 = Bx^2 + (A+C)x + A$$
The polynomials on both sides of the above equation are identical，so we equate coefficients of like powers of x obtaining，
$$\begin{cases} B=1 \\ A+C=-1 \\ A=1 \end{cases}$$
Solve the equations，we get
$$A=1, B=1, C=-2$$
That is，
$$\frac{x^2-x+1}{x(x+1)} = \frac{1}{x} + \frac{x-2}{x+1}$$
Hence，
$$\begin{aligned}\int \frac{x^2-x+1}{x(x+1)}\mathrm{d}x &= \int \left(\frac{1}{x} + \frac{x-2}{x+1}\right)\mathrm{d}x \\ &= \int \frac{1}{x}\mathrm{d}x + \int \frac{x-2}{x+1}\mathrm{d}x \\ &= \int \frac{1}{x}\mathrm{d}x + 1 - \int \frac{3}{x+1}\mathrm{d}x \\ &= \ln|x| + x - 3\ln|x+1| + C\end{aligned}$$

Example 7.49 Evaluate $\int \frac{3x+2}{x(x+1)^2}\mathrm{d}x$.

Solution：

The partial fraction decomposition has the from
$$\frac{3x+2}{x(x+1)^2} = \frac{A}{x} + \frac{Bx+C}{(x+1)^2}$$
To find the values of the undetermined coefficients A、B、C，multiply through by $x(x+1)^2$，we clear fractions and get，
$$3x+2 = A(x+1)^2 + x(Bx+C)$$
Simplify and group the terms：
$$3x+2 = (A+B)x^2 + (2A+C)x + A$$
The polynomials on both sides of the above equation are identical，so we equate coefficients of like powers of x obtaining，
$$\begin{cases} A+B=0 \\ 2A+C=3 \\ A=2 \end{cases}$$

Solve the equations, we get
$$A=2, B=-2, C=-1$$
That is,
$$\frac{3x+2}{x(x+1)^2} = \frac{2}{x} - \frac{2x+1}{(x+1)^2}$$
And let,
$$\frac{2x+1}{(x+1)^2} = \frac{D}{x+1} + \frac{E}{(x+1)^2}$$
Then,
$$D=2, E=-1$$
So
$$\frac{3x+2}{x(x+1)^2} = \frac{2}{x} - \frac{2}{x+1} + \frac{1}{(x+1)^2}$$
Hence,
$$\begin{aligned}\int \frac{3x+2}{x(x+1)^2}dx &= \int \left[\frac{2}{x} - \frac{2}{x+1} + \frac{1}{(x+1)^2}\right]dx \\ &= \int \frac{2}{x}dx - \int \frac{2}{x+1}dx + \int \frac{1}{(x+1)^2}dx \\ &= 2\ln|x| - 2\ln|x+1| - \frac{1}{x+1} + C\end{aligned}$$

7.6 不定积分的应用 Applications of Indefinite Integral

AP 微积分上，不定积分应用有：初值问题（Initial Value Problems）和运动问题（Motion Problems）。

7.6.1 初值问题 Initial Value Problems

已知函数的导数以及在某个特殊点 x_0 处的值 y_0，求 x 的函数 y 的问题称为初值问题（Initial Value Problems）。

Example 7.50 Find the curve whose slope at the point (x,y) is $2x^2$ if the curve is required to pass through the point $(1,-2)$.

Solution：

Since the curve whose slope at the point (x,y) is $2x^2$, then
$$\frac{dy}{dx} = 2x^2$$
The function y is an antiderivative of $f(x)=2x^2$, so
$$y = \int 2x^2 dx = \frac{2}{3}x^3 + C$$
The initial condition $y(1)=-2$, then
$$-2 = \frac{2}{3} \times 1^3 + C$$

$$C = -\frac{8}{3}$$

The curve we want is $y = \frac{2}{3}x^3 - \frac{8}{3}$.

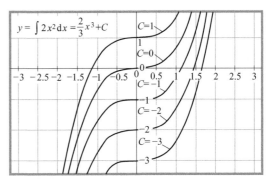

Figure 7.6.1

【方法总结】函数 $f(x)$ 不定积分的一般解 $F(x)+C$，这些解的个数有无穷多个，见 Figure 7.6.1，对应每一个 C 的值都有一个解，通过求满足初始条件（The Initial Condition）$y_0(x_0) = y_0$ 的特殊解，来解初值问题．初值问题运用广泛，特别在数学建模中．

7.6.2 运动问题 Motion Problems

我们前面讲过由位置（Position）求速度（Velocity）和加速度（Acceleration）的运动问题，见 6.3.2 节，由速度求解物体的位置则应用不定积分．

Example 7.51 The velocity of particle moving along a line is given by $\dfrac{ds}{dt} = 9.8t + 5$ at time t. If $s(0) = 10$. Find its position when $t = 3$.

Solution：Since
$$v(t) = \frac{ds}{dt} = 9.8t + 5$$

Then,
$$s = \int (9.8t + 5)dt = 4.9t^2 + 5t + C$$

Since $s(0) = 10$, $4.9 \times 0^2 + 5 \times 0 + C = 10$, then
$$C = 10$$

And that the position function is
$$s = 4.9t^2 + 5t + 10$$

Plug in $t = 3$ to find the position at that time：
$$s(3) = 4.9 \times 3^2 + 5 \times 3 + 10 = 69.1$$

7.7 习题 Practice Exercises

(1) Evaluate the following integrals.

1) $\int 3x \, dx$；

2) $\int 2\sqrt{x} \, dx$；

3) $\int (3x^2 + 2x + 1) \, dx$；

4) $\int (x+1)(x-1) \, dx$；

5) $\int 3\sin x \, dx$.

(2) Find $f(x)$ if $f'(x)=2x$ and $f(1)=2$.

(3) Find $\int e^{2x} f'(e^x) dx$, if $\ln x - 3$ is an antiderivative of the function $f(x)$.

(4) Evaluate the following integrals using U-substitution.

1) $\int (2x+1)^5 dx$;

2) $\int x e^{x^2+3} dx$;

3) $\int \sin 5x \, dx$;

4) $\int \dfrac{x}{1+2x^2} dx$;

5) $\int \dfrac{x}{(1+x^2)^2} dx$.

(5) Evaluate the following integrals.

1) $\int \dfrac{\tan^{-1} x}{1+x^2} dx$;

2) $\int \tan^2 2x \, dx$;

3) $\int \cos^2 x \sin x \, dx$;

4) $\int \cos^4 x \, dx$;

5) $\int \cos x \, e^{\sin x} dx$.

(6)# Evaluate the following integrals using Integration by Parts.

1) $\int x \cos x \, dx$;

2) $\int x^3 \ln x \, dx$;

3) $\int \arcsin x \, dx$;

4) $\int x^2 e^{-x} dx$;

5) $\int e^x \cos x \, dx$.

(7)# Evaluate the following integrals.

1) $\int \dfrac{7x-12}{(x-1)(x-2)} dx$;

2) $\int \dfrac{x+9}{(x+2)^2} dx$;

3) $\int \dfrac{x+2}{x(x+1)^2} dx$.

(8) The velocity of particle moving along a line is given by $\dfrac{ds}{dt} = 9.8t - 5$ at time t. If $s(0) = 20$. Find its position when $t = 5$.

第 8 章

定积分
The Definite Integral

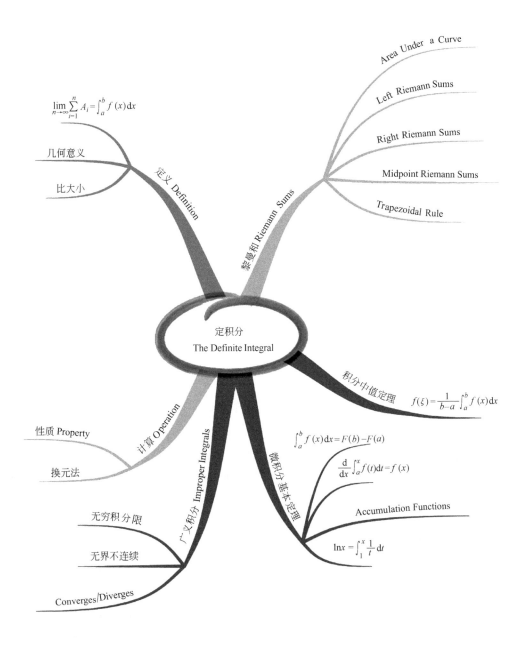

8.1 黎曼和与梯形法则 Riemann Sums and Trapezoid Rule

初等数学背景下，求曲线下的面积（Area Under a Curve）相当困难．但是，在微积分背景下，黎曼和（Riemann Sums）与梯形法则（Trapezoidal Rule）给出了巧妙的方法，求曲线下的面积变得简单．并且，定积分的概念也从计算曲线下的面积引出．

8.1.1 曲线下的面积 Area Under a Curve

How to find the shaded area under the curve of $y=f(x)$ from $x=a$ to $x=b$?

在直角坐标系中，由曲线 $f(x)=\dfrac{1}{5}(x-1)^2+1$，直线 $x=a$，$x=b$ 及 x 轴所围成的面积，叫做曲线下的面积（Area Under a Curve），记为 A，见 Figure 8.1.1.

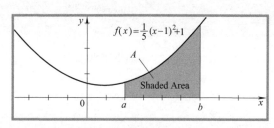

Figure 8.1.1

8.1.2 黎曼和 Riemann Sums

The theory of limits of finite approximations was made precise by the German mathematician Bernhard Riemann.

对于求曲线下的面积（Area Under a Curve）问题，德国数学家黎曼（Bernhard Riemann）给出了一个巧妙的方法，并衍生出了黎曼和（Riemann Sums）．

黎曼的巧妙方法大致是这样：

1）将曲线下的不规则图像近似切割成等宽的一个个小矩形（Rectangle）；
2）测量所有小矩形的面积，累加所有小矩形的面积，得到一个面积和；
3）使用面积和估算曲线下的面积 A；
4）将小矩形切割得再小些，重复上述过程，使得估算值更为准确．

以下，运用黎曼的巧妙方法，在直角坐标系中，估算曲线 $f(x)=\dfrac{1}{5}(x-1)^2+1$，直线 $x=1$，$x=3$ 及 x 轴所围成的曲线下面积 A．

1）把闭区间 $[1,3]$ 分为 2 个子区间（Subinterval），见 Figure 8.1.2(a).

原曲边形被近似切割成等宽（Width）的 2 个小矩形，小矩形的宽为 1，高（Height）

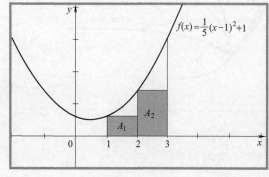

Figure 8.1.2(a)

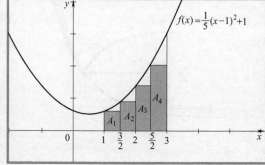

Figure 8.1.2(b)

分别为 $f(1)$ 和 $f(2)$，面积和 A_1+A_2：

$$A \approx A_1+A_2 = 1 \times f(1) + 1 \times f(2) = 1 + \frac{6}{5} = \frac{11}{5}$$

即：我们认为上述曲线下的面积 A 估计为 $\frac{11}{5}$.

2) 把闭区间 $[1,3]$ 分为 4 个子区间，见 Figure 8.1.2(b)，原曲边形被近似切割成等宽的 4 个小矩形，小矩形的宽为 $\frac{1}{2}$，高分别为 $f(1)$、$f\left(\frac{3}{2}\right)$、$f(2)$、$f\left(\frac{5}{2}\right)$，小矩形的面积和 $A_1+A_2+A_3+A_4$：

$$A \approx A_1+A_2+A_3+A_4 = \frac{1}{2} \times f(1) + \frac{1}{2} \times f\left(\frac{3}{2}\right) + \frac{1}{2} \times f(2) + \frac{1}{2} \times f\left(\frac{5}{2}\right)$$

$$= \frac{1}{2} + \frac{21}{40} + \frac{3}{5} + \frac{29}{40} = \frac{47}{20}$$

那么，我们估计上述曲线下的面积 A 为 $\frac{47}{20}$，比 $\frac{11}{5}$ 更精确一点.

为了使得估算更为准确，把闭区间 $[1,3]$ 分为 8 个子区间，见 Figure 8.1.2(c)，那么小矩形的面积和为 $A_1+A_2+A_3+\cdots+A_8$，随着小矩形切割得越小，小矩形的面积和越接近曲线下的面积，估算更为准确.

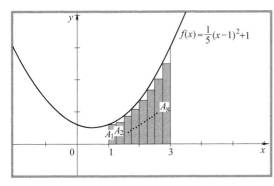

Figure 8.1.2(c)

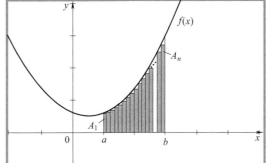

Figure 8.1.2(d)

一般而言，为了估算曲线 $f(x)$，直线 $x=a$，$x=b$ 及 x 轴所围成的面积 A，把闭区间 $[a,b]$ 划分为 n 个子区间，见 Figure 8.1.2(d)，则曲边形分割成 n 个小曲边形. 用 n 个小矩形估算 n 个小曲边形，我们使用 A_i 表述第 i 个小矩形的面积，n 个小矩形的面积和：

$$\sum_{i=1}^{n} A_i = A_1 + A_2 + \cdots + A_n$$

我们称 $\sum_{i=1}^{n} A_i$ 为黎曼和（Riemann Sums）.

8.1.3 左黎曼和 Left Riemann Sums

If we use the function value of the left point of the interval, the sum $\sum_{i=1}^{n} A_i$ is called a Left Riemann Sum $L(n)$.

由黎曼和的定义可知，见 Figure 8.1.3：

$$A_i = \frac{b-a}{n} y_i$$

所以：

$$L(n) = \sum_{i=1}^{n} \left(\frac{b-a}{n} \right) y_i \quad \text{or} \quad L(n) = \left(\frac{b-a}{n} \right)(y_1 + y_2 + \cdots + y_n)$$

其中，n 为区间 $[a, b]$ 子区间的个数。

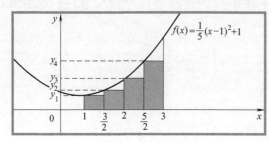

Figure 8.1.3

Example 8.1 Approximation the area under the curve $y = \frac{1}{5}(x-1)^2 + 1$ from $x = 1$ to $x = 3$ using $L(4)$.

Solution：

Draw four rectangles that look like Figure 8.1.3：

The width of each rectangle is

$$\frac{b-a}{n} = \frac{3-1}{4} = \frac{1}{2}$$

The heights of the rectangle are：

$$y_1 = f(1) = 1, \quad y_2 = f\left(\frac{3}{2}\right) = \frac{21}{20}, \quad y_3 = f(2) = \frac{6}{5} \text{ and } y_4 = f\left(\frac{5}{2}\right) = \frac{29}{20}$$

Therefore, the area is：

$$L(4) = \frac{1}{2} \times 1 + \frac{1}{2} \times \frac{21}{20} + \frac{1}{2} \times \frac{6}{5} + \frac{1}{2} \times \frac{29}{20} = \frac{47}{20} = 2.35$$

8.1.4 右黎曼和 Right Riemann Sums

If we use the function value of the right point of the interval, the sum $\sum_{i=1}^{n} A_i$ is called a Right Riemann Sum $R(n)$.

由黎曼和的定义可知，见 Figure 8.1.4：

$$A_i = \frac{b-a}{n} y_{i+1}$$

所以：

$$R(n) = \sum_{i=1}^{n} \left(\frac{b-a}{n} \right) y_{i+1} \quad \text{or} \quad R(n) = \left(\frac{b-a}{n} \right)(y_2 + y_3 + \cdots + y_{n+1})$$

其中，n 为区间 $[a, b]$ 子区间的个数。

Example 8.2 Approximation the area under the curve $y = \frac{1}{5}(x-1)^2 + 1$ from $x = 1$ to $x = 3$ using $R(4)$.

Solution：

Draw four rectangles that look like Figure 8.1.4：

The width of each rectangle is
$$\frac{b-a}{n}=\frac{3-1}{4}=\frac{1}{2}$$

The heights of the rectangle are:
$$y_2=f\left(\frac{3}{2}\right)=\frac{21}{20},\ y_3=f(2)=\frac{6}{5},$$
$$y_4=f\left(\frac{5}{2}\right)=\frac{29}{20}\text{ and }y_5=f(3)=\frac{9}{5}$$

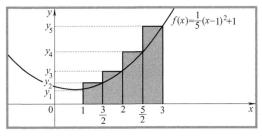

Figure 8.1.4

Therefore, the area is:
$$R(4)=\frac{1}{2}\times\frac{21}{20}+\frac{1}{2}\times\frac{6}{5}+\frac{1}{2}\times\frac{29}{20}+\frac{1}{2}\times\frac{9}{5}=\frac{11}{4}=2.75$$

8.1.5 中黎曼和 Midpoint Riemann Sums

If we use the function value of the midpoint of the interval, the sum $\sum_{i=1}^{n}A_i$ is called a Midpoint Riemann Sum $M(n)$.

由黎曼和的定义可知，见 Figure 8.1.5:
$$A_i=\frac{b-a}{n}y_{\frac{2i-1}{2}}$$

所以:
$$M(n)=\sum_{i=1}^{n}\left(\frac{b-a}{n}\right)y_{\frac{2i-1}{2}}\quad\text{or}\quad M(n)=\left(\frac{b-a}{n}\right)\left(y_{\frac{3}{2}}+y_{\frac{5}{2}}+\cdots+y_{\frac{2n-1}{2}}\right)$$

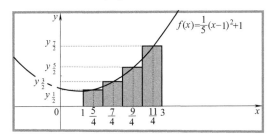

Figure 8.1.5

其中，n 为区间 $[a,b]$ 子区间的个数.

Example 8.3 Approximation the area under the curve $y=\frac{1}{5}(x-1)^2+1$ from $x=1$ to $x=3$ using $M(4)$.

Solution:

Draw four rectangles that look like Figure 8.1.5:

The width of each rectangle is
$$\frac{b-a}{n}=\frac{3-1}{4}=\frac{1}{2}$$

The heights of the rectangle are:
$$y_{\frac{1}{2}}=f\left(\frac{5}{4}\right)=\frac{81}{80},\ y_{\frac{3}{2}}=f\left(\frac{7}{4}\right)=\frac{89}{80},\ y_{\frac{5}{2}}=f\left(\frac{9}{4}\right)=\frac{21}{16}\text{ and }y_{\frac{7}{2}}=f\left(\frac{11}{4}\right)=\frac{129}{80}$$

Therefore, the area is:
$$M(4)=\frac{1}{2}\times\frac{81}{80}+\frac{1}{2}\times\frac{89}{80}+\frac{1}{2}\times\frac{21}{16}+\frac{1}{2}\times\frac{129}{80}=\frac{101}{40}=2.525$$

8.1.6 梯形法则 Trapezoidal Rule

To approximate the shaded area under the curve of $y=f(x)$ from a to b, we add the areas of the trapezoids made by joining the ends of the segments to the x-axis. We denote the bases of the trapezoids by y_0, y_1, y_2, $\cdots$, y_n, and the heights by $h_i = \dfrac{b-a}{n}$ [See Figure 8.1.6(a)]:

$$T(n) = \left(\frac{b-a}{2n}\right)(y_0 + 2y_1 + \cdots + 2y_{n-1} + y_n)$$

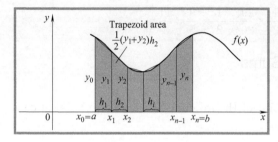

Figure 8.1.6(a)

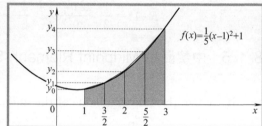

Figure 8.1.6(b)

梯形法则又叫辛普森法则（Simpson's Rule）。黎曼和是将曲线下的不规则图像近似切割成等宽的小矩形（Rectangle），而梯形法则是将曲线下的面积近似切割成等宽的小梯形，梯形的高为$h_i = \dfrac{b-a}{n}$，上底和下底分别为y_{i-1}、y_i，见 Figure 8.1.6(a)，其中任意一个梯形的面积为：

$$A_i = \frac{1}{2}\left(\frac{b-a}{n}\right)(y_{i-1} + y_i)$$

将所有小梯形累加得到梯形法则：

$$T(n) = \left(\frac{b-a}{2n}\right)(y_0 + 2y_1 + \cdots + 2y_{n-1} + y_n)$$

Example 8.4 Approximation the area under the curve $y = \dfrac{1}{5}(x-1)^2 + 1$ from $x=1$ to $x=3$ using $T(4)$.

Solution:

Draw four trapezoids that look like Figure 8.1.6(b):

The height of each trapezoids is

$$\frac{b-a}{n} = \frac{3-1}{4} = \frac{1}{2}$$

The bases of the trapezoids are:

$$y_0 = f(1) = 1,\ y_1 = f\left(\frac{3}{2}\right) = \frac{21}{20},\ y_2 = f(2) = \frac{6}{5},\ y_3 = f\left(\frac{5}{2}\right) = \frac{29}{20},\ y_4 = f(3) = \frac{9}{5}$$

Therefore, the area is:

$$T(4) = \frac{1}{2} \times \frac{1}{2}\left(1 + 2 \times \frac{21}{20} + 2 \times \frac{6}{5} + 2 \times \frac{29}{20} + \frac{9}{5}\right) = \frac{51}{20} = 2.55$$

8.2 定积分的定义 Definition of the Definite Integral

几何上的面积、长度和体积问题；物理上的速度、距离和变力做功问题是导致定积分（Definite Integral）出现的主要背景. 在黎曼和（Riemann Sum）的基础上，理解定积分的概念，我们马上就要触及微积分的基本定理.

8.2.1 定积分的定义 Definition of the Definite Integral

Suppose $f(x)$ be defined on the closed interval $[a, b]$ and the sum $\sum_{i=1}^{n} A_i$ is Riemann Sum. If $\lim_{n \to \infty} \sum_{i=1}^{n} A_i$ exists, we say $f(x)$ is integrable on $[a, b]$.

$$\lim_{n \to \infty} \sum_{i=1}^{n} A_i = \int_a^b f(x) \, dx$$

Which is read as "the integral from a to b of $f(x)$ x dee x". It has names：

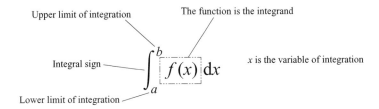

Integral of $f(x)$ from a to b

简而言之，定积分是黎曼和（Riemann Sum）的极限. 我带领同学们再把整个过程回顾一下：

假设函数 $f(x)$ 在闭区间 $[a, b]$ 上连续，求曲线 $f(x)$ 和直线 $x=a$，$x=b$ 及 x 轴所围成的面积 A，见 Figure 8.2.1(a).

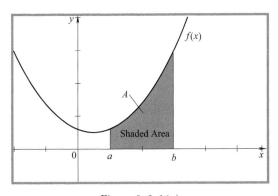

Figure 8.2.1(a)

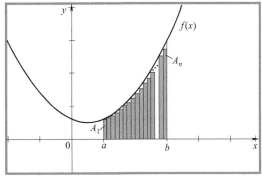

Figure 8.2.1(b)

在开区间 (a, b) 内任意插入 $n-1$ 个分点，见 Figure 8.2.1(b)：

$$a = x_0 < x_1 < x_2 < \cdots < x_{i-1} < x_i < \cdots < x_{n-1} < x_n = b$$

把区间 $[a,b]$ 划分成 n 个子区间（Subinterval）：

$$[x_0, x_1), (x_1, x_2], (x_2, x_3], \cdots, (x_{i-1}, x_i], \cdots, (x_{n-1}, x_n]$$

并记 $\Delta x_i = x_i - x_{i-1}, i = 1, 2, \cdots, n$. 此外，在每个子区间上分别任意取点 $\xi_1, \xi_2, \cdots, \xi_i, \cdots, \xi_n$，求积分和：$\sum_{i=1}^{n} f(\xi_i) \Delta x_i$，这个和就是黎曼和（Riemann Sum）.

如果对 $[a,b]$ 的任意划分和在各个子区间上任意取点，极限（Limit）

$$\lim_{\Delta x \to 0} \sum_{i=1}^{n} f(\xi_i) \Delta x_i$$

（其中 $\Delta x = \max\limits_{1 \leqslant i \leqslant n} \{\Delta x_i\}$）总存在且相等，则称 $f(x)$ 在 $[a,b]$ 上可积（Integrable），且称这个极限值为 $f(x)$ 在 $[a,b]$ 上的定积分（Definite Integral），记为：

$$\int_a^b f(x) \mathrm{d}x$$

其中：a 叫做积分下限（Lower Limit of Integral）；b 叫做积分上限（Upper Limit of Integral）；区间 $[a,b]$ 叫做积分区间（Interval of Integration）；函数 $f(x)$ 叫做被积函数（Integrand）；x 叫做积分变量（Variable of Integration）；$f(x)\mathrm{d}x$ 叫做被积表达式；$\int$ 叫做积分号（Sign of Integration）.

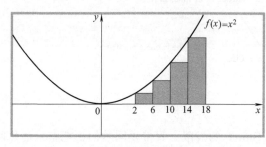

Figure 8.2.1(c)

Example 8.5 Approximate $\int_2^{18} x^2 \mathrm{d}x$ using four left endpoint rectangles with equal width.

Solution：

Draw four trapezoids that look like Figure 8.2.1(c)：

The width of each rectangle is

$$\frac{b-a}{n} = \frac{18-2}{4} = 4$$

The heights of the rectangle are：

$y_1 = f(2) = 4, y_2 = f(6) = 36, y_3 = f(10) = 100$ and $y_4 = f(14) = 196$

Therefore, the area is：

$$\int_2^{12} x^2 \mathrm{d}x \approx L(4) = 4 \times (4 + 36 + 100 + 196) = 1344$$

Note：定积分（Definite Integral），积分后得到的值是确定的，是一个数，不是一个函数.

Example 8.6 If we are given the following table of values for x and $f(x)$：

x	3	6	9	12	15
$f(x)$	4	8	11	17	19

Use Trapezoidal Rule with 4 subintervals indicated by the data in the table to approximate $\int_3^{15} f(x) \mathrm{d}x$.

Solution：

The height of each trapezoids is 3，and the bases of the trapezoids are：
$$y_0=4, y_1=8, y_2=11, y_3=17, y_4=19$$

Therefore：
$$\int_3^{15} f(x)\mathrm{d}x \approx T(4) = \frac{1}{2} \times 4(4+2\times 8+2\times 11+2\times 17+19)=190$$

8.2.2 定积分的几何意义 The Geometric Interpretation of Definite Integral

Definite Integral as an Area Under a Curve：If $y=f(x)$ is nonnegative and integrable over a closed interval $[a,b]$, then the integral of $f(x)$ from a to b is the area under the curve $y=f(x)$ over $[a,b]$. See Figure 8.2.2(a).

$$A = \int_a^b f(x)\mathrm{d}x$$

假设函数 $f(x)$ 是 $[a,b]$ 上的可积函数，$\int_a^b f(x)\mathrm{d}x$ 的几何意义是：曲线 $y=f(x)$ 下，直线 $x=a$ 到直线 $x=b$ 与 x 轴所围成的面积 A. 并且，我们规定，在 x 轴上方的面积为正的，x 下方的面积为负的.

Example 8.7　Using $M(4)$, find that the approximate area of the shaded region Figure 8.2.2(b).

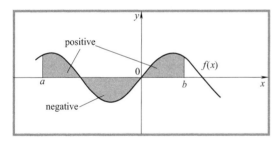

Figure 8.2.2(a)

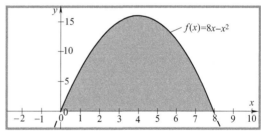

Figure 8.2.2(b)

Solution：

The area of the shaded region is $\int_0^8 f(x)\mathrm{d}x$.

On $[0,8]$ with $n=4$, and the width of each rectangle is
$$\frac{b-a}{n} = \frac{8-0}{4} = 2$$

The heights of the rectangle are：
$$y_{\frac{1}{2}}=f(1)=7, y_{\frac{3}{2}}=f(3)=15, y_{\frac{5}{2}}=f(5)=15 \text{ and } y_{\frac{7}{2}}=f(7)=7$$

Therefore，the area is：
$$\int_0^8 f(x)\mathrm{d}x \approx M(4) = 2 \times (7+15+15+7) = 88$$

8.2.3 近似求和比大小 Comparing Approximating Sums

前面我们已经讲过，黎曼和 $L(n)$、$M(n)$、$R(x)$ 和梯形法则 $T(n)$ 可以估算

$\int_a^b f(x)\mathrm{d}x$ 的大小，那么，它们之间有什么样的大小关系呢？

1) If $f(x)$ is an increasing function on $[a, b]$, then $L(n) \leqslant \int_a^b f(x)\mathrm{d}x \leqslant R(n)$; while if $f(x)$ is decreasing, then $R(n) \leqslant \int_a^b f(x)\mathrm{d}x \leqslant L(n)$.

受函数的单调性影响，当函数 $f(x)$ 单调递增（Increasing）时，见 Figure 8.2.3(a)，实线（阴影面积）表示左黎曼和 $L(4)$，虚线包围的面积表示右黎曼和 $R(4)$，显然，$L(4) \leqslant \int_a^b f(x)\mathrm{d}x \leqslant R(4)$；

类似地，当函数 $f(x)$ 单调递减（Decreasing）时，见 Figure 8.2.3(b)，实线（阴影面积）表示左黎曼和 $R(4)$，虚线包围的面积表示右黎曼和 $L(4)$，显然 $R(4) \leqslant \int_a^b f(x)\mathrm{d}x \leqslant L(4)$.

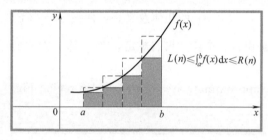

Figure 8.2.3(a)

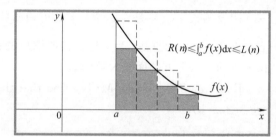

Figure 8.2.3(b)

2) If the graph of $f(x)$ is concave up, then $M(n) \leqslant \int_a^b f(x)\mathrm{d}x \leqslant T(n)$; while if the graph of $f(x)$ is concave down, then $T(n) \leqslant \int_a^b f(x)\mathrm{d}x \leqslant M(n)$.

受函数的凹凸性影响，见 Figure 8.2.3(c)，显然，当函数 $f(x)$ 上凹（Concave Up）时，梯形法则估计的面积大于实际面积；当函数 (x) 下凹（Concave Down）时，实际面积大于梯形法则估计的面积.

Figure 8.2.3(d)，当函数 $f(x)$ 下凹时，中黎曼和面积 $ABCD$，其中 $\triangle AEG \cong \triangle DEF$，将 $\triangle AEG$ 面积割补到 $\triangle DEF$，比覆盖了实际面积还有剩余，即 $\int_a^b f(x)\mathrm{d}x \leqslant M(n)$；同理，当函数 $f(x)$ 上凹时，$M(n) \leqslant \int_a^b f(x)\mathrm{d}x$. 函数凹向的判定见本书 6.2.2 节.

Example 8.8 Write an inequality including $L(n)$, $M(n)$, $R(n)$, $T(n)$, and $\int_a^b f(x)\mathrm{d}x$ for the graph of $f(x)$ shown in Figure 8.2.3(e).

Solution：

Since $f(x)$ increase on $[a, b]$ and is concave down, the inequality is

$$L(n) \leqslant T(n) \leqslant \int_a^b f(x)\mathrm{d}x \leqslant M(n) \leqslant R(n)$$

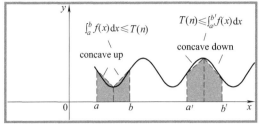

Figure 8.2.3(c)　　　　　　　　　　Figure 8.2.3(d)

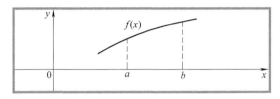

Figure 8.2.3(e)

8.3　微积分基本定理 The Fundamental Theorem of Calculus

微积分基本定理是微积分中最基本的公式，证明了微分与积分是可逆运算，是联系微分学与积分学的桥梁，同时在理论上标志着微积分完整体系的形成．

8.3.1　牛顿-莱布尼茨公式 Newton-Leibniz Formula

If $f(x)$ is continuous on $[a,b]$, then
$$\int_a^b f(x)\mathrm{d}x = F(b) - F(a)$$

Where $F(x)$ is any antiderivative of $f(x)$, that is, a function such $F'(x) = f(x)$.

牛顿-莱布尼茨公式也称为微积分基本定理（The Fundamental Theorem of Calculus）．它表明：一个连续函数在区间 $[a,b]$ 上的定积分等于它的任意一个原函数在区间 $[a,b]$ 上的增量；它揭示了定积分与不定积分之间的联系．

Example 8.9　Find $\int_0^1 x^2 \mathrm{d}x$．

Solution：
$$\int_0^1 x^2 \mathrm{d}x = \frac{x^3}{3}\bigg|_0^1 = \frac{1^3}{3} - \frac{0^3}{3} = \frac{1}{3}$$

Example 8.10　Find $\int_2^5 (x^3 + x)\mathrm{d}x$．

Solution：
$$\int_2^5 (x^3 + x)\mathrm{d}x = \left(\frac{x^4}{4} + \frac{x^2}{2}\right)\bigg|_2^5 = \left(\frac{5^4}{4} + \frac{5^2}{2}\right) - \left(\frac{2^4}{4} + \frac{2^2}{2}\right) = \frac{651}{4}$$

Example 8.11　Find $\int_0^{\frac{\pi}{2}} \cos x \, \mathrm{d}x$．

Solution:
$$\int_0^{\frac{\pi}{2}} \cos x \, dx = \sin x \Big|_0^{\frac{\pi}{2}} = \sin\left(\frac{\pi}{2}\right) - \sin 0 = 1$$

Example 8.12　Find $\int_{-2}^{-1} \frac{1}{x} dx$.

Solution:
$$\int_{-2}^{-1} \frac{1}{x} dx = (\ln |x|)\Big|_{-2}^{-1} = \ln|-1| - \ln|-2| = -\ln 2$$

Example 8.13　Find $\int_{-1}^{\sqrt{3}} \frac{1}{1+x^2} dx$.

Solution:
$$\int_{-1}^{\sqrt{3}} \frac{1}{1+x^2} dx = \arctan x \Big|_{-1}^{\sqrt{3}} = \arctan\sqrt{3} - \arctan(-1)$$
$$= \frac{\pi}{3} - \left(-\frac{\pi}{4}\right) = \frac{7}{12}\pi$$

8.3.2　微积分第二基本定理 The Second Fundamental Theorem of Calculus

If $f(x)$ is continuous on $[a, b]$, then derivative of the function $F(x) = \int_a^x f(t) dt$ is:
$$\frac{dF}{dx} = \frac{d}{dx}\int_a^x f(t) dt = f(x), \quad x \in [a, b]$$

假设函数 $f(x)$ 在区间 $[a,b]$ 上连续（Continuous），函数 $F(x) = \int_a^x f(t) dt$，那么，函数 $F(x)$ 的导数为：
$$\frac{dF}{dx} = \frac{d}{dx}\int_a^x f(t) dt = f(x)$$

$\int_a^x f(t) dt$ 称为变上限定积分（The Integral With A Variable Upper Limit）。微积分第二基本定理指出，连续函数 $f(x)$ 取变上限 x 的定积分，然后求导，其结果还原为 $f(x)$ 本身，再次揭示了不定积分和微分互为逆运算。

Example 8.14　Find $\frac{d}{dx}\int_2^x \sin t \, dt$.

Solution:
By the second fundamental theorem of calculus,
$$\frac{dy}{dx}\int_2^x \sin t \, dt = \sin x$$

Example 8.15　Find $\frac{d}{dx}\int_5^x \sin t \, dt$.

Solution:
By the second fundamental theorem of calculus,
$$\frac{dy}{dx}\int_5^x \sin t \, dt = \sin x$$

Note：根据微积分第二基本定理，本题和上一题答案相同．

Example 8.16 Find $\dfrac{d}{dx}\displaystyle\int_5^x (5t^3 - 2t^2)dt$.

Solution：

By the second fundamental theorem of calculus,

$$\frac{dy}{dx}\int_5^x (5t^3 - 2t^2)dt = 5x^3 - 2x^2$$

Example 8.17 Find $\dfrac{d}{dx}\displaystyle\int_5^{x^2} (5t^3 - 2t^2)dt$.

Solution：

Let $u = x^2$, then $du = 2x\,dx$

$$dx = \frac{du}{2x}$$

$$\frac{d}{dx}\int_5^{x^2}(5t^3 - 2t^2)dt = 2x\,\frac{d}{du}\int_5^u (5t^3 - 2t^2)dt$$

By the second fundamental theorem of calculus,

$$\frac{d}{du}\int_5^u (5t^3 - 2t^2)dt = 5u^3 - 2u^2$$

Therefor,

$$\frac{d}{dx}\int_5^{x^2}(5t^3 - 2t^2)dt = 2x[5(x^2)^3 - 2(x^2)^2] = 10x^7 - 4x^5$$

Example 8.18 Find $\dfrac{d}{dx}\displaystyle\int_5^{x^2} \sin(\ln t)dt$.

Solution：

Let $u = x^2$, then $du = 2x\,dx$

$$\frac{d}{dx}\int_5^{x^2}\sin(\ln t)dt = 2x\,\frac{d}{du}\int_5^u \sin(\ln t)dt$$

By the second fundamental theorem of calculus,

$$\frac{d}{du}\int_5^u \sin(\ln t)dt = \sin(\ln u)$$

Therefor,

$$\frac{d}{dx}\int_5^{x^2}\sin(\ln t)dt = 2x\sin(\ln x^2)$$

8.3.3 累积函数 Accumulation Functions

The function $F(x) = \displaystyle\int_a^x f(t)dt$ is named accumulation function.

前面我们讲，假设函数 $f(x)$ 在 $[a,b]$ 上可积，$\displaystyle\int_a^x f(t)dt$ 称为变上限定积分（The Integral With A Variable Upper Limit），则如此确定的函数 $F(x) = \displaystyle\int_a^x f(t)dt$ 称为累积函数（Accumulation Functions）. 累积函数表示在曲线 $f(x)$ 下，从 $x = a$ 到变量 $x = x$ 的

面积,见 Figure 8.3.3 (a)。

Example 8.19 Suppose we have the function $F(x) = \int_1^x (t-1)^2 dt$. Let's evaluate this as x increases from 1 to 5.

Solution:

Let's find $F(2)$. Graphically, we are looking for the area under the curve $y = (x-1)^2$ from $x=1$ to $x=2$. See Figure 8.3.3(b):

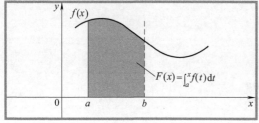

Figure 8.3.3(a)   Figure 8.3.3(b)

If we evaluate the integral, we get: $F(2) = \int_1^2 (x-1)^2 dx = \dfrac{1}{3}$

As in the previous example, let's make a table of values of the accumulation function for different values of x:

x	2	3	4	5
$F(x)$	$\dfrac{1}{3}$	$\dfrac{8}{3}$	$\dfrac{27}{3}$	$\dfrac{64}{3}$

We can see that the values of $F(x)$ will increase as x increases from 1 to 5.

Example 8.20 Let $f(x)$ be the function given by $F(x) = \int_1^x (t^2 - 4) dt$ on $[-3, 3]$. At which values of x does $F(x)$ attain a relative minimum?

Solution:

Find the derivative of the function
$$F'(x) = \frac{d}{dx} \int_1^x (t^2 - 4) dt = x^2 - 4$$

Set the derivative equal to zero and solve for x:
$$x^2 - 4 = 0, \quad x = \pm 2$$

Take the second derivative of the function:
$$F''(x) = (x^2 - 4)' = 2x$$

Since $F''(2) = 4 > 0$ and $F''(-2) = -4 < 0$,

then, the function $F(x)$ attain a relative minimum at $x = 2$.

8.3.4 lnx 的定积分定义 lnx as A Definite Integral

The natural logarithm of $x(x > 0)$, as a definite integral:

$$\ln x = \int_1^x \frac{1}{t} dt$$

自然对数 $\ln x$ 在物理学、生物学等自然学科有重要的意义．它表示曲线 $y = \frac{1}{t}$（$t > 0$）、t 轴、直线 $t = 1$ 和直线 $t = x$（$x > 1$）所围成的面积，可以看成累积函数（Accumulation Functions）．见 Figure 8.3.4(a)．

根据微积分基本定理和 $\ln 1 = 0$，我们也可以大致得到上述结论：

$$\int_1^x \frac{1}{t} dt = \ln x - \ln 1 = \ln x$$

Example 8.21 Show that $\frac{5}{6} < \ln 3 < \frac{3}{2}$.

Solution:

Using the definition of $\ln x$ by definite integral:

$$\ln 3 = \int_1^3 \frac{1}{x} dx$$

Interpret as the area under $y = \frac{1}{x}$, above the x-axis, and bounded at the left by $x = 1$ and the right by $x = 3$. See Figure 8.3.4(b).

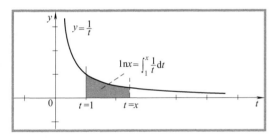
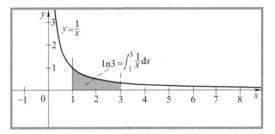

Figure 8.3.4(a)   Figure 8.3.4(b)

The area of the shaded region is $\int_1^3 \frac{1}{x} dx$. On $[1, 3]$ with $n = 2$, and the width of each rectangle is

$$\frac{b-a}{n} = \frac{3-1}{2} = 1$$

The heights of the rectangle are:

$$y_1 = f(1) = 1,\ y_2 = f(2) = \frac{1}{2},\ y_3 = f(3) = \frac{1}{3}$$

Therefore,

$$L(2) = \frac{3-1}{2}\left(\frac{1}{1} + \frac{1}{2}\right) = \frac{3}{2} \text{ and } R(2) = \frac{3-1}{2} \times \left(\frac{1}{2} + \frac{1}{3}\right) = \frac{5}{6}$$

Since $y = \frac{1}{x}$ decreasing on $[1, 3]$, then $R(2) < \int_1^3 \frac{1}{x} dx < L(2)$

Thus,

$$\frac{5}{6} < \ln 3 < \frac{3}{2}$$

8.4 定积分的性质 Properties of Definite Integral

$f(x)$ 和 $g(x)$ 同时为可积（Integrable）函数，则定积分有如下性质．

8.4.1 零区间和数乘 Zero Width Interval and Constant Multiple

1) Zero Width Interval, see Figure 8.4.1(a):

$$\int_a^a f(x)\,\mathrm{d}x = 0$$

2) Constant Multiple, see Figure 8.4.1(b):

$$\int_a^b kf(x)\,\mathrm{d}x = k\int_a^b f(x)\,\mathrm{d}x,\ k \text{ is a constant}$$

$$\int_a^b kf(x)\,\mathrm{d}x = k\int_a^b f(x)\,\mathrm{d}x$$

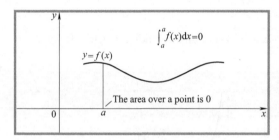

Figure 8.4.1(a)　　　　　　　　　Figure 8.4.1(b)

Example 8.22　Find $\int_3^3 \dfrac{\sin^2(e^x)}{1-x^2}\,\mathrm{d}x$.

Solution:

Since $\int_a^a f(x)\,\mathrm{d}x = 0$, then

$$\int_3^3 \frac{\sin^2(e^x)}{1-x^2}\,\mathrm{d}x = 0$$

Example 8.23　Find $\int_1^4 2x\sqrt{x}\,\mathrm{d}x$.

Solution:

$$\int_1^4 2x\sqrt{x}\,\mathrm{d}x = 2\int_1^4 x^{\frac{3}{2}}\,\mathrm{d}x = 2 \times \frac{2}{5}x^{\frac{5}{2}}\bigg|_1^4 = \frac{124}{5}$$

8.4.2 加减法和可加性 Sum/Difference and Additivity

1) Sum/Difference, see Figure 8.4.2(a):

$$\int_a^b [f(x) \pm g(x)]\,\mathrm{d}x = \int_a^b f(x)\,\mathrm{d}x \pm \int_a^b g(x)\,\mathrm{d}x$$

2) Additivity, see Figure 8.4.2(b):

$$\int_a^b f(x)\,\mathrm{d}x = \int_a^c f(x)\,\mathrm{d}x + \int_c^b f(x)\,\mathrm{d}x,\ a < c < b$$

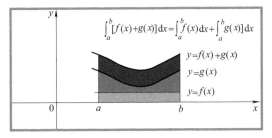

Figure 8.4.2(a)

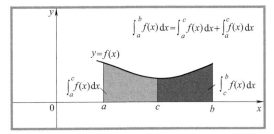

Figure 8.4.2(b)

Example 8.24 Find $\int_0^3 (x^2 + x + 1)\,dx$.

Solution:

$$\int_0^3 (x^2 + x + 1)\,dx = \int_0^3 x^2\,dx + \int_0^3 x\,dx + \int_0^3 dx$$
$$= \frac{x^3}{3}\Big|_0^3 + \frac{x^2}{2}\Big|_0^3 + x\Big|_0^3 = \frac{33}{2}$$

Example 8.25 Find $\int_1^3 |x - 2|\,dx$.

Solution:

$$\int_1^3 |x - 2|\,dx = \int_1^2 |x - 2|\,dx + \int_2^3 |x - 2|\,dx$$
$$= \int_1^2 -(x - 2)\,dx + \int_2^3 (x - 2)\,dx$$
$$= -\frac{(x-2)^2}{2}\Big|_1^2 + \frac{(x-2)^2}{2}\Big|_2^3 = 1$$

8.4.3 积分的次序 Order of Integration

$$\int_a^b f(x)\,dx = -\int_b^a f(x)\,dx$$

Example 8.26 Find $\int_2^0 x e^{x^2}\,dx$.

Solution:
Since

$$\int x e^{x^2}\,dx = \frac{1}{2}\int e^{x^2}\,d(x^2) = \frac{1}{2}e^{x^2} + C$$

Then,

$$\int_2^0 x e^{x^2}\,dx = -\int_0^2 x e^{x^2}\,dx = -\frac{1}{2}e^{x^2}\Big|_0^2 = \frac{1}{2}(1 - e^4)$$

8.4.4 对称函数积分 Definite Integrals of Symmetric Functions

1) If $f(x)$ is an odd function, then

$$\int_{-a}^a f(x)\,dx = 0$$

2) If $f(x)$ is an even function, then

$$\int_{-a}^a f(x)\,dx = 2\int_0^a f(x)\,dx$$

Example 8.27 Find $\int_{-2}^{2} \sin x \, dx$.

Solution：

Since $y = \sin x$ is an odd function, see Figure 8.4.4(a), then

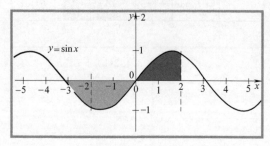
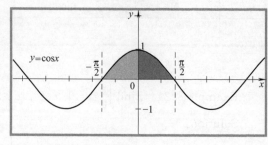

Figure 8.4.4(a)   Figure 8.4.4(b)

$$\int_{-2}^{2} \sin x \, dx = 0$$

Example 8.28 Find $\int_{-\pi/2}^{\pi/2} \cos x \, dx$.

Solution：

Since $y = \cos x$ is an even function, see Figure 8.4.4(b), then

$$\int_{-\pi/2}^{\pi/2} \cos x \, dx = 2 \int_{0}^{\pi/2} \cos x \, dx = 2 \sin x \Big|_{0}^{\pi/2} = 2$$

Example 8.29 Find $\int_{-\pi}^{\pi} (\sin x + \cos x) \, dx$.

Solution：

$$\int_{-\pi}^{\pi} (\sin x + \cos x) \, dx = \int_{-\pi}^{\pi} \sin x \, dx + \int_{-\pi}^{\pi} \cos x \, dx$$

$$= 0 + 2 \int_{0}^{\pi} \cos x \, dx$$

$$= 0 + 2 \sin x \Big|_{0}^{\pi} = 0$$

8.5 积分中值定理 The Mean Value Theorem for Integrals

If $f(x)$ is continuous on $[a, b]$, there is a number ξ between a and b, such that

$$\int_{a}^{b} f(x) \, dx = f(\xi)(b - a).$$

Then, the average value of $f(x)$ on $[a, b]$ is

$$f(\xi) = \frac{1}{b-a} \int_{a}^{b} f(x) \, dx, \quad a < \xi < b$$

前面我们讲过，微分中值定理（The Mean Value Theorem for Derivatives），见本书 6.4 节．对应地，积分中值定理也是 AP 微积分考试的一个热点内容．

积分中值定理的几何意义：曲线 $y = f(x)$，直线 $x = a$，$x = b$ 与 x 轴所围成的曲边梯形的面积，等于以区间 $[a, b]$ 为底，以这个区间内的某一点处曲线的纵坐标 $f(\xi)$ 为高的矩形面积，见 Figure 8.5(a)．

Example 8.30 Find the average value of $f(x)=3x^2$ from $x=1$ to $x=3$.

Solution:

The function $f(x)=3x^2$ is continuous on $[1,3]$ and the average value of $f(x)$ is

$$f(\xi) = \frac{1}{b-a}\int_a^b f(x)\,\mathrm{d}x = \frac{1}{3-1}\int_1^3 3x^2\,\mathrm{d}x$$
$$= \frac{1}{2}\times 3\times \frac{x^3}{3}\bigg|_1^3 = 13$$

Example 8.31 The graph in Figure 8.5(b) shows the speed $v(t)$ of a car, in miles per hour, at 5-minute intervals during a 1-hour period. Estimate the average speed of the car during the interval $20 \leqslant t \leqslant 50$.

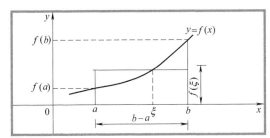

Figure 8.5(a)　　　　　　　　Figure 8.5(b)

Solution:

Approximate the distance from $t=20$ to $t=30$ by the area under the curve, roughly, by the sum of the areas of a trapezoid:

$$\frac{1}{2}\times\frac{1}{12}(40+2\times 50+60)=\frac{25}{3}(\mathrm{m})$$

Similarly, Approximate the distance from $t=40$ to $t=50$:

$$\frac{1}{2}\times\frac{1}{12}(60+2\times 50+30)=\frac{95}{12}(\mathrm{m})$$

Approximate the distance from $t=30$ to $t=40$ by the area under the curve, roughly, by the sum of the areas of a rectangle:

$$\frac{1}{12}\times(60+60)=10(\mathrm{m})$$

Then, approximate the distance from $t=20$ to $t=50$:

$$\frac{25}{3}+\frac{95}{12}+10=\frac{105}{4}(\mathrm{m})$$

The average speed equals the distance traveled divided by the time. Thus the averge speed from $t=20$ to $t=50$:

$$\frac{\frac{105}{4}\mathrm{m}}{(50-20)\mathrm{min}}=\frac{7\mathrm{m}}{8\mathrm{min}}=\frac{105}{2}\mathrm{m/h}$$

Example 8.32 Find the average value of $f(x)=\sqrt{16-x^2}$ on $[-4,4]$.

Solution:

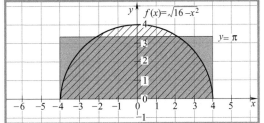

Figure 8.5(c)

Recognize $f(x) = \sqrt{16-x^2}$ as a function whose graph is the upper semicircle of radius 4 centered at the origin, see Figure 8.5(c).

The area between the semicircle and the x-axis from -4 to 4 can be computed using the geometry formula

$$A = \frac{1}{2}\pi r^2 = \frac{1}{2}\pi \times 4^2 = 8\pi$$

Because $f(x)$ is nonnegative, the area is also the value of the integral of $f(x)$ from -4 to 4,

$$\int_{-4}^{4} \sqrt{16-x^2}\,dx = 8\pi$$

Therefore, the average value of $f(x)$ is

$$f(\xi) = \frac{1}{4-(-4)} \int_{-4}^{4} \sqrt{16-x^2}\,dx = \frac{1}{8} \times 8\pi = \pi$$

8.6 定积分的计算 The Operations of Definite Integrate

有了牛顿-莱布尼茨公式（Newton-Leibniz Formula）后，一般的定积分计算便转化成了不定积分的计算．类似于不定积分（The Indefinite Integral），定积分也有换元法和分部积分法，这两个基本方法对定积分的计算很重要．

8.6.1 定积分的换元法 Substitution in Definite Integrals

If $\varphi'(x)$ is continuous on the interval $[a, b]$ and $f(x)$ is continuous on the range of $\varphi(x)$, then

$$\int_a^b f[\varphi(x)]\varphi'(x)\,dx = \int_{\varphi(a)}^{\varphi(b)} f(u)\,du$$

Where $u = \varphi(x)$.

【方法总结】应用定积分的换元法有两点值得同学们注意：

1) 用 $u = \varphi(x)$ 把原来的变量 x 替换成新变量 u 后，积分上限和积分下限也要替换成相应于新变量 u 的积分限．

2) 求出 $f(u)$ 的一个原函数（Antiderivative）后，不必类似于不定积分，再替换成原来的变量 x 的函数，只要把新变量 u 的上、下限分别代入，做差得出结果即可．

Example 8.33　Find $\int_{-1}^{1} 3x^2 \sqrt{x^3+1}\,dx$.

Solution：

Let $u = x^3 + 1$, $du = 3x^2\,dx$. To find the new limits of integration：

When $x = -1$, $u = (-1)^3 + 1 = 0$; when $x = 1$, $u = (1)^3 + 1 = 2$

Therefore,

$$\int_{-1}^{1} 3x^2 \sqrt{x^3+1}\,dx = \int_0^2 \sqrt{u}\,du$$

Since

$$\int_0^2 \sqrt{u}\,du = \frac{2}{3} u^{\frac{3}{2}} \Big|_0^2 = \frac{2}{3}(2^{\frac{3}{2}} - 0^{\frac{3}{2}}) = \frac{4\sqrt{2}}{3}$$

Then,
$$\int_{-1}^{1} 3x^2 \sqrt{x^3+1}\,dx = \frac{4\sqrt{2}}{3}$$

Example 8.34 Find $\int_{1}^{2} \frac{1}{(3-5x)^2}\,dx$.

Solution:

Let $u = 3-5x$, $du = -5dx$, so $dx = -\frac{1}{5}du$. To find the new limits of integration:
When $x=1$, $u=3-5\times 1=-2$; when $x=1$, $u=3-5\times 2=-7$.
Therefore,
$$\int_{1}^{2} \frac{1}{(3-5x)^2}\,dx = -\frac{1}{5}\int_{-2}^{-7}\frac{1}{u^2}du = \frac{1}{5}\int_{-7}^{-2}\frac{1}{u^2}du$$

Since
$$\frac{1}{5}\int_{-7}^{-2}\frac{1}{u^2}du = \frac{1}{5}\left(-\frac{1}{u}\right)\Big|_{-7}^{-2} = \frac{1}{5}\times\left(\frac{1}{2}-\frac{1}{7}\right) = \frac{1}{14}$$

Then,
$$\int_{1}^{2}\frac{1}{(3-5x)^2}dx = \frac{1}{14}$$

Example 8.35 Find $\int_{\pi/4}^{\pi/2} \cot x \csc^2 x\,dx$.

Solution:

Let $u = \cot x$, $du = -\csc^2 x\,dx$. To find the new limits of integration:
When $x=\frac{\pi}{4}$, $u=\cot\frac{\pi}{4}=1$; when $x=\frac{\pi}{2}$, $u=\cot\frac{\pi}{2}=0$

Therefor,
$$\int_{\pi/4}^{\pi/2} \cot x \csc^2 x\,dx = \int_{1}^{0} u(-du) = \int_{0}^{1} u\,du$$

Since
$$\int_{0}^{1} u\,du = \frac{u^2}{2}\Big|_{0}^{1} = \frac{1^2}{2}-\frac{0^2}{2}=\frac{1}{2}$$

Then,
$$\int_{\pi/4}^{\pi/2}\cot x\csc^2 x\,dx = \frac{1}{2}$$

8.6.2# 定积分的分部积分法 Integration by Parts Formula for Definite Integrals

If $u=u(x)$ and $v=v(x)$ are continuous on $[a,b]$, then
$$\int_{a}^{b} u\,dv = uv\Big|_{a}^{b} - \int_{a}^{b} v\,du$$

定积分的分部积分公式可由不定积分（The Indefinite Integral）的分部积分法和牛顿-莱布尼茨公式推导得到．

【方法总结】上述公式表明：

1) 当右边的定积分比较容易计算时，利用定积分的分部积分公式可算出左边的定积分；

2）原函数（Antiderivative）已经积出的部分可以先用上、下限代入计算．

Example 8.36 Find $\int_0^4 x e^{-x} dx$.

Solution：

Let $u = x$, $dv = e^{-x} dx$, then $du = dx$, $v = -e^{-x}$

$$\int_0^4 x e^{-x} dx = -x e^{-x} \Big|_0^4 - \int_0^4 (-e^{-x}) dx$$
$$= -4e^{-4} - 0 + \int_0^4 e^{-x} dx$$
$$= -4e^{-4} - e^{-x} \Big|_0^4 = -4e^{-4} - e^{-4} + e^0$$
$$= 1 - 5e^{-4} \approx 0.91$$

Note：AP 微积分考试中，有部分题需要使用计算器求定积分，同学们需要掌握指定型号的图形计算器的相关操作方法．

Example 8.37 Find $\int_0^1 \arctan x \, dx$.

Solution：

Let $u = \arctan x$, $dv = dx$, then $du = \dfrac{1}{1+x^2} dx$, $v = x$

$$\int_0^1 \arctan x \, dx = x \arctan x \Big|_0^1 - \int_0^1 \dfrac{x}{1+x^2} dx$$
$$= 1 \times \arctan 1 - 0 \times \arctan 0 - \int_0^1 \dfrac{x}{1+x^2} dx$$
$$= \dfrac{\pi}{4} - \int_0^1 \dfrac{x}{1+x^2} dx$$

Let $t = 1 + x^2$, $dt = 2x \, dx$, To find the new limits of integration：
When $x = 0$, $t = 1$; when $x = 1$, $t = 2$

$$\int_0^1 \dfrac{x}{1+x^2} dx = \dfrac{1}{2} \int_1^2 \dfrac{1}{t} dt = \dfrac{1}{2} \ln|t| \Big|_1^2 = \dfrac{1}{2}(\ln 2 - \ln 1) = \dfrac{1}{2} \ln 2$$

Therefore,

$$\int_0^1 \arctan x \, dx = \dfrac{\pi}{4} - \int_0^1 \dfrac{x}{1+x^2} dx = \dfrac{\pi}{4} - \dfrac{\ln 2}{2}$$

8.7# 广义积分 Improper Integrals

前面我们探讨的定积分（Definite Integral）前提是有限积分限和有界函数，接下来，我们会遇到无穷积分限或无界不连续函数的积分，这两类积分叫做广义积分（Improper Integrals）．

8.7.1# 无穷积分限的积分 Integration on an Infinite Interval

1）If $f(x)$ is continuous on $[a, +\infty)$, then

$$\int_a^{+\infty} f(x)\mathrm{d}x = \lim_{b\to+\infty} \int_a^b f(x)\mathrm{d}x$$

2) If $f(x)$ is continuous on $(-\infty, b]$, then

$$\int_{-\infty}^b f(x)\mathrm{d}x = \lim_{a\to-\infty} \int_a^b f(x)\mathrm{d}x$$

3) If $f(x)$ is continuous on $(-\infty, +\infty)$, then

$$\int_{-\infty}^{+\infty} f(x)\mathrm{d}x = \int_{-\infty}^c f(x)\mathrm{d}x + \int_c^{+\infty} f(x)\mathrm{d}x$$

where c is any real number.

式 1) 和式 2) 中，如果极限（Limit）是有限的，广义积分收敛（Converges）且极限是广义积分的值；如果极限不存在，广义积分发散（Diverges）。

式 3) 中，如果等式右端的两个积分都收敛（Converges），则广义积分收敛；否则发散（Diverges）；并且，c 的选择不影响广义积分的敛散性（Convergence or Divergence），我们可以选择合适的 c 求解广义积分.

Example 8.38 Find $\int_1^{+\infty} \frac{1}{\sqrt{x}}\mathrm{d}x$

Solution：

$$\int_1^{+\infty} \frac{1}{\sqrt{x}}\mathrm{d}x = \lim_{b\to+\infty} \int_1^b \frac{1}{\sqrt{x}}\mathrm{d}x = \lim_{b\to+\infty} 2\sqrt{x}\Big|_1^b$$
$$= \lim_{b\to+\infty} (2\sqrt{b}-1) = +\infty$$

The limit does not exist as a finite number and so the improper integral $\int_1^{+\infty} \frac{1}{\sqrt{x}}\mathrm{d}x$ is diverges.

Example 8.39 Find $\int_1^{+\infty} \frac{1}{x}\mathrm{d}x$

Solution：

$$\int_1^{+\infty} \frac{1}{x}\mathrm{d}x = \lim_{b\to+\infty} \int_1^b \frac{1}{x}\mathrm{d}x = \lim_{b\to+\infty} \ln|x|\Big|_1^b$$
$$= \lim_{b\to+\infty} (\ln b - \ln 1) = \lim_{b\to+\infty} \ln b = +\infty$$

The limit does not exist as a finite number and so the improper integral $\int_1^{+\infty} \frac{1}{x}\mathrm{d}x$ is diverges.

Example 8.40 Find $\int_1^{+\infty} \frac{1}{x^2}\mathrm{d}x$.

Solution：

$$\int_1^{+\infty} \frac{1}{x^2}\mathrm{d}x = \lim_{b\to+\infty} \int_1^b \frac{1}{x^2}\mathrm{d}x = \lim_{b\to+\infty} \left(-\frac{1}{x}\right)\Big|_1^b$$
$$= \lim_{b\to+\infty} \left(-\frac{1}{b}+1\right) = 1$$

Thus, the improper integral converges and the area has finite value 1.

【方法总结】广义积分 $\int_1^{+\infty} \frac{1}{x^p}\mathrm{d}x$，如果 $p>1$，积分收敛到值 $\frac{1}{p-1}$；如果 $p\leqslant 1$，积

分发散. 后续级数（Series）章节还将深入讲解.

Example 8.41 Find $\int_{-\infty}^{0} x e^x \, dx$.

Solution:

$$\int_{-\infty}^{0} x e^x \, dx = \lim_{a \to -\infty} \int_{a}^{0} x e^x \, dx$$

Let $u = x$, $dv = e^x$, then $du = dx$, $v = e^x$:

$$\int_{a}^{0} x e^x \, dx = x e^x \Big|_{a}^{0} - \int_{a}^{0} e^x \, dx = -a e^a - 1 + e^a$$

By Hospital's Rule:

$$\lim_{a \to -\infty} a e^a = \lim_{a \to -\infty} \frac{a}{e^{-a}} = \lim_{a \to -\infty} \frac{1}{-e^{-a}} = \lim_{a \to +\infty} (-e^a) = 0$$

Therefore,

$$\int_{-\infty}^{0} x e^x \, dx = \lim_{a \to -\infty} (-a e^a - 1 + e^a) = 0 - 1 + 0 = -1$$

Thus, the improper integral converges and the area has finite value -1.

Example 8.42 Find $\int_{-\infty}^{+\infty} \frac{1}{9 + x^2} \, dx$.

Solution:

It's convenient to choose $c = 0$:

$$\int_{-\infty}^{+\infty} \frac{1}{9 + x^2} \, dx = \int_{-\infty}^{0} \frac{1}{9 + x^2} \, dx + \int_{0}^{+\infty} \frac{1}{9 + x^2} \, dx$$

Find the integrals on the right side separately:

$$\int_{-\infty}^{0} \frac{1}{9 + x^2} \, dx = \lim_{a \to -\infty} \int_{a}^{0} \frac{1}{9 + x^2} \, dx = \lim_{a \to -\infty} \frac{1}{3} \arctan \frac{x}{3} \Big|_{a}^{0}$$

$$= \frac{1}{3} \lim_{a \to -\infty} \left(\arctan \frac{0}{3} - \arctan \frac{a}{3} \right) = \frac{1}{3} \left[0 - \left(-\frac{\pi}{2} \right) \right] = \frac{\pi}{6}$$

$$\int_{0}^{+\infty} \frac{1}{9 + x^2} \, dx = \lim_{b \to +\infty} \int_{0}^{b} \frac{1}{9 + x^2} \, dx = \lim_{b \to +\infty} \frac{1}{3} \arctan \frac{x}{3} \Big|_{0}^{b}$$

$$= \frac{1}{3} \lim_{b \to +\infty} \left(\arctan \frac{b}{3} - \arctan \frac{0}{3} \right) = \frac{1}{3} \left(\frac{\pi}{2} - 0 \right) = \frac{\pi}{6}$$

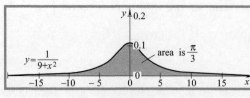

Figure 8.7.1

Since both integrals are convergent, the given integral is convergent:

$$\int_{-\infty}^{+\infty} \frac{1}{9 + x^2} \, dx = \frac{\pi}{6} + \frac{\pi}{6} = \frac{\pi}{3}$$

Since $\frac{1}{9 + x^2} > 0$, the given improper integral can be interpreted as the area of the infinite region that lies under the cure $y = \frac{1}{9 + x^2}$ and above the x-axis, see Figure 8.7.1.

8.7.2# 无界不连续的积分 Integrands with Infinite Discontinuities

1) If $f(x)$ is continuous on $(a, b]$ and is discontinuous at a then

$$\int_a^b f(x)\mathrm{d}x = \lim_{c \to a^+} \int_c^b f(x)\mathrm{d}x$$

2) If $f(x)$ is continuous on $[a, b)$ and is discontinuous at a then

$$\int_a^b f(x)\mathrm{d}x = \lim_{c \to b^-} \int_a^c f(x)\mathrm{d}x$$

3) If $f(x)$ is discontinuous at c, where $a < c < b$, and continuous on $[a, c) \cup (c, b]$, then

$$\int_a^b f(x)\mathrm{d}x = \int_a^c f(x)\mathrm{d}x + \int_c^b f(x)\mathrm{d}x$$

式 1) 和式 2) 中，如果极限（Limit）是有限的，广义积分收敛（Converges）且极限是广义积分的值；如果极限不存在，广义积分发散（Diverges）。

式 3) 中，如果等式右端的两个积分都收敛（Converges），则广义积分收敛；否则发散（Diverges）。

Example 8.43 Find $\int_3^4 \dfrac{1}{\sqrt{x-3}} \mathrm{d}x$.

Solution：

$$\begin{aligned}\int_3^4 \dfrac{1}{\sqrt{x-3}}\mathrm{d}x &= \lim_{c \to 3^+} \int_c^4 \dfrac{1}{\sqrt{x-3}}\mathrm{d}x \\ &= \lim_{c \to 3^+} 2\sqrt{x-3}\,\Big|_c^4 \\ &= \lim_{c \to 3^+} 2(\sqrt{4-3} - \sqrt{c-3}) = 2\end{aligned}$$

Example 8.44 Determine whether $\int_0^{\frac{\pi}{2}} \tan x\,\mathrm{d}x$ converges or diverges.

Solution：

$$\begin{aligned}\int_0^{\frac{\pi}{2}} \tan x\,\mathrm{d}x &= \lim_{c \to \frac{\pi}{2}^-} \int_0^c \tan x\,\mathrm{d}x \\ &= \lim_{c \to \frac{\pi}{2}^-} (\ln|\sec x|)\Big|_0^c \\ &= \lim_{c \to \frac{\pi}{2}^-} (\ln\sec c - \ln 1) = +\infty\end{aligned}$$

Thus, the given improper integral is divergent.

Example 8.45 Find $\int_0^4 \dfrac{1}{x-2}\mathrm{d}x$.

Solution：

$$\int_0^4 \dfrac{1}{x-2}\mathrm{d}x = \int_0^2 \dfrac{1}{x-2}\mathrm{d}x + \int_2^4 \dfrac{1}{x-2}\mathrm{d}x$$

Where

$$\begin{aligned}\int_0^2 \dfrac{1}{x-2}\mathrm{d}x &= \lim_{c \to 2^-} \int_0^c \dfrac{1}{x-2}\mathrm{d}x = \lim_{c \to 2^-} (\ln|x-2|)\Big|_0^c \\ &= \lim_{c \to 2^-} (\ln|c-2| - \ln 2) = -\infty\end{aligned}$$

Thus, $\int_0^2 \dfrac{1}{x-2}\mathrm{d}x$ is divergent. So, $\int_0^4 \dfrac{1}{x-2}\mathrm{d}x$ is divergent.

8.7.3# 收敛和发散判别法 Test for Converges and Diverges

Let $f(x)$ and $g(x)$ be continuous on $[a, \infty)$ with $0 \leqslant f(x) \leqslant g(x)$ for all $x \geqslant a$, then

1) $\int_a^\infty f(x) \mathrm{d}x$ converges if $\int_a^\infty g(x) \mathrm{d}x$ converges;

2) $\int_a^\infty g(x) \mathrm{d}x$ diverges if $\int_a^\infty f(x) \mathrm{d}x$ diverges.

上述是直接比较判别法，其极限形式，Limit Comparison Test：

If the positive functions $f(x)$ and $g(x)$ be continuous on $[a, \infty)$ and if

$$\lim_{n \to \infty} \frac{f(x)}{g(x)} = L, \ 0 < L < \infty$$

Then, $\int_a^\infty f(x) \mathrm{d}x$ and $\int_a^\infty g(x) \mathrm{d}x$ both converge or both diverge.

在实际中，遇到不能直接求出广义积分（Improper Integrals）的值时，首先尝试确定它收敛（Converges）或发散（Diverges）. 其主要判别法是直接比较判别法和极限比较判别法. 如果收敛，还可以用数值方法逼近它的值.

Example 8.46   Does $\int_1^\infty \dfrac{\sin^2 x}{x^2} \mathrm{d}x$ converge or diverge?

Solution：

$$\frac{\sin^2 x}{x^2} \leqslant \frac{1}{x^2}, \text{ on } [1, \infty)$$

And

$$\int_1^\infty \frac{1}{x^2} \mathrm{d}x = \lim_{a \to \infty} \int_1^a \frac{1}{x^2} \mathrm{d}x = \lim_{a \to \infty} \left(-\frac{1}{x} \bigg|_1^a \right) = \lim_{a \to \infty} \left(-\frac{1}{a} + \frac{1}{1} \right) = 1$$

Since $\int_1^\infty \dfrac{1}{x^2} \mathrm{d}x$ converge and $\dfrac{\sin^2 x}{x^2} \leqslant \dfrac{1}{x^2}$, $\int_1^\infty \dfrac{\sin^2 x}{x^2} \mathrm{d}x$ converge by the Comparison Test.

Example 8.47   Does $\int_1^\infty \mathrm{e}^{-x^2} \mathrm{d}x$ converge or diverge?

Solution：

Since $x \geqslant 1$, then $x \leqslant x^2$, so that $-x^2 \leqslant -x$ and $\mathrm{e}^{-x^2} \leqslant \mathrm{e}^{-x}$ on $[1, \infty)$.

$$\int_1^\infty \frac{1}{\mathrm{e}^x} \mathrm{d}x = \lim_{a \to \infty} \int_1^a \frac{1}{\mathrm{e}^x} \mathrm{d}x = \lim_{a \to \infty} \left(-\frac{1}{\mathrm{e}^x} \bigg|_1^a \right) = \frac{1}{\mathrm{e}}$$

Since $\int_1^\infty \mathrm{e}^{-x} \mathrm{d}x$ converge and $\mathrm{e}^{-x^2} \leqslant \mathrm{e}^{-x}$, $\int_1^\infty \mathrm{e}^{-x^2} \mathrm{d}x$ converge by the Comparison Test.

Example 8.48   Does $\int_1^\infty \dfrac{2}{\mathrm{e}^x - 5} \mathrm{d}x$ converge or diverge?

Solution：

Let $f(x) = \dfrac{2}{\mathrm{e}^x - 5}$, $g(x) = \dfrac{1}{\mathrm{e}^x}$

$$\frac{f(x)}{g(x)} = \frac{\dfrac{2}{\mathrm{e}^x - 5}}{\dfrac{1}{\mathrm{e}^x}} = \frac{2 \mathrm{e}^x}{\mathrm{e}^x - 5}$$

$$\lim_{n\to\infty}\frac{f(x)}{g(x)}=\lim_{n\to\infty}\frac{2\mathrm{e}^x}{\mathrm{e}^x-5}=2>0$$

And

$$\int_1^\infty \frac{1}{\mathrm{e}^x}\mathrm{d}x=\lim_{a\to\infty}\int_1^a \frac{1}{\mathrm{e}^x}\mathrm{d}x=\lim_{a\to\infty}\left(-\frac{1}{\mathrm{e}^x}\bigg|_1^a\right)=\frac{1}{\mathrm{e}}$$

Therefore, $\int_1^\infty \frac{1}{\mathrm{e}^x}\mathrm{d}x$ converge, $\int_1^\infty \frac{2}{\mathrm{e}^x-5}\mathrm{d}x$ also converge by the Limit Comparison Test.

8.8 习题 Practice Exercises

(1) Approximation the area under the curve $y=(x-1)^2+1$ from $x=1$ to $x=5$ using $L(4)$.

(2) Write an inequality including $L(n)$, $M(n)$, $R(n)$, $T(n)$, and $\int_a^b f(x)\mathrm{d}x$ for the graph of $f(x)$ shown in Figure 8.8.

(3) Evaluate the following integrals.

1) $\int_2^5 (x^3+x^2+x)\mathrm{d}x$;

2) $\int_0^{\frac{\pi}{2}} \sin x\,\mathrm{d}x$;

3) $\int_2^3 \frac{1}{3x-2}\mathrm{d}x$;

4) $\int_{-4}^4 \frac{1}{16+x^2}\mathrm{d}x$;

5) $\int_1^e \frac{\ln x}{x}\mathrm{d}x$.

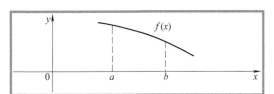

Figure 8.8

(4) Find $\dfrac{\mathrm{d}}{\mathrm{d}x}\int_5^x (5t^6-2t^3)\mathrm{d}t$.

(5) Find $\dfrac{\mathrm{d}}{\mathrm{d}x}\int_5^{x^2} (t^3-1)\mathrm{d}t$.

(6) Let $f(x)$ be the function given by $F(x)=\int_1^x (t^2-2t-3)\mathrm{d}t$ on $[-3,3]$. At which values of x does $F(x)$ attain a relative minimum?

(7) Find the average value of $f(x)=2x^2+x+2$ from $x=1$ to $x=4$.

(8)♯ Evaluate the following integrals.

1) $\int_1^{+\infty} \frac{1}{x^3}\mathrm{d}x$;

2) $\int_{-\infty}^0 x\mathrm{e}^{x^2}\mathrm{d}x$;

3) $\int_0^2 \frac{1}{x-1}\mathrm{d}x$.

第 9 章

积分的应用
Applications of Integral

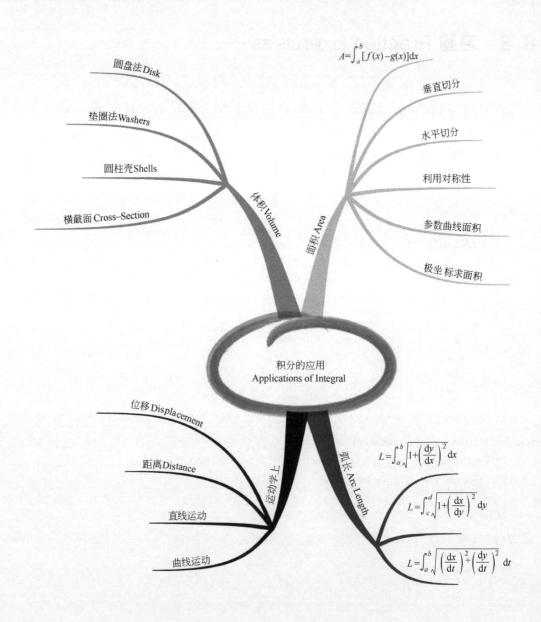

9.1 面积 Area

前面我们已经讲过，定积分 $\int_a^b f(x)\mathrm{d}x$ 的几何意义是：在区间 $[a,b]$ 上，曲线 $y=f(x)$ 下的面积．本节，我们将运用定积分（Definite Integral）求解一些更为复杂的平面图形面积．

9.1.1 两条曲线之间的面积 The Area Between Two Curves

If $f(x)$ and $g(x)$ are continuous with $f(x) \geqslant g(x)$ throughout $[a,b]$, then the area of region between the curves $y=f(x)$ and $y=g(x)$ from a to b is the integral of $[f(x)-g(x)]$ from a to b:

$$A = \int_a^b [f(x)-g(x)]\mathrm{d}x$$

【方法总结】求两条曲线之间的面积一般有 3 个步骤：

1) 画出曲线的图形并画出一个典型的矩形，这样能帮助同学们显示出两条曲线的上下位置关系，形象地帮你求出积分限；
2) 求积分限和写出 $[f(x)-g(x)]$ 的表达式；
3) 从 a 到 b 积分 $[f(x)-g(x)]$，得到所求面积．

9.1.2 垂直切分 Vertical Slices

对于类似于 Figure 9.1.2(a)，在区间 $[a,b]$ 上，一条函数曲线总是在另外一条曲线的上方，一般采用垂直切分求两条曲线之间的面积．把两条曲线之间的面积垂直切分成无穷多个小长方形，每一个长方形长为 (y_2-y_1)，高为 $\mathrm{d}x$，面积为 $(y_2-y_1)\mathrm{d}x$，则两条曲线之间的面积为：

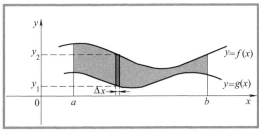

Figure 9.1.2(a)

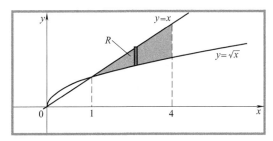

Figure 9.1.2(b)

$$\int_a^b (y_2-y_1)\mathrm{d}x = \int_a^b [f(x)-g(x)]\mathrm{d}x$$

Example 9.1 Find the area of the region R bounded by the curve $y=x$ and the curve $y=\sqrt{x}$ on the interval $[1,4]$.

Solution：

Make a sketch of the region, see Figure 9.1.2(b)：

The area between the curves is

$$\int_1^4 (x - \sqrt{x})\,dx$$

Then, evaluate it:

$$\int_1^4 (x - \sqrt{x})\,dx = \left(\frac{x^2}{2} - \frac{2}{3}x^{\frac{3}{2}}\right)\bigg|_1^4$$
$$= \left[\left(\frac{4^2}{2} - \frac{2}{3} \times 4^{\frac{3}{2}}\right) - \left(\frac{1^2}{2} - \frac{2}{3} \times 1^{\frac{3}{2}}\right)\right] = \frac{17}{6}$$

Example 9.2 Find the area of the region enclosed by the parabola $y = x^2$ and the line $y = 2 - x$.

Solution:

Make a sketch of the region, see Figure 9.1.2(c):

Find the points of intersection:

$$x^2 = 2 - x$$

Solve for x:

$$x = -2,\ x = 1$$

The limits of integration $a = -2$, $b = 1$. The area between the curves is

$$\int_{-2}^1 [(2-x) - x^2]\,dx = \int_{-2}^1 (-x^2 - x + 2)\,dx$$
$$= \left(-\frac{x^3}{3} - \frac{x^2}{2} + 2x\right)\bigg|_{-2}^1$$
$$= \frac{9}{2}$$

Note：本题求两条相交曲线之间的面积．

Example 9.3 Find the area of the region R bounded by the curve $y = \sin x$ and the curve $y = \cos x$ on the interval $[0, \pi]$.

Solution:

Make a sketch of the region, see Figure 9.1.2(d):

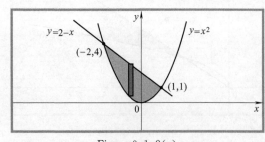

Figure 9.1.2(c)

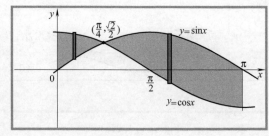

Figure 9.1.2(d)

Find the points of intersection:

$$\sin x = \cos x,\ x \in [0, \pi]$$

Solve for x:

$$x = \frac{\pi}{4}$$

Notice that $\cos x$ is on top between 0 and $\frac{\pi}{4}$, then $\sin x$ is on top between $\frac{\pi}{4}$ and π.

The area between the curves is

$$\int_0^{\pi/4} (\cos x - \sin x) dx + \int_{\pi/4}^{\pi} (\sin x - \cos x) dx$$
$$= (\sin x + \cos x)\Big|_0^{\pi/4} + (-\cos x - \sin x)\Big|_{\pi/4}^{\pi}$$
$$= \left(\sin\frac{\pi}{4} + \cos\frac{\pi}{4}\right) - (\sin 0 + \cos 0) + (-\cos\pi - \sin\pi) - \left(-\cos\frac{\pi}{4} - \sin\frac{\pi}{4}\right)$$
$$= \left(\frac{\sqrt{2}}{2} + \frac{\sqrt{2}}{2}\right) - (0+1) + (1-0) - \left(-\frac{\sqrt{2}}{2} - \frac{\sqrt{2}}{2}\right)$$
$$= 2\sqrt{2}$$

Note：如果所求面积边界曲线的表达式在一个或多个点改变，应分割区域为对应不同表达式的子区域，并且对每个子区域应用定积分．

9.1.3 水平切分 Horizontal Slices

类似于 Figure 9.1.3(a)，在区间 $[c,d]$ 上，一条函数曲线总是在另外一条曲线的左方，一般采用水平切分求两条曲线之间的面积．把两条曲线之间的面积水平切分成无穷多个小长方形，每一个长方形长为 $(x_2 - x_1)$，高为 dy，面积为 $(x_2 - x_1)dy$，则两条曲线之间的面积为：

$$\int_c^d (x_2 - x_1) dy = \int_c^d [f(y) - g(y)] dy$$

Example 9.4 Find the area of the region R between the curve $x = \sqrt{2y}$ and the line $x = y$ from $y = 0$ to $y = 2$.

Solution：

Make a sketch of the region, see Figure 9.1.3(b)：

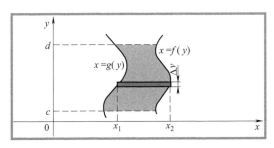
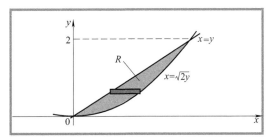

Figure 9.1.3(a)　　　　　　　　　Figure 9.1.3(b)

The area between the curves is：

$$\int_0^2 (\sqrt{2y} - y) dy = \left(\sqrt{2}\frac{y^{\frac{3}{2}}}{\frac{3}{2}} - \frac{y^2}{2}\right)\Big|_0^2$$
$$= \left(\sqrt{2}\frac{2^{\frac{3}{2}}}{\frac{3}{2}} - \frac{2^2}{2}\right) - 0 = \frac{2}{3}$$

Example 9.5 Find the area of the region between the curve $x = y^2 - 5y$ and the line $x = y$.

第 9 章　积分的应用 Applications of Integral

Solution:

Make a sketch of the region, see Figure 9.1.3(c):

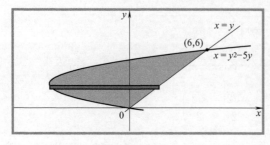

Figure 9.1.3(c)　　　　　　　　Figure 9.1.3(d)

Find the points of intersection:
$$y^2 - 5y = y$$

Solve for y:
$$y = 0, \quad y = 6$$

The limits of integration $c = 0$, $d = 6$. The area between the curves is
$$\int_0^6 [y - (y^2 - 5y)]dy = \int_0^6 (-y^2 + 6y)dy$$
$$= \left(-\frac{y^3}{3} + 6 \times \frac{y^2}{2}\right)\Big|_0^6 = 36$$

Example 9.6 Find the area of the region between the curve $x = y^3 - 4y$ and the line $x = 0$.

Solution:

Make a sketch of the region, see Figure 9.1.3(d):

Find the points of intersection:
$$y^3 - 4y = 0$$

Solve for y:
$$y = 0, \quad y = -2, \quad y = 2$$

Notice that $x = y^3 - 4y$ is to the right of y-axis from $y = -2$ and $y = 0$, then $x = y^3 - 4y$ is to the left of y-axis from $y = 0$ and $y = 2$. The area between the curves is
$$\int_{-2}^0 (y^3 - 4y)dy + \int_0^2 (4y - y^3)dy$$
$$= \left(\frac{y^4}{4} - 4 \times \frac{y^2}{2}\right)\Big|_{-2}^0 + \left(4 \times \frac{y^2}{2} - \frac{y^4}{4}\right)\Big|_0^2$$
$$= 4 + 4 = 8$$

9.1.4 利用对称性 Using Symmetry

我们发现：部分曲线围成的面积或关于 x 轴、y 轴对称，或关于原点（Origin）对称，可以利用面积的对称性简化求积分．前面我们讲过对称函数的定积分，见本书 8.4.4 节，它们之间既有联系，又有区别．

Example 9.7 Find the area of the region between the curve $y = \cos 2x$ and the x-axis on the interval $\left[-\dfrac{\pi}{4},\dfrac{\pi}{4}\right]$.

Solution:

Make a sketch of the region, see Figure 9.1.4(a):

The area bounded by the x-axis and this arch of cosine curve is symmetric to the y-axis:

$$\int_{-\pi/4}^{\pi/4}\cos2x\,\mathrm{d}x = 2\int_{0}^{\pi/4}\cos2x\,\mathrm{d}x$$
$$= 2\times\dfrac{1}{2}\sin2x\Big|_{0}^{\pi/4} = \sin\left(2\times\dfrac{\pi}{4}\right)-\sin0 = 1$$

Note: 曲线 $y=\cos2x$ 和 x 轴围成的面积关于 y 轴对称，因此，其面积等于两倍 y 轴右半部分的面积.

Example 9.8 Find the area of the region enclosed by the parabola $x=y^2-2$ and the y-axis.

Solution:

Make a sketch of the region, see Figure 9.1.4(b):

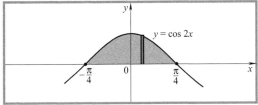

Figure 9.1.4(a)

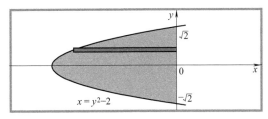

Figure 9.1.4(b)

Find the points of intersection:
$$y^2-2=0$$

Solve for y:
$$y=-\sqrt{2},\ y=\sqrt{2}$$

The area bounded by the parabola and y-axis is symmetric to the x-axis:

$$2\int_{0}^{\sqrt{2}}(2-y^2)\mathrm{d}y = 2\left(2y-\dfrac{y^3}{3}\right)\Big|_{0}^{\sqrt{2}}$$
$$= 2\left(2\sqrt{2}-\dfrac{(\sqrt{2})^3}{3}\right)-0 = \dfrac{8\sqrt{2}}{3}$$

Note: 曲线 $x=y^2-2$ 和 y 轴围成的面积关于 x 轴对称，因此，其面积等于两倍 x 轴右上半部分的面积. 再如椭圆，既关于 x 轴对称，又关于 y 轴对称，其面积等于 4 倍第一象限上的面积，见 Example 9.9.

9.1.5# 参数曲线有界的区域 Region Bounded by a Parametric Curve

The equations $\begin{cases} x=f(t) \\ y=g(t) \end{cases}$, for t in I, are parametric equations for C with parameter t, then to evaluate $\int_{a}^{b}y\,\mathrm{d}x$, we express y, $\mathrm{d}x$, and the limits a and b in terms of t and $\mathrm{d}t$, then integrate. Remember that we define dx to $x'(t)dt$.

根据定积分的换元法，见本书 8.6.1 节，求参数曲线的积分 $\int_a^b y\,dx$，可以把被积函数 y、微分因子 dx 以及积分限 a 和 b 都化为变量 t 和 dt 的表达式，然后再积分．

Example 9.9 Find the area enclosed by the ellipse with parametric equations
$\begin{cases} x = 3\cos t \\ y = 2\sin t \end{cases}$.

Solution：

Make a sketch of the region, see Figure 9.1.5(a)：

The ellipse is symmetric to both axes, the area inside the ellipse is four times the area in the first quadrant：

$$4\int_0^3 y\,dx$$

Since $x = 3\cos t$, then $dx = -3\sin t\,dt$. When $x = 3$, $3\cos t = 3$, $t = 0$; $x = 0$, $3\cos t = 0$, $t = \dfrac{\pi}{2}$, therefore,

$$4\int_0^3 y\,dx = 4\int_{\pi/2}^0 2\sin t(-3\sin t)\,dt$$
$$= -24\int_{\pi/2}^0 \sin^2 t\,dt = -24\int_{\pi/2}^0 \frac{1-\cos 2t}{2}\,dt$$
$$= -24\int_{\pi/2}^0 \left(\frac{1}{2} - \frac{1}{2}\cos 2t\right)dt$$
$$= -24\left(\frac{t}{2} - \frac{1}{2}\times\frac{1}{2}\sin 2t\right)\Big|_{\pi/2}^0 = 0 - (-6\pi) = 6\pi$$

Example 9.10 Find the area enclosed by one arch of the cycloid with parametric equations $\begin{cases} x = t - \sin t \\ y = 1 - \cos t \end{cases}$.

Solution：

Sketch the graph using graphing calculator, see Figure 9.1.5(b)：

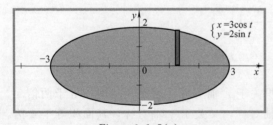

Figure 9.1.5(a)

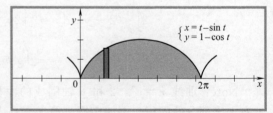

Figure 9.1.5(b)

The area enclosed by one arch of the cycloid is

$$\int_0^{2\pi} y\,dx$$

Since $x = t - \sin t$, then $dx = (1 - \cos t)dt$. When $x = 0$, $t - \sin t = 0$,
$$t = 0 \text{ and } t = 2\pi$$

Therefore,

$$\int_0^{2\pi} y\,dx = \int_0^{2\pi}(1-\cos t)(1-\cos t)\,dt = \int_0^{2\pi}(1-2\cos t+\cos^2 t)\,dt$$
$$= \int_0^{2\pi}\left(1-2\cos t+\frac{\cos 2t+1}{2}\right)$$
$$= \int_0^{2\pi}\left(\frac{3}{2}-2\cos t+\frac{1}{2}\cos 2t\right)dt$$
$$= \left(\frac{3t}{2}-2\sin t+\frac{1}{2}\times\frac{1}{2}\sin 2t\right)\Big|_0^{2\pi} = 3\pi$$

Note：本题需要结合图形计算器，能使用图形计算器绘制函数图像是 AP 微积分考试要求.

9.1.6# 极坐标求面积 Find Area in Polar Coordinates

If the area A bounded by the polar curve $r=f(\theta)$ and the rays $\theta=\alpha$ and $\theta=\beta$ is
$$A = \int_\alpha^\beta \frac{1}{2} r^2\,d\theta$$

假设曲线 $r=f(\theta)$ 在区间 $[\alpha,\beta]$ 连续，见 Figure 9.1.6(a)，求 $r=f(\theta)$ 和 $\theta=\alpha$、$\theta=\beta$ 围成的面积. 当 $\Delta\theta$ 足够小时，曲线面积 ΔA 近似为小扇形的面积，即：

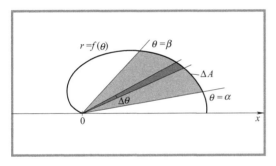

Figure 9.1.6(a)   Figure 9.1.6(b)

$$\Delta A = \frac{1}{2}rl = \frac{1}{2}r(r\Delta\theta) = \frac{1}{2}r^2\Delta\theta$$

根据我们讲过的黎曼和（Riemann Sum）可以得到：
$$A = \lim_{n\to\infty}\sum_{i=1}^n \frac{1}{2} r_i^2 \Delta\theta$$

写成定积分：$A = \int_\alpha^\beta \frac{1}{2} r^2\,d\theta$.

Example 9.11   Find the area inside both the circle $r=3\cos\theta$ and the cardioid $r=1+\cos\theta$.

Solution：

Sketch the graph using graphing calculator, see Figure 9.1.6(b), where one half of the required area is shaded.

Since $3\cos\theta=1+\cos\theta$ when $\theta=\frac{\pi}{3}$ or $\theta=-\frac{\pi}{3}$.

The area of the cardioid swept out by a radius as θ varies from 0 to $\frac{\pi}{3}$：

$$\int_\alpha^\beta \frac{1}{2} r^2 d\theta = \int_0^{\pi/3} \frac{1}{2}(1+\cos\theta)^2 d\theta$$

$$= \int_0^{\pi/3} \frac{1}{2}\left(1 + 2\cos\theta + \frac{\cos 2\theta + 1}{2}\right) d\theta$$

$$= \frac{1}{2}\left(\frac{3}{2}\theta + 2\sin\theta + \frac{1}{2} \times \frac{1}{2}\sin 2\theta\right)\bigg|_0^{\pi/3}$$

$$= \frac{\pi}{4} + \frac{9\sqrt{3}}{16}$$

The area of the circle swept out by θ as it varies from $\frac{\pi}{3}$ to $\frac{\pi}{2}$:

$$\int_\alpha^\beta \frac{1}{2} r^2 d\theta = \int_{\pi/3}^{\pi/2} \frac{1}{2}(3\cos\theta)^2 d\theta = \int_{\pi/3}^{\pi/2} \frac{9}{4}(1+\cos 2\theta) d\theta$$

$$= \frac{9}{4}\left(\theta + \frac{1}{2}\sin 2\theta\right)\bigg|_{\pi/3}^{\pi/2} = \frac{3\pi}{8} - \frac{9\sqrt{3}}{16}$$

The desired area is twice the sum of two parts:

$$A = 2\left[\int_0^{\pi/3} \frac{1}{2}(1+\cos\theta)^2 d\theta + \int_{\pi/3}^{\pi/2} \frac{1}{2}(3\cos\theta)^2 d\theta\right] = \frac{5\pi}{4}$$

Example 9.12 Find the shaded area A of the polar curve $r = 4 - 2\cos\theta$ for $0 \leqslant \theta \leqslant \pi$, see Figure 9.1.6(c).

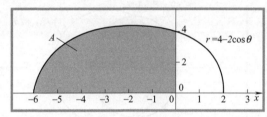

Figure 9.1.6(c)

Solution:

Let $r = 4$ and $r = 6$, since $r = 4 - 2\cos\theta$, then $\theta = \frac{\pi}{2}$ and $\theta = \pi$.

The shaded area A of the polar curve $r = 4 - 2\cos\theta$ swept out by θ as it varies from $\frac{\pi}{2}$ to π.

Consequently,

$$A = \int_{\pi/2}^{\pi} \frac{1}{2}(4-2\cos\theta)^2 d\theta = \int_{\pi/2}^{\pi} \frac{1}{2}(16 - 16\cos\theta + 4\cos^2\theta) d\theta$$

$$= \int_{\pi/2}^{\pi} \frac{1}{2}[16 - 16\cos\theta + 2(\cos 2\theta + 1)] d\theta$$

$$= \int_{\pi/2}^{\pi} \frac{1}{2}(18 - 16\cos\theta + 2\cos 2\theta) d\theta$$

$$= \frac{1}{2}(18\theta - 16\sin\theta + \sin 2\theta)\bigg|_{\pi/2}^{\pi}$$

$$= \frac{1}{2}(18\pi - 16\sin\pi + \sin 2\pi) - \frac{1}{2}\left(18 \times \frac{\pi}{2} - 16\sin\frac{\pi}{2} + \sin\pi\right)$$

$$= \frac{9}{2}\pi + 8$$

Note：本题注意到两点．①极坐标中角的规定同三角函数中角的规定一致，故 $r=6$ 而非 $r=-6$；②极坐标下的面积由两条直线和极坐标曲线围成，两条直线分别由极坐标曲线的起点和终点与极点连线得到，并以此确定积分限．

9.2 体积 Volume

应用定积分求体积有两大类型：①求旋转体（Solids of Revolution）的体积，其解法有圆盘法（Disks）、垫圈法（Washers）和圆柱壳法（Shells）；②已知横截面（Cross-Sections）的体积．

旋转体是由一个平面图形绕着平面内一条直线旋转一周而得到的立体，这条直线称为旋转的轴（Axis）．比如说：由矩形（Rectangle）绕它的一条边旋转一周可以得到圆柱（Cylinder）；直角三角形（Right-angled Triangle）绕它的直角边、直角梯形（Right-angled Trapezoid）绕它的直角腰、半圆（Semicircle）绕它的直径旋转一周，分别可以得到圆锥（Cone）、圆台（Circular Truncated Cone）和球体（Sphere）．

如果一个立体不是旋转体，但已知其垂直于某一定轴的各个横截面，那么这个立体的体积也可以运用定积分计算．

9.2.1 旋转体：圆盘法 Solids of Revolution: Disks

1) If the solid of revolution V is obtained by revolving the plane region R bounded by $y=f(x)$ on the interval $[a,b]$ like one shown in Figure 9.2.1(a), is revolved about the x-axis. We need only observe that the cross-sectional area $A(x)$ is the area of a disk of radius $f(x)$, the distance of the planar region's boundary from the axis of revolution. The area is then

$$A(x)=\pi\,(radius)^2=\pi\,[f(x)]^2$$

The volume of the solid is：

$$V=\int_a^b \pi\,[f(x)]^2\,\mathrm{d}x$$

如上述过程，见 Figure 9.2.1(b)，曲线 $y=f(x)$ 与 $x=a$，$x=b$ 和 x 轴围成的区域绕 x 轴旋转一周得到旋转体（Solids of Revolution），其垂直于 x 轴的横截面是圆盘（Disk），圆盘的面积 $A(x)$ 是变化的，鉴于此，这个方法称为圆盘法．

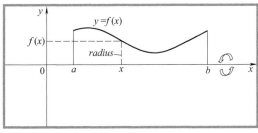

Figure 9.2.1(a)

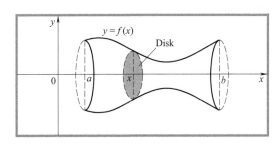

Figure 9.2.1(b)

2) If the solid of revolution V is obtained by revolving the plane region R bounded by $x=g(y)$, $y=c$, $y=d$ and y-axis like one shown in Figure 9.2.1(c), is revolved about the y-axis. The area is then

$$A(y)=\pi\,(radius)^2=\pi\,[g(y)]^2$$

The volume of the solid is:

$$V = \int_c^d \pi [g(y)]^2 \, dy$$

如上述过程，见 Figure 9.2.1(d)，曲线 $x=g(x)$ 与 $y=c$，$y=d$ 和 y 轴围成的区域绕 y 轴旋转一周得到旋转体（Solids of Revolution），其垂直于 y 轴的横截面是圆盘，圆盘的面积 $A(y)$ 是变化的，这是绕 y 轴旋转的圆盘法.

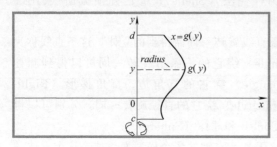

Figure 9.2.1(c)

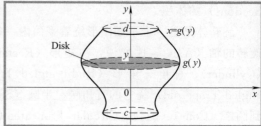

Figure 9.2.1(d)

Example 9.13 The region between the curve $y=x$, $0 \leqslant x \leqslant 4$, and the x-axis is revolved about the x-axis to generate a solid. Find its volume.

Solution:

Make a sketch of the region, a typical radius, and the generated solid, see Figure 9.2.1(e):

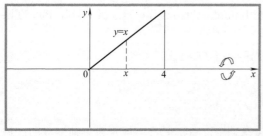

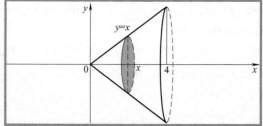

Figure 9.2.1(e)

The volume is

$$\int_0^4 \pi x^2 \, dx = \pi \left. \frac{x^3}{3} \right|_0^4 = \frac{64\pi}{3}$$

Example 9.14 Find the volume of the solid that results when the region between the curve $y=2\sqrt{x}$ and the x-axis, from $x=0$ to $x=3$, is revolved about the x-axis.

Solution:

Make a sketch of the region, a typical radius, and the generated solid, see Figure 9.2.1(f):

The volume is

$$\int_0^3 \pi (2\sqrt{x})^2 \, dx = \int_0^3 4\pi x \, dx = 4\pi \left. \frac{x^2}{2} \right|_0^3 = 18\pi$$

Example 9.15 Find the volume of the solid that results when the region between the curve $x=\dfrac{y^2}{4}$ and the y-axis, from $y=0$ to $y=3$, is revolved about the y-axis.

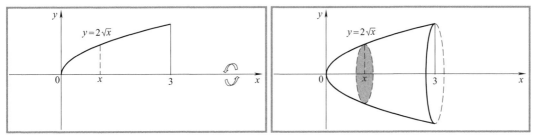

Figure 9.2.1(f)

Solution:

Make a sketch of the region, a typical radius, and the generated solid, see Figure 9.2.1(g):

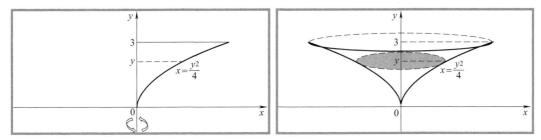

Figure 9.2.1(g)

The volume is

$$\int_0^3 \pi \left(\frac{y^2}{4}\right)^2 dy = \frac{\pi}{16}\int_0^3 y^4 dy = \frac{\pi}{16} \times \frac{y^5}{5}\bigg|_0^3 = \frac{243\pi}{80}$$

Example 9.16 Find the volume of a sphere of radius r.

Solution:

Let the region bounded by a semicircle with center O and radius r.

The sphere of radius r that results when the semicircle's diameter is revolved about the x-axis, see Figure 9.2.1(h).

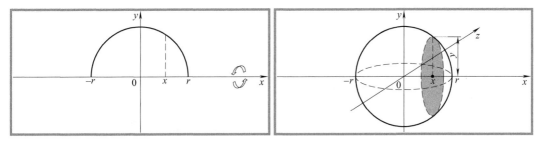

Figure 9.2.1(h)

The cross-sectional area $A(x)$ is the area of a disk of radius y:

$$A(x) = \pi y^2$$

The volume of the sphere is:

$$V = \int_{-r}^{r} \pi y^2 dx$$

第9章 积分的应用 Applications of Integral

Since the equation of the circle is:
$$x^2 + y^2 = r^2$$
Hence,
$$V = \int_{-r}^{r} \pi y^2 \, dx = \pi \int_{-r}^{r} (r^2 - x^2) \, dx$$
$$= \pi \left(r^2 x - \frac{x^3}{3} \right) \bigg|_{-r}^{r} = \frac{4}{3} \pi r^3$$

【方法总结】运用圆盘法（Disks）一般有 3 个步骤：
1) 画出区域草图并确定半径函数 $f(x)$；
2) 平方 $f(x)$ 并乘以 π；
3) 定积分（Definite Integral）求体积．

9.2.2 旋转体：垫圈法 Solids of Revolution：Washers

If the solid of revolution V is obtained by revolving the plane region R bounded above by $y = f(x)$ and below by $y = g(x)$ on the interval $[a, b]$ like one shown in Figure 9.2.2 (a), is revolved about the x-axis. Then each washer has an area:
$$A(x) = \pi [f(x)^2 - g(x)^2]$$
The volume of the solid is:
$$V = \int_a^b \pi [f(x)^2 - g(x)^2] \, dx$$

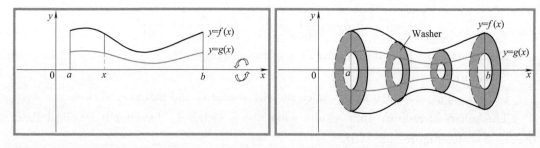

Figure 9.2.2(a)

Similarly, shown in Figure 9.2.2(b), the volume of the solid is:
$$V = \int_c^d \pi [f(y)^2 - g(y)^2] \, dy$$

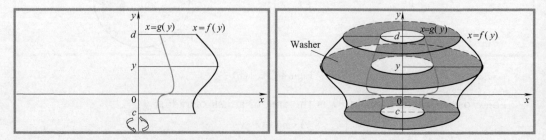

Figure 9.2.2(b)

如果产生旋转体（Solids of Revolution）的区域不通过旋转轴，那么该旋转体就含有洞，见 Figure 9.2.2(a)、Figure 9.2.2(b)，其垂直于旋转轴（x 轴或 y 轴）的横截面不是圆盘而是垫圈（Washer），这个方法称为垫圈法．

Example 9.17 Find the volume of the solid that results when the region bounded by the curve $y=x$ and $y=x^3$, from $x=0$ to $x=1$, is revolved about the x-axis.

Solution：

Make a sketch of the region and the generated solid，see Figure 9.2.2(c)：

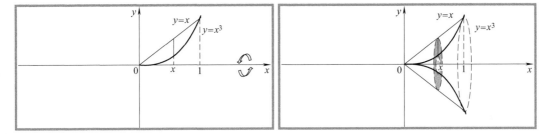

Figure 9.2.2(c)

The top curve is $y=x$ and the bottom curve is $y=x^3$，then the volume is：

$$V=\int_0^1 \pi[x^2-(x^3)^2]\mathrm{d}x = \pi\int_0^1(x^2-x^6)\mathrm{d}x$$
$$=\pi\left(\frac{x^3}{3}-\frac{x^7}{7}\right)\bigg|_0^1=\frac{4\pi}{21}$$

Example 9.18 Find the volume of the solid that results when the region bounded by the curve $y=x$ and $y=x^3$, from $y=0$ to $y=1$, is revolved about the y-axis.

Solution：

Make a sketch of the region and the generated solid，see Figure 9.2.2(d)：

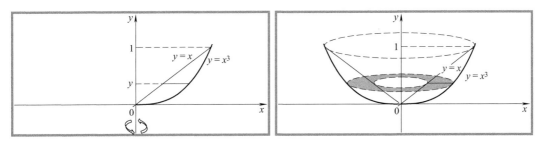

Figure 9.2.2(d)

The curve $y=x^3$ on the outside and the curve $y=x$ on the inside，then the volume is：

$$V=\int_0^1 \pi[(\sqrt[3]{y})^2-y^2]\mathrm{d}y = \pi\int_0^1(y^{\frac{2}{3}}-y^2)\mathrm{d}y$$
$$=\pi\left(\frac{y^{\frac{5}{3}}}{\frac{5}{3}}-\frac{y^3}{3}\right)\bigg|_0^1=\frac{4}{15}\pi$$

Example 9.19 Find the volume of the solid that results when the region bounded the curve $y=x$ and $y=x^3$, from $x=0$ to $x=1$, is revolved about the line $y=-1$.

Solution：

Make a sketch of the region and the generated solid, see Figure 9.2.2(e)：

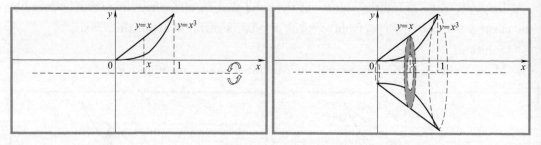

Figure 9.2.2(e)

The top curve is $y=x$ and the bottom curve is $y=x^3$, then the volume is：

$$V=\int_0^1 \pi[(x+1)^2-(x^3+1)^2]dx = \pi\int_0^1(x^2+2x-x^6-2x^3)dx$$
$$=\pi\left(\frac{x^3}{3}+2\times\frac{x^2}{2}-\frac{x^7}{7}-2\times\frac{x^4}{4}\right)\Bigg|_0^1=\frac{29\pi}{42}$$

Example 9.20 Find the volume of the solid that results when the region bounded by the curve $y=x$ and $y=x^3$, from $y=0$ to $y=1$, is revolved about the line $x=-1$.

Solution：

Make a sketch of the region and the generated solid, see Figure 9.2.2(f)：

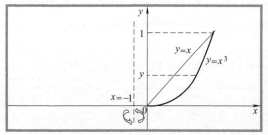 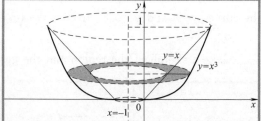

Figure 9.2.2(f)

The curve $y=x^3$ on the outside and the curve $y=x$ on the inside, then the volume is：

$$V=\int_0^1\pi[(\sqrt[3]{y}+1)^2-(y+1)^2]dy=\pi\int_0^1(y^{\frac{2}{3}}+2y^{\frac{1}{3}}-y^2-2y)dy$$
$$=\pi\left(\frac{y^{\frac{5}{3}}}{\frac{5}{3}}+2\times\frac{y^{\frac{4}{3}}}{\frac{4}{3}}-\frac{y^3}{3}-2\times\frac{y^2}{2}\right)\Bigg|_0^1=\frac{23\pi}{30}$$

【方法总结】运用垫圈法（Washers）一般有3个步骤：

1）描绘要旋转区域的草图并画一条垂直于旋转轴的线段横穿它，当这个区域绕旋转轴旋转一周时，这条线段产生一个垫圈；

2）通过曲线和交点确定积分限；

3）确定垫圈的内外半径，积分求体积．

9.2.3 旋转体：圆柱壳 Solids of Revolution: Cylindrical Shells

The volume of the solid generated by revolving the region between the x-axis and the graph of a continuous function $y=f(x)\geqslant 0$, $0\leqslant a\leqslant x\leqslant b$, about the y-axis is

$$V=\int_a^b 2\pi \begin{pmatrix} shell \\ radius \end{pmatrix}\begin{pmatrix} shell \\ height \end{pmatrix} \mathrm{d}x$$

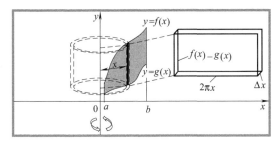 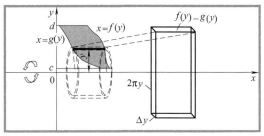

Figure 9.2.3(a)　　　　　　　　　Figure 9.2.3(b)

圆柱壳法是求旋转体（Solids of Revolution）体积的另一种方法，类似于树的年轮一样从旋转轴向外增长．

求曲线 $y=f(x)$ 和 $y=g(x)$ 在区间 $[a,b]$ 上围成的区域绕 y 轴旋转得到旋转体体积．把被旋转区域垂直 x 轴切分成无穷多个小长方形，见 Figure 9.2.3(a)，每个小长方形绕 y 轴旋转一圈形成高为 $[f(x)-g(x)]$，底面半径为 x，壁厚为 Δx 的空心圆柱，沿圆柱的母线方向切开，空心圆柱为一个长方体，体积为 $2\pi[f(x)-g(x)]\Delta x$，由黎曼和（Riemann Sums）得到：

$$V=\int_a^b 2\pi x[f(x)-g(x)]\mathrm{d}x$$

同理，见 Figure 9.2.3(b) 曲线 $x=f(y)$ 和 $x=g(y)$ 在区间 $[c,d]$ 上围成的区域绕 y 轴旋转得到旋转体体积：

$$V=\int_c^d 2\pi y[f(y)-g(y)]\mathrm{d}y$$

Example 9.21 Find the volume of the region that when bounded by the curve $y=x^2$, the x-axis, and the line $x=3$ is revolved about y-axis.

Solution：

Make a sketch of the region and the generated solid, see Figure 9.2.3(c)：

The volume is：

$$V=\int_0^3 2\pi x(x^2-0)\mathrm{d}x=2\pi\int_0^3 x^3\mathrm{d}x=2\pi\left.\frac{x^4}{4}\right|_0^3=\frac{81\pi}{2}$$

Note：本题也可用垫圈法（Washers）解答，答案是一样的，请同学们自行验证．

Example 9.22 Find the volume of the region that when bounded by the curve $y=x^2$, the x-axis, and the line $x=3$ is revolved about the line $x=-1$.

Solution：

Make a sketch of the region and the generated solid, see Figure 9.2.3(d)：

The volume is：

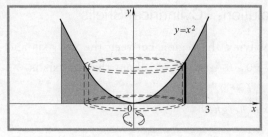
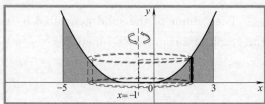

Figure 9.2.3(c)   Figure 9.2.3(d)

$$V = \int_0^3 2\pi(x+1)(x^2-0)dx = 2\pi\int_0^3 (x^3+x^2)dx$$
$$= 2\pi\left(\frac{x^4}{4}+\frac{x^3}{3}\right)\Big|_0^3 = \frac{117\pi}{2}$$

Example 9.23 Find the volume of the region that results when the region bounded by the curve $x=y^2$ and the line $x=y$, from $y=0$ to $y=1$, is revolved about the line x-axis.

Solution:

Make a sketch of the region and the generated solid, see Figure 9.2.3(e):

The volume is:

$$V = \int_0^1 2\pi y(y-y^2)dy = 2\pi\int_0^1 (y^2-y^3)dy$$
$$= 2\pi\left(\frac{y^3}{3}-\frac{y^4}{4}\right)\Big|_0^1 = \frac{\pi}{12}$$

Example 9.24 Find the volume of the region that results when the region bounded by the curve $x=y^2$ and the line $x=y$, from $y=0$ to $y=1$, is revolved about the line $y=-1$.

Solution:

Make a sketch of the region and the generated solid, see Figure 9.2.3(f):

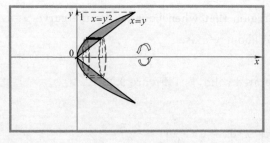
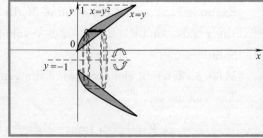

Figure 9.2.3(e)   Figure 9.2.3(f)

The volume is:

$$V = \int_0^1 2\pi(y+1)(y-y^2)dy = 2\pi\int_0^1 (y-y^3)dy$$
$$= 2\pi\left(\frac{y^2}{2}-\frac{y^4}{4}\right)\Big|_0^1 = \frac{\pi}{2}$$

【方法总结】运用圆柱壳法（Cylindrical Shells）一般有 3 个步骤：

1) 描绘要旋转区域的草图并画一条平行于旋转轴的线段横穿它，线段的高度为圆柱壳高度，线段到旋转轴的距离为圆柱壳半径，线段的宽度为圆柱壳厚度；

2) 针对圆柱壳厚度变量确定积分限；

3) 写出体积的积分，积分求体积．

9.2.4　已知横截面的体积 Volumes of solids with Known Cross-Sections

The volume of a solid of known integrable cross-sectional area $A(x)$ from $x=a$ to $x=b$ is the integral of A from a to b：

$$V = \int_a^b A(x) \mathrm{d}x$$

假设 $A(x)$ 是连续（Continuous）函数，表示过点 x 且垂直 x 轴的立体横截面 (Cross-Sections) 面积．此时，取 x 为积分变量 (Variable of Integration)，它的变化区间为 $[a, b]$．见 Figure 9.2.4(a)，在任一小区间 $[x, x+\Delta x]$ 上，薄片的体积近似于底面积为 $A(x)$、高为 Δx 的扁柱体，即积分元素：

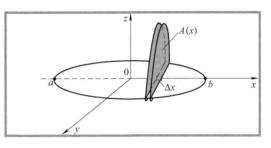

Figure 9.2.4(a)

$$\mathrm{d}V = A(x)\Delta x$$

所以，已知从 $x=a$ 到 $x=b$ 横截面积 $A(x)$ 的立体，如果 $A(x)$ 可积，那么它的体积是：

$$V = \int_a^b A(x) \mathrm{d}x$$

Example 9.25　Find the volume of the solid whose base is the region between the semicircle $y = \sqrt{9-x^2}$ and x-axis, and whose cross-sections perpendicular to the x-axis are equilateral triangle with a side on the base.

Solution：

Make a sketch of the region, see Figure 9.2.4(b)：

The side of the equilateral triangle is $\sqrt{9-x^2}$. The area of a cross-section is

$$A(x) = \frac{\sqrt{3}}{4}(\sqrt{9-x^2})^2 = \frac{\sqrt{3}}{4}(9-x^2)$$

The volume is：

$$V = \int_{-3}^{3} \frac{\sqrt{3}}{4}(9-x^2)\mathrm{d}x = \frac{\sqrt{3}}{4}\left(9x - \frac{x^3}{3}\right)\Big|_{-3}^{3} = 9\sqrt{3}$$

Example 9.26　Find the volume of the solid whose base is the region between $y = 4 - x^2$ and x-axis, and whose cross-sections perpendicular to the x-axis are squares with a side on the base.

Solution：

Make a sketch of the region, see Figure 9.2.4(c)：

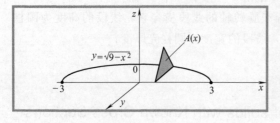
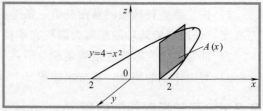

Figure 9.2.4(b)　　　　　　　　Figure 9.2.4(c)

The side of the square is $4-x^2$. The area of a cross-section is
$$A(x)=(4-x^2)^2$$

The volume is：
$$V=\int_{-2}^{2}(4-x^2)^2\,dx=\int_{-2}^{2}(16-8x^2+x^4)\,dx$$
$$=\left(16x-8\times\frac{x^3}{3}+\frac{x^5}{5}\right)\bigg|_{-2}^{2}=\frac{512}{15}$$

【方法总结】运用横截面（Cross-Sections）求体积一般有 3 个步骤：
1) 画一个该立体及其典型横截面的草图；
2) 确定 $A(x)$ 的公式和积分限；
3) 积分求体积．

9.3[#]　弧长 Arc Length

定积分求曲线的弧长分两种情况：光滑曲线下求弧长和参数曲线下求弧长．

9.3.1[#]　光滑曲线的弧长公式 Formula for the Length of $y=f(x)$

1) If $f(x)$ is continuously differentiable on the closed interval $[a,b]$, the length of the curve $y=f(x)$ from $x=a$ to $x=b$ is

$$L=\int_{a}^{b}\sqrt{1+\left(\frac{dy}{dx}\right)^2}\,dx$$

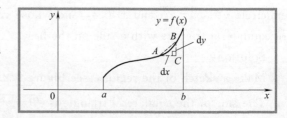

Figure 9.3.1

曲线 $y=f(x)$ 是一条连续光滑的曲线，见 Figure 9.3.1，我们分割区间 $a\leqslant x\leqslant b$ 成很短的子区间，使得其上的曲线弧近似是直的．即：曲线弧长 $\widehat{AB}\approx AB$，$AC=dx$，$BC=dy$，那么 $\widehat{AB}\approx\sqrt{(AC)^2+(BC)^2}$

$$\widehat{AB}\approx\sqrt{(dx)^2+(dy)^2}=\sqrt{(dx)^2\left[1+\left(\frac{dy}{dx}\right)^2\right]}=\sqrt{1+\left(\frac{dy}{dx}\right)^2}\,dx$$

根据黎曼和（Riemann Sums），从而得到光滑曲线的弧长公式．

Example 9.27　Find the length of the curve $y=2x\sqrt{x}$ from $x=0$ to $x=1$.

Solution:

Find the derivative of the function:

$$\frac{dy}{dx} = (2x^{\frac{3}{2}})' = 2 \times \frac{3}{2} x^{\frac{1}{2}} = 3\sqrt{x}$$

The length of the curve from $x=0$ to $x=1$ is

$$L = \int_0^1 \sqrt{1 + (3\sqrt{x})^2} \, dx = \int_0^1 \sqrt{1 + 9x} \, dx$$

$$= \frac{2}{27}(1+9x)^{\frac{3}{2}} \Big|_0^1 = \frac{20\sqrt{10} - 2}{27}$$

2) If $g(y)$ is continuously differentiable on the closed interval $[c, d]$, the length of the curve $x = g(y)$ from $y = c$ to $y = d$ is

$$L = \int_c^d \sqrt{1 + \left(\frac{dx}{dy}\right)^2} \, dy$$

以上，类似于曲线 $y = f(x)$ 的弧长公式，由黎曼和可得曲线 $x = g(y)$ 的弧长公式.

Example 9.28 Find the length of the curve $x = \frac{y^3}{3} + \frac{1}{4y}$ from $y = 1$ to $y = 2$.

Solution:

Find the derivative of the function:

$$\frac{dx}{dy} = y^2 - \frac{1}{4y^2}$$

The length of the curve from $y = 1$ to $y = 2$ is

$$L = \int_1^2 \sqrt{1 + \left(y^2 - \frac{1}{4y^2}\right)^2} \, dy = \int_1^2 \sqrt{\left(y^2 + \frac{1}{4y^2}\right)^2} \, dy$$

$$= \int_1^2 \left(y^2 + \frac{1}{4y^2}\right) dy = \left(\frac{y^3}{3} - \frac{1}{4y}\right) \Big|_1^2 = \frac{59}{24}$$

9.3.2# 参数曲线的长度 Length of a Parametric Curve

If a smooth curve C is defined parametrically by $\begin{cases} x = f(t) \\ y = g(t) \end{cases}$, $a \leqslant t \leqslant b$, the length of C is

$$L = \int_a^b \sqrt{\left(\frac{dx}{dt}\right)^2 + \left(\frac{dy}{dt}\right)^2} \, dt$$

如果曲线 C 用参数方程 $\begin{cases} x = f(t) \\ y = g(t) \end{cases}$, $a \leqslant t \leqslant b$ 表示. 结合参数方程求导 $\frac{dy}{dx} = \frac{\frac{dy}{dt}}{\frac{dx}{dt}}$，有：

$$\sqrt{1 + \left(\frac{dy}{dx}\right)^2} = \sqrt{1 + \left(\frac{\frac{dy}{dt}}{\frac{dx}{dt}}\right)^2} = \sqrt{\frac{\left(\frac{dx}{dt}\right)^2 + \left(\frac{dy}{dt}\right)^2}{\left(\frac{dx}{dt}\right)^2}} = \sqrt{\left(\frac{dx}{dt}\right)^2 + \left(\frac{dy}{dt}\right)^2} \frac{dt}{dx}$$

代入光滑曲线的弧长公式，我们可以得到：

$$L = \int_a^b \sqrt{1 + \left(\frac{dy}{dx}\right)^2} \, dx = \int_a^b \sqrt{\left(\frac{dx}{dt}\right)^2 + \left(\frac{dy}{dt}\right)^2} \, \frac{dt}{dx} \times dx = \int_a^b \sqrt{\left(\frac{dx}{dt}\right)^2 + \left(\frac{dy}{dt}\right)^2} \, dt$$

以上，我们得到参数方程形式下的弧长公式．

Example 9.29 Find the length of the curve $x = t^3$, $y = \frac{3t^2}{2}$ from $t = 0$ to $t = \sqrt{3}$.

Solution:

Find the derivative of the function:

$$\frac{dx}{dt} = 3t^2, \quad \frac{dy}{dt} = 3t$$

The length of the curve from $t = 0$ to $t = \sqrt{3}$ is

$$L = \int_0^{\sqrt{3}} \sqrt{(3t^2)^2 + (3t)^2} \, dt = \int_0^{\sqrt{3}} 3t\sqrt{1+t^2} \, dt = (1+t^2)^{\frac{3}{2}} \Big|_0^{\sqrt{3}} = 7$$

9.4 位移和距离 Displacement and Distance

微分（Differential）应用章节讲了部分运动学上的应用，涉及运动学上的位移和距离问题，应用积分来解答．

9.4.1 直线运动 Motion Along a Straight Line

The object P has velocity $v(t)$, where $v(t)$ is a continuous function, then the displacement in the object's position from $t = a$ to $t = b$ is:

$$\int_a^b v(t) \, dt$$

The distance traveled by the object during the time interval $[a, b]$ is the definite integral of its speed:

$$\int_a^b |v(t)| \, dt$$

运动问题是 AP 微积分考试的经典题型．已知位置函数求速度（Velocity）或加速度（Acceleration），应用微分求解，见本书 6.3.2 节；若已知速度函数，通过定积分可求解物体某段时间经过的位移或者运动距离．

Example 9.30 If a girl moves along a straight line with velocity $v(t) = 3t^2 - 2t + 5$, find the displacement traveled between $t = 1$ and $t = 5$.

Solution:

$$\int_1^5 v(t) \, dt = \int_1^5 (3t^2 - 2t + 5) \, dt = (t^3 - t^2 + 5t) \Big|_1^5 = 120$$

Example 9.31 A particle moves along the x-axis so that its velocity at time t is given by $v(t) = t^2 - 4t + 3$. Find the total distance covered between $t = 0$ and $t = 5$.

Solution:

$$\int_0^5 |v(t)| \, dt = \int_0^5 |t^2 - 4t + 3| \, dt$$

Since $v(t) = t^2 - 4t + 3 = (t-1)(t-3)$, we see that:

t	[0,1]	[1,3]	[3,5]
$v(t)$	$v(t)>0$	$v(t)<0$	$v(t)>0$

Then,

$$\int_0^5 |t^2 - 4t + 3| \, dt$$
$$= \int_0^1 (t^2 - 4t + 3) \, dt + \int_1^3 -(t^2 - 4t + 3) \, dt + \int_3^5 (t^2 - 4t + 3) \, dt$$
$$= \left(\frac{t^3}{3} - 2t^2 + 3t\right)\Big|_0^1 - \left(\frac{t^3}{3} - 2t^2 + 3t\right)\Big|_1^3 + \left(\frac{t^3}{3} - 2t^2 + 3t\right)\Big|_3^5 = \frac{28}{3}$$

9.4.2[#] 曲线运动 Motion Along a Curve

If the object P moves along a curve defined parametrical by velocity vector $\overrightarrow{v(t)} = \dfrac{d\vec{r}}{dt} = \left(\dfrac{dx}{dt}, \dfrac{dy}{dt}\right)$, then the distance the object travels from time from $t=a$ to $t=b$ is:

$$\int_a^b |v(t)| \, dt = \int_a^b \sqrt{\left(\frac{dx}{dt}\right)^2 + \left(\frac{dy}{dt}\right)^2} \, dt$$

曲线运动下的求解速度或加速度问题,见本书 6.3.3 节. 已知物体 P 速度向量,求解 P 在某一段时间的运动距离等同于求解参数曲线的长度.

Example 9.32 A particle moves along a curve so that its velocity at time t is given by $v(t) = \left\langle 2t + 2, \dfrac{1}{(t+1)^2} \right\rangle$. Find the total distance covered between $t=0$ and $t=1$.

Solution:

Since $v(t) = \left\langle 2t + 2, \dfrac{1}{(t+1)^2} \right\rangle$, then

$$\frac{dx}{dt} = 2t + 2 \quad \text{and} \quad \frac{dy}{dt} = \frac{1}{(t+1)^2}$$

On the interval [0, 1], the distance traveled by the particle is

$$\int_0^1 \sqrt{(2t+2)^2 + \left(\frac{1}{(t+1)^2}\right)^2} \, dt$$

Use calculator to calculate integral

$$\int_0^1 \sqrt{(2t+2)^2 + \left(\frac{1}{(t+1)^2}\right)^2} \, dt \approx 3.057$$

9.5 习题 Practice Exercises

(1) Find the area of the region R bounded by the curve $y = x^2 + 1$ and the line $y = 0$ on the interval $[-2, 1]$.

(2) Find the area of the region enclosed by the parabola $y = x^2 + 1$ and the line $y = 1 - 2x$.

(3) Find the area of the region between the curve $x = y^3 - 3y$ and the line $x = y$.

(4) In the first quadrant, find the area of the region bounded above by the x-axis and below by the

curves of $y=\sin x$ and $y=\cos x$.

(5) The area A enclosed by the four-leaved rose $r=\cos 2\theta$ equals, to three decimal places.

(6) The region between the curve $y=x^2$, $0 \leqslant x \leqslant 4$, and the x-axis is revolved about the x-axis to generate a solid. Find its volume.

(7) Find the volume of the solid that results when the region bounded by the curve $y=x$ and $y=x^2$, from $y=0$ to $y=1$, is revolved about the line $x=-1$.

(8) Use the method of cylindrical shells to find the volume of the solid that results when the region bounded by $y=\sqrt{x}$, $y=2x-1$, and $x=0$ is revolved around the y-axis.

(9) Find the volume of the solid whose base is the region between the semicircle $y=\sqrt{9-x^2}$ and x-axis, and whose cross-sections perpendicular to the x-axis are squares with a side on the base.

(10) Find the length of the curve $y=\dfrac{x^3}{12}+\dfrac{1}{x}$ from $x=1$ to $x=3$.

第 10 章

微分方程
Differential Equations

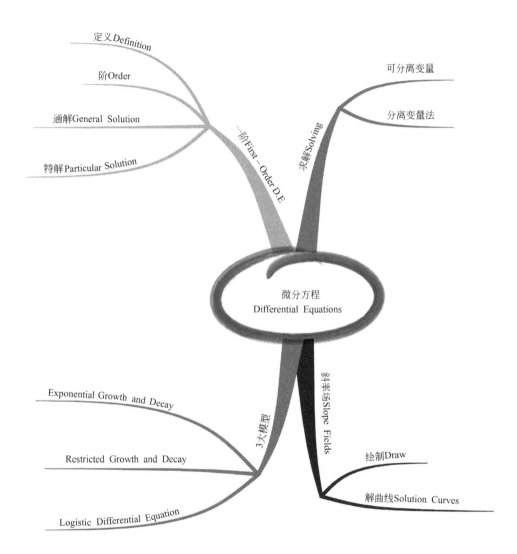

10.1 一阶微分方程 First-Order Differential Equations

微分方程（Differential Equation）是伴随着微积分学一起发展起来的数学分支，应用十分广泛，可以解决许多与导数有关的问题．本节我们认识下微分方程的基本概念．

10.1.1 基本定义 Basic Definitions

A first-order differential equation is an equation

$$\frac{dy}{dx} = f(x, y)$$

in which $f(x, y)$ is a function of two variables defined on a region in the xy-plane.

一阶微分方程表示未知函数（Function）、未知函数的导数（Derivative）和自变量（Independent Variable）之间的关系，简记为"D. E."．

10.1.2 微分方程的阶 D. E. Order

The order of a differential equation is the order of the highest derivative that occurs in the equation.

微分方程中所出现的未知函数的最高阶导数（Highest Derivative）的阶数，称为微分方程的阶（The Order of a Differential Equation）．

例如：$\frac{dy}{dx} + x = e^x$ 及 $\left(\frac{dy}{dx}\right)^2 - y = \sin x$ 为一阶微分方程；$\frac{d^2y}{dx^2} + 2\frac{dy}{dx} - 3y = x^2$ 为二阶微分方程．

10.1.3 通解和特解 General Solution and Particular Solution

A solution of equation $\frac{dy}{dx} = f(x, y)$ is a differentiable function $y = y(x)$ defined on an interval I of x-values such that $\frac{dy}{dx} = f(x, y)$ on that interval.

一个函数以及它的导数代入微分方程（Differential Equations），使两端恒等，则称此函数为微分方程的解（Solution）．

如果微分方程的解中含有任意常数，且独立任意常数的个数等于微分方程的阶数（D. E. Order），这样的解为通解（General Solution）；由给定的条件代入通解，确定任意常数的值而得到的解为特解（Particular Solution）；用来确定特解的条件称为初始条件（Initial Condition）．

Example 10.1 Solve the differential equations $\frac{dy}{dx} = -\frac{x}{y}$, given the initial condition $y(0) = 3$.

Solution：

We rewrite the equation as $g(y)dy = f(x)dx$：

$$y\,dy = -x\,dx$$

Integrate both side of the equation：

$$\int y \, dy = \int (-x) \, dx$$

$$\frac{1}{2}y^2 + C_1 = -\frac{1}{2}x^2 + C_2$$

Let $C = 2(C_2 - C_1)$, then

$$x^2 + y^2 = C$$

Since $y(0) = 3$, we get: $0 + 3^2 = C$, the particular solution is

$$x^2 + y^2 = 9$$

10.2 求解可分离变量微分方程 Solving Separable D. E.

求微分方程的解是本章重点之一，AP 微积分主要讨论一阶可分离变量微分方程.

10.2.1 可分离变量微分方程 Separable Differential Equations

The differential equation $\frac{dy}{dx} = f(x, y)$ is separable if $f(x, y)$ can be expressed as a product of a function of x and a function of y.

一阶可分离变量的微分方程可以表示成一个 x 的函数和一个 y 的函数的乘积，那么，原微分方程可写成如下形式：

$$g(y) \, dy = f(x) \, dx$$

一端只含 y 的函数和 dy，另一端只含 x 的函数和 dx.

Example 10.2　If $\frac{dy}{dx} = \frac{x}{y^2}$ and $y(2) = 3$, find an equation for y in terms of x.

Solution: Separate variables,

$$y^2 \, dy = x \, dx$$

Integrate both side of the equation:

$$\int y^2 \, dy = \int x \, dx$$

$$\frac{y^3}{3} = \frac{x^2}{2} + C$$

Solve for the constant: $\frac{3^3}{3} = \frac{2^2}{2} + C$, so $C = 7$. The solution is

$$y = \sqrt[3]{\frac{3}{2}x^2 + 21}$$

10.2.2 分离变量法 Separating the Variables

【方法总结】运用分离变量法（Separating the Variables）求一阶可分离变量微分方程的解一般有 3 个步骤：

1) 将所求微分方程分离变量化为：$g(y) \, dy = f(x) \, dx$；
2) 同时对两边积分（Integrate）；
3) 解得通解（General Solution）.

Example 10.3 Solve the differential equation $\dfrac{dy}{dx} = \dfrac{x}{y+\sin y}$.

Solution: Separate variables,
$$(y+\sin y)dy = x\,dx$$

Integrate both side of the equation:
$$\int (y+\sin y)dy = \int x\,dx$$
$$\frac{y^2}{2} - \cos y = \frac{x^2}{2} + C$$

Example 10.4 If $\dfrac{dy}{dx} = \sqrt{xy}$ and $x=0$ when $y=4$, find y when $x=9$.

Solution: Separate variables,
$$\frac{dy}{\sqrt{y}} = \sqrt{x}\,dx$$

Integrate both side of the equation:
$$\int \frac{dy}{\sqrt{y}} = \int \sqrt{x}\,dx$$
$$2\sqrt{y} = \frac{2}{3}x^{\frac{3}{2}} + C$$

Using $y=4$ and $x=0$, we get $2\times\sqrt{4} = \dfrac{2}{3}\times 0^{\frac{3}{2}} + C$, so $C=4$. Then
$$2\sqrt{y} = \frac{2}{3}x^{\frac{3}{2}} + 4$$

When $x=9$, we find that $y=121$.

Example 10.5 If $\dfrac{dy}{dx} = 2xy$ and $y(0)=4$, find an equation for y in terms of x.

Solution: Separate variables,
$$\frac{1}{y}dy = 2x\,dx$$

Integrate both side of the equation:
$$\int \frac{1}{y}dy = \int 2x\,dx$$
$$\ln y = x^2 + C$$

Putting both sides into exponential form:
$$y = e^{x^2 + C} = e^{x^2} \times e^C$$

Since e^C is a constant, the equation becomes:
$$y = Ce^{x^2}$$

Using $y(0)=4$, we get $4 = Ce^0 = C$, so $C=4$. Then
$$y = 4e^{x^2}$$

Example 10.6 If the acceleration of a particle is given by $a(t)=G$, G is a constant, find the equation of velocity $v(t)$ and displacement $h(t)$ when it begins at $t=0$ with $v(0)=a$ and $h(0)=b$.

Solution: Since $a(t) = G$, then
$$v(t) = \int a(t) dt = \int G dt = Gt + C_0$$

Plug $v(0) = a$ into the above equation, we get
$$C_0 = a$$

Hence,
$$v(t) = Gt + a$$

Similarly,
$$h(t) = \int v(t) dt = \int (Gt + a) dt = \frac{1}{2} Gt^2 + at + C_1$$

Plug $h(0) = b$ into the above equation, we get
$$C_1 = b$$

Therefore,
$$h(t) = \frac{1}{2} Gt^2 + at + b$$

Example 10.7 If the acceleration of a particle is given by $a(t) = 20t^3 - 6t$, $s(t)$ denotes its position and $s(-1) = 2$, $s(1) = 4$, find the equation of velocity $v(t)$.

Solution:

Since $a(t) = 20t^3 - 6t$, then
$$v(t) = \int a(t) dt = \int (20t^3 - 6t) dt = 5t^4 - 3t^2 + C_1$$

Similarly,
$$s(t) = \int v(t) dt = \int (5t^4 - 3t^2 + C_1) dt = t^5 - t^3 + C_1 t + C_2$$

Since $s(-1) = 2$ and $s(1) = 4$, then
$$\begin{cases} s(-1) = -1 + 1 - C_1 + C_2 = 2 \\ s(1) = 1 - 1 + C_1 + C_2 = 4 \end{cases}$$

So,
$$C_1 = 1$$

Therefore,
$$v(t) = 5t^4 - 3t^2 + 1$$

10.3 斜率场 Slope Fields

已知微分方程，一根简短的切线表示平面上每一个点的斜率，绘制形成斜率场．通过斜率场，可以了解微分方程解的曲线．

10.3.1 绘制斜率场 Draw the SlopeField

We can picture the slopes graphically by drawing short line segments of slope $f(x, y)$ at selected points (x, y) in the region of the xy-plane that constitutes the domain of $f(x)$. The resulting picture is called a slope fields.

函数 $y=f(x)$ 在定义域（Domain）内每一个点 (x,y) 的斜率用一根简短的切线刻画出来，形成的图形区域是斜率场（Slope Fields），见 Figure 10.3.1(a)，表示微分方程 $\dfrac{dy}{dx}=2x$ 斜率场.

Example 10.8 Given $\dfrac{dy}{dx}=y$, sketch the slope field of the function.

Solution：

The slope of the curve at $y=1$ is 1. The slope of the curve at $y=2$ is 2. The slope of the curve at the origin is 0. The slope of the curve at $y=-1$ is -1. The slope of the curve at $x=-2$ is -2. See Figure 10.3.1(b).

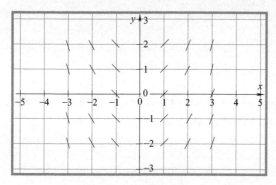

Figure 10.3.1(a)　　　　　　　　　　Figure 10.3.1(b)

Example 10.9 Given $\dfrac{dy}{dx}=\dfrac{x}{y}$, sketch the slope field of the function.

Solution：

Set up a table of values for $\dfrac{dy}{dx}$ selected values of x and y.

y \ x	-3	-2	-1	0	1	2	3
-3	1	$\dfrac{2}{3}$	$\dfrac{1}{3}$	0	$-\dfrac{1}{3}$	$-\dfrac{2}{3}$	-1
-2	$\dfrac{3}{2}$	1	$\dfrac{1}{2}$	0	$-\dfrac{1}{2}$	-1	$-\dfrac{3}{2}$
-1	3	2	1	0	-1	-2	-3
1	-3	-2	-1	0	1	2	3
2	$-\dfrac{3}{2}$	-1	$-\dfrac{1}{2}$	0	$\dfrac{1}{2}$	1	$\dfrac{3}{2}$
3	-1	$-\dfrac{2}{3}$	$-\dfrac{1}{3}$	0	$\dfrac{1}{3}$	$\dfrac{2}{3}$	1

Draw short tangent line segments with the given slopes at variouspoints. See Figure 10.3.1(c).

Example 10.10 Given $\dfrac{dy}{dx}=x-y$, sketch the slope field of the function.

Solution:

Set up a table of values for $\dfrac{dy}{dx}$ selected values of x and y.

y \ x ($x-y$)	−3	−2	−1	0	1	2	3
−3	0	1	2	3	4	5	6
−2	−1	0	1	2	3	4	5
−1	−2	−1	0	1	2	3	4
0	−3	−2	−1	0	1	2	3
1	−4	−3	−2	−1	0	1	2
2	−5	−4	−3	−2	−1	0	1
3	−6	−5	−4	−3	−2	−1	0

Draw short tangent line segments with the given slopes at various points. See Figure 10.3.1(d).

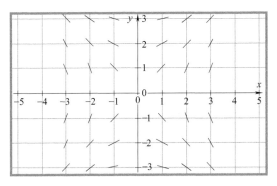
Figure 10.3.1(c)

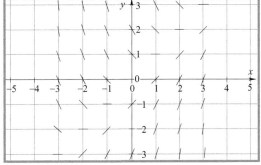
Figure 10.3.1(d)

10.3.2 看出解曲线 Viewing Solution Curves

Each time we specify an initial condition $y(x_0)=y_0$ for the solution of a differential equation $\dfrac{dy}{dx}=f(x,y)$, the solution curve is required to pass through the point (x_0,y_0) and to have slope $f(x_0,y_0)$ there.

斜率场中的直线是微分方程（Differential Equation）求解得到的函数 $y=f(x)$ 在点 (x,y) 处的一小段切线（Tangent Line），见 Figure 10.3.2(a)，微分方程 $\dfrac{dy}{dx}=2x$ 的一个解为 $y=x^2-2$。

Example 10.11 Given $\dfrac{dy}{dx}=x$, sketch the slope field of the function.

Solution:

The slope of the curve at $x=1$ is 1. The slope of the curve at $x=2$ is 2. The slope of the curve at $x=0$ is 0. The slope of the curve at $x=-1$ is -1. The slope of the curve at $x=-2$ is -2. These different slopes by drawing small segments of the tangent lines at those point. See Figure 10.3.2(b).

In fact, $y=\frac{1}{2}x^2+C$ is a solution of the d.e. for every constant C, since $y'=\left(\frac{1}{2}x^2+C\right)'=x$. We see the curve of $y=\frac{1}{2}x^2-2$ with slope segments drawn in Figure 10.3.2(c).

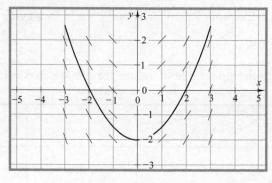

Figure 10.3.2(a)

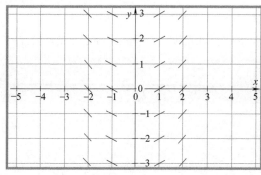

Figure 10.3.2(b)　　　　　　　　Figure 10.3.2(c)

Note：斜率场是直观表达微分方程求解之后函数 $y=f(x)$ 在点 (x,y) 的斜率情况，图形上的小段切线参考 30°、45°、60°的 $\tan\theta$ 值绘制，允许部分误差．

10.4　指数增长与衰减 Exponential Growth and Decay

指数增长与衰减是微分方程的常见应用之一，常见于细菌的繁殖、货币投资和同位素衰减等场景．

10.4.1　基本定义 Basic Definitions

If $y=f(t)$ is the value of a quantity y at time t and if the rate of change of y with respect to t is proportional to its size $f(t)$ at any time, then

$$\frac{dy}{dt}=ky$$

当一个变量 y 在一个既定的时间周期 $t\in D$ 里，其增长百分比是一个常量 k 时，这个量 y 就会显示出指数增长/衰减（Exponential Growth/Decay）．

Example 10.12　Solve the differential equation $\frac{dy}{dt}=ky$．

Solution：Separate variables，
$$\frac{1}{y}dy = k\,dt$$
Integrate both side of the equation：
$$\int \frac{1}{y}dy = \int k\,dt$$
$$\ln y = kt + C$$
Putting both sides into exponential form：
$$y = e^{kt+C} = e^{kt} \times e^{C}$$
Since $C_0 = e^C$ is a constant，the equation becomes：
$$y = C_0 e^{kt}$$

Example 10.13 The colony of bacteria is growing at a rate proportional to its size. If the growth rate per minute is 2% of the current size，how long will it take for the size to double?

Solution：

If the size at time t is S，then we are given that：
$$\frac{dS}{dt} = 0.02P$$
So，
$$S = S_0 e^{0.02t}$$
where S_0 is the initial size. We seek t when $S = 2S_0$
$$2S_0 = S_0 e^{0.02t}$$
$$2 = e^{0.02t}, \ \ln 2 = 0.02t, \ t = \frac{\ln 2}{0.02} \approx 34.6574 (\text{minute}) \ .$$

10.4.2 指数增长与衰变的应用 Applications of Exponential Growth and Decay

当一个变量（Variable）从一个时期以固定比例（Proportional）增长时，指数增长（Exponential Growth）就发生了．即：
$$\frac{dy}{dt} = ky \ \text{且} \ y = C_0 e^{kt}$$

当 $k > 0$ 时，$y = f(t)$ 单调递增，即函数 $f(t)$ 呈指数增长（Exponential Growth），见 Figure 10.4.2(a)；

当 $k < 0$ 时，$y = f(t)$ 单调递减，即函数 $f(t)$ 呈指数衰减（Exponential Decay），见 Figure 10.4.2(b)．

假如 $t = 0$ 时，给定初始值（Initial Amount）y_0，有：$y_0 = C_0 e^{k \times 0} = C_0 \times 1 = C_0$．即：$C_0$ 为初始值，k 为增长的百分比，t 为时间．下面归纳一下 AP 微积分考试中常见的指数增长与衰变应用．

1）指数增长的应用 Applications of Exponential Growth

① 细菌（Bacteria）的繁殖；

② 食物充足条件下，人类（Humans）、果蝇（Fruit Flies）、啮齿类动物（Rodents）

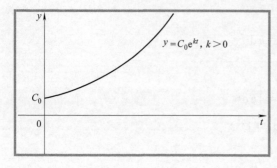

Figure 10.4.2(a)　　　　　　　　　Figure 10.4.2(b)

的繁殖；

③ 能够赚取复利的货币（Money）投资；

④ 诸如气（Gas）、油（Oil）、电（Electricity）或贵重金属（Valuables Metals）等商品的需求.

2）指数衰减的应用 Applications of Exponential Decay

① 同位素（Isotopes）的衰变；

② 一定数额货币（Money）的贬值速度；

③ 血液中药物（Drug）浓度的衰减；

④ 光束穿过水减弱的强度（Intensity）.

Example 10.14　The fruit flies in a certain culture increase continuously at a rate proportional to the number present. If the number triples in 10 days, how many will there be in 30 days?

Solution：

We let P be the number at time t and P_0 the number initially.

Then，
$$P = P_0 e^{kt}$$

Since $P = 3P_0$ when $t = 10$, we see that：
$$3P_0 = P_0 e^{10k}$$

and that
$$k = \frac{1}{10}\ln 3$$

Then,
$$P = P_0 e^{\frac{t\ln 3}{10}}$$

When $t = 30$,
$$P = P_0 e^{3\ln 3} = P_0 e^{\ln 3^3} = P_0 e^{\ln 27} = 27 P_0$$

Example 10.15　Uranium-235 decays at a rate proportional to the quantity present. Its half-life is 708 years. How long will it take for one quarter of a given of uranium-235 to decay?

Solution：

If $U(t)$ is the amount present at time t, then it satisfies the equation

$$U(t) = U_0 e^{kt}$$

where U_0 is the initial amount and k is the factor of proportionality.

Since it is given that $U = \frac{1}{2} U_0$ when $t = 708$, we see that:

$$\frac{1}{2} U_0 = U_0 e^{708k}$$

$$\frac{1}{2} = e^{708k}$$

$$k = \frac{\ln \frac{1}{2}}{708} \approx -0.000979$$

Then,

$$U(t) = U_0 e^{-0.000979 t}$$

When one quarter of U_0 has decayed, three quarters of the initial amount remains.

$$U(t) = \frac{3}{4} U_0 = U_0 e^{-0.000979 t}$$

$$\frac{3}{4} = e^{-0.000979 t}$$

$$t = \frac{\ln \frac{3}{4}}{-0.000979} \approx 294 \, (\text{years})$$

10.5 约束增长与衰减 Restricted Growth and Decay

约束增长与衰减是微分方程的另一个常见应用,如牛顿冷却定律就是其典型应用.

10.5.1 基本定义 Basic Definitions

If $y = f(t)$ is the value of a quantity y at time t and if the rate of change of y with respect to t is proportional to its size $f(t)$ and a fixed constant A at any time, then

$$\frac{dy}{dt} = k(A - y) \text{ or } \frac{dy}{dt} = -k(y - A)$$

where $k > 0$ and $A > 0$.

1) 当一个变量 y 在一个既定的时间周期 $t \in D$ 里,其增长百分比是一个常量 k 时,并且其增长会受到一个固定常数约束时,这个量 y 就会显示出约束增长 (Restricted Growth). 用 D.E. 表示:

$$\frac{dy}{dt} = k(A - y), A - y > 0$$

2) 当一个变量 y 在一个既定的时间周期 $t \in D$ 里,其衰减百分比是一个常量 k ($k > 0$) 时,并且其衰减会受到一个固定常数约束时,这个量 y 就会显示出约束衰减 (Restricted Decay). 用 D.E. 表示:

$$\frac{dy}{dt} = -k(y - A), y - A > 0$$

Example 10.16 Solve the differential equation $\dfrac{dy}{dt}=k(A-y)$, $A-y>0$.

Solution: Separate variables,
$$\dfrac{1}{A-y}dy=k\,dt$$

Integrate both side of the equation:
$$\int\dfrac{1}{A-y}dy=\int k\,dt$$

Since $A-y>0$, then
$$-\ln(A-y)=kt+C$$
$$\ln(A-y)=-kt-C$$

Putting both sides into exponential form:
$$A-y=e^{-kt-C}=e^{-kt}\times e^{-C}$$

Since $C_0=e^{-C}$ is a constant, the equation becomes:
$$y=A-C_0 e^{-kt}$$

Example 10.17 Solve the differential equation $\dfrac{dy}{dt}=-k(y-A)$, $y-A>0$.

Solution: Separate variables,
$$\dfrac{1}{y-A}dy=-k\,dt$$

Integrate both side of the equation:
$$\int\dfrac{1}{y-A}dy=-\int k\,dt$$

Since $y-A>0$, then
$$\ln(y-A)=-kt+C$$

Putting both sides into exponential form:
$$y-A=e^{-kt+C}=e^{-kt}\times e^{C}$$

Since $C_0=e^{C}$ is a constant, the equation becomes:
$$y=A+C_0 e^{-kt}$$

10.5.2 约束增长与衰减的应用 Applications of Restricted Growth and Decay

与指数增长（Exponential Growth）比较而言，约束增长的变量（Variable）在一个时期内不单受到固定比例（Proportional）影响，还受到固定常数（Fixed Constant）影响. 分两种情况：

1) 当 $A-y>0$ 时，变量 y 显示出约束增长，即：
$$\dfrac{dy}{dt}=k(A-y) \text{ 且 } y=A-C_0 e^{-kt}$$

其中 A 为约束常数，$A-C_0$ 为初始值，k 为增长的百分比，t 为时间，$y=f(t)$ 单调递增，见 Figure 10.5.2(a)；

2) 当 $y-A>0$ 时，变量 y 显示出约束衰减，即：
$$\dfrac{dy}{dt}=-k(y-A) \text{ 且 } y=A+C_0 e^{-kt}$$

其中 A 为约束常数，$A+C_0$ 为初始值（Initial Amount），k 为增长的百分比，t 为时间，$y=f(t)$ 单调递减，见 Figure 10.5.2(b)。

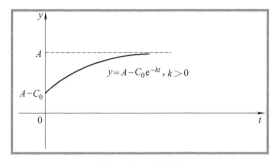

Figure 10.5.2(a)

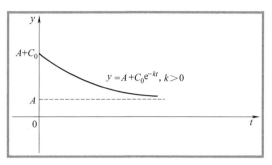

Figure 10.5.2(b)

举例常见的约束增长的应用：

① 一个高温物体和环境存在温度差，根据牛顿冷却定律（Newton's Law of Cooling），物体降温的速度和温度差成正比，显示出约束衰减（Restricted Decay）。而牛顿加热定律（Newton's Law of Heating）告诉我们，低温物体升温的速度和物体与环境温度差成正比，显示出约束增长（Restricted Growth）。

② 由于空气摩擦，物体下落的速度不是不加约束地增加，而是会接近物体下落的极限速度 L。其加速度与极限速度和物体速度之间的差成正比，显示出约束增长。

③ 轮胎内部压力和外部固定压力之间存在压力差，如果轮胎有一个小泄漏，其内部空气压力下降的速度和压力差成正比，显示出约束衰减。

Example 10.18 According to Newton's law of heating, a cold object warms up at a rate proportional to the difference between its own temperature and that of its environment. If you put a roast at 50°F into an oven of 500°F, and if, after 20 minutes, the temperature of the roast is 200°F. What is its temperature after 60 minutes?

Solution：

If $H(t)$ is the temperature of the roast at time t, then
$$\frac{\mathrm{d}H(t)}{\mathrm{d}t}=k[500-H(t)]$$
$$H(t)=500-C_0\mathrm{e}^{-kt}$$

Since $H(0)=50°F$, we have
$$50=500-C_0$$
$$C_0=450$$

Then,
$$H(t)=500-450\mathrm{e}^{-kt}$$

Also, $H(20)=200°F$, so
$$200=500-450\mathrm{e}^{-k\times 20}$$

and
$$\mathrm{e}^{-k}\approx 0.9799$$
$$H(t)=500-450\times 0.9799^t$$

Consequently,

$$H(60)=500-450\times 0.9799^{60}\approx 366.918(°F)$$

Example 10.19 According to Newton's law of cooling, a hot object cools at a rate proportional to the difference between its own temperature and that of its environment. If a roast at room temperature 88°F is put into a 15°F freezer, and if, after 20 minutes, the temperature of the roast is 30°F. How long will it take for the temperature of the roast to fall to 20°F?

Solution:

If $H(t)$ is the temperature of the roast at time t, then

$$\frac{dH(t)}{dt}=-k[H(t)-15]$$

$$H(t)=15+C_0 e^{-kt}$$

Since $H(0)=88°F$, we have

$$88=15+C_0$$
$$C_0=73$$

Then,

$$H(t)=15+73e^{-kt}$$

Also, $H(20)=30°F$, so

$$30=15+73e^{-k\times 20}$$

and

$$e^{-k}\approx 0.9239$$
$$H(t)=15+73\times 0.9239^t$$

We must find t when $H(t)=20$:

$$20=15+73\times 0.9239^t$$

$$0.9239^t=\frac{5}{73}$$

$$t\times \ln 0.9239=\ln \frac{5}{73}$$

$$t\approx 33.872(\text{minute})$$

10.6# 逻辑斯谛微分方程 Logistic Differential Equation

逻辑斯谛微分方程是具有现实意义的微分方程应用,当一个物种迁入到一个新生态系统中后,其数量发生的变化往往遵循逻辑斯谛微分方程.

10.6.1# 基本定义 Basic Definitions

If $y=f(t)$ is the value of a quantity y at time t, and if the rate of change of y may be proportional both to amount of y and to the difference between a fixed constant A and its amount. Then,

$$\frac{dy}{dt}=ky(A-y)$$

where $k>0$ and $A>0$.

当一个变量 y 在一个既定的时间周期 $t\in D$ 里，其增长百分比是一个常量 k 时，并且其增长会受到一个固定常数约束且受自身影响时，得到上述微分方程，称为逻辑斯谛微分方程，$y=f(t)$ 显示出 Logistic 增长（Logistic Growth）。

Example 10.20 Solve the differential equation $\dfrac{\mathrm{d}y}{\mathrm{d}t}=ky(A-y)$.

Solution: Separate variables,

$$\frac{1}{y(A-y)}\mathrm{d}y=k\,\mathrm{d}t$$

Integrate both side of the equation:

$$\int\frac{1}{y(A-y)}\mathrm{d}y=\int k\,\mathrm{d}t$$

The rational function of indefinite integral. Since

$$\frac{1}{y(A-y)}=\frac{1}{A}\left(\frac{1}{y}+\frac{1}{A-y}\right)$$

Then,

$$\int\frac{1}{y(A-y)}\mathrm{d}y=\frac{1}{A}\int\left(\frac{1}{y}+\frac{1}{A-y}\right)\mathrm{d}y=\int k\,\mathrm{d}t$$

That is,

$$\frac{1}{A}\left(\int\frac{1}{y}\mathrm{d}y+\int\frac{1}{A-y}\mathrm{d}y\right)=\int k\,\mathrm{d}t$$

$$\frac{1}{A}[\ln|y|-\ln|A-y|]=kt+C$$

$$\ln|y|-\ln|A-y|=Akt+AC$$

Hence,

$$\ln\left|\frac{A-y}{y}\right|=-Akt-AC$$

$$\left|\frac{A-y}{y}\right|=\mathrm{e}^{-Akt-AC}=\mathrm{e}^{-Akt}\times\mathrm{e}^{-AC}$$

Since $C_0=\pm\mathrm{e}^{-AC}$ is a constant, the equation becomes:

$$\frac{A-y}{y}=C_0\mathrm{e}^{-Akt}$$

$$\frac{A}{y}-1=C_0\mathrm{e}^{-Akt}$$

$$y=\frac{A}{1+C_0\mathrm{e}^{-Akt}}$$

10.6.2# Logistic 增长的应用 Applications of Logistic Growth

相比较而言，指数增长（Exponential Growth）是无限增长，Logistic 增长是有限增长。Logistic 微分方程的增长百分比 k 是固定常量，并且，其增长还会受到另一个固定常数 A 约束和自身的影响。即：

$$\frac{\mathrm{d}y}{\mathrm{d}t}=ky(A-y) \text{ 且 } y=\frac{A}{1+C_0\mathrm{e}^{-Akt}}$$

其中，$k>0$，$A>0$，A 表示承载能力（carrying capacity），k 表示增长的百分比，t 表示时间，$\dfrac{A}{1+C_0}$ 表示初始值（Initial Amount），见 Figure 10.6.2 (a).

Figure 10.6.2 上曲线 y 称为 Logistic 曲线（Logistic Curve）. 根据 Logistic 曲线，我们知道：

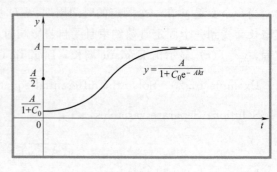

Figure 10.6.2

① 承载能力 A 是 y 能取到的极大值，当 $t\to+\infty$ 时取得；

② 在 Logistic 增长前期，y 的增长速度缓慢上升，$\dfrac{d^2 y}{dt^2}>0$；

③ 当 $y=\dfrac{A}{2}$，Logistic 增长增长速度达到最大；

④ Logistic 增长后期，y 的增长速度缓慢下降，y 接近于承载能力 A 时，增长速度接近于 0.

常见的 Logistic 增长应用：

① 某些疾病的传播；

② 谣言、时尚信息或者新的宗教的传播等；

③ 细胞在培养皿中的繁殖，蚂蚁在小容器中的繁殖，鱼在小鱼塘里的繁殖，果蝇在有限食物条件下的繁殖，还有酵母细胞的繁殖等；

④ 化学反应中，在催化剂的作用下，物质变成新物质的速度等.

Example 10.21 Because of limited food and space, an ant population cannot exceed 3000. It grows at a rate proportional both to the existing population and to the attainable additional population. If there were 200 ants 3 months ago, and 1 month ago the population was 1000, about how many ants are there now?

Solution：

If $y(t)$ is the ant population at time t, then

$$\dfrac{dy}{dt}=ky(3000-y)$$

$$y(t)=\dfrac{3000}{1+C_0 e^{-3000kt}}$$

Using initial condition $y(0)=200$, we have

$$\dfrac{3000}{1+C_0}=200$$

$$C_0=14$$

Then,

$$y(t)=\dfrac{3000}{1+14e^{-3000kt}}$$

Also, $y(2)=1000$, so

$$\frac{3000}{1+14e^{-3000k \times 2}} = 1000$$

and

$$1+14e^{-3000k \times 2} = 3$$

$$e^{-3000k} = \frac{\sqrt{7}}{7}$$

Then the particular solution is

$$y(t) = \frac{3000}{1+14 \times \left(\frac{\sqrt{7}}{7}\right)^t}$$

Thus,

$$y(3) = \frac{3000}{1+14 \times \left(\frac{\sqrt{7}}{7}\right)^3} \approx 1708.5$$

10.7 习题 Practice Exercises

(1) Solve the differential equations $\frac{dy}{dx} = 2x+1$, given the initial condition $y(0) = 3$.

(2) If $\frac{dy}{dx} = \frac{y}{\sqrt{x}}$ and $y(1) = 4$, find an equation for y in terms of x.

(3) If $f''(x) = x-1$, $f'(1) = 0$ and $f(1) = 4$, find $f(x)$.

(4) If the acceleration of a car is given by $a(t) = 3t^2 - 6t$, $s(t)$ denotes its position and $s(-1) = 3$, $s(1) = 5$, find the equation of velocity $v(t)$.

(5) Given $\frac{dy}{dx} = x+y$, sketch the slope field of the function.

(6) Suppose you deposit \$821 in a bank account that pay 4% compounded continuously. How much money will you have 10 years later?

(7) Determine $\frac{dy}{dx}$ for the curve defined by $y \sin x = 1$.

(8) Advertisers generally assume that the rate at which people hear about a product is proportional to the number of people who have not yet heart about it. Suppose that the size of a community is 20000, that to begin with no one has heart about a product, but that after 10 days 2000 people know about it. How long will it take for 5000 people to have heard of it?

(9) Write the solution of the initial-value problem

$$\frac{dP}{dt} = 0.08P\left(1 - \frac{P}{1000}\right), \quad P(0) = 100$$

Use it find the population size $P(50)$.

第 11 章

无穷级数
Infinite Series

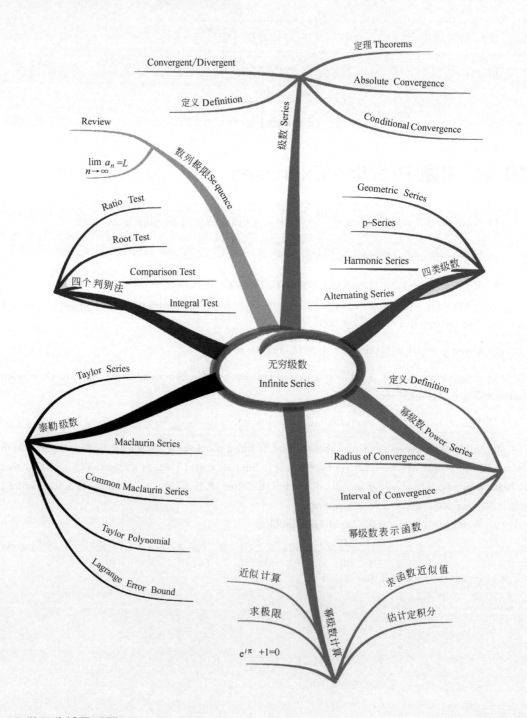

11.1 数列的极限 The Limit of The Sequence

高中阶段，我们学习了数列，特别是等差数列（Arithmetic Sequence）和等比数列（Geometric Sequence）的递推公式、通项公式以及前 n 项和公式．本节，我们回顾一下数列并学习数列的极限．

11.1.1 回顾数列 Review of the sequence

A sequence can be thought of as a list of numbers written in a definite order：
$$a_1, a_2, a_3, a_4, \cdots, a_n, \cdots$$
The number a_1 is called the first term, and in general a_n is the nth term.

1) 等差数列（Arithmetic Sequence）．如果一个数列从第二项起，每一项与它的前一项的差都等于同一个常数 d，这个数列叫做等差数列．常数 d 叫做等差数列的公差（Common Difference）．

递推公式（Definition）：$a_n = a_{n-1} + d$；

通项公式（The nth Term）：$a_n = a_1 + (n-1)d$；

前 n 项和公式（Formulas for S_n）：$S_n = na_1 + \dfrac{n(n-1)}{2}d$，$n \in N^*$；

等差中项（Arithmetic Mean）：若三个数 a、b、c 成等差数列，即 $2b = a + c$，则称 b 为 a 和 c 的等差中项．

2) 等比数列（Geometric Sequence）．如果一个数列从第二项起，每一项与它的前一项的比值都等于同一个常数 $q(q \neq 0)$，这个数列叫做等比数列．这个常数 q 叫做等比数列的公比（Common Ratio）．其中，$a_1 \neq 0$，a_n 中的每一项均不为 0．

递推公式（Definition）：$a_n = a_{n-1} q (n \geqslant 2, a_{n-1} \neq 0, q \neq 0)$；

通项公式（The nth Term）：$a_n = a_1 q^{n-1}$；

前 n 项和公式（Formulas for S_n）：$S_n = \begin{cases} \dfrac{a_1(1-q^n)}{1-q}, & q \neq 1 \\ na_1, & q = 1 \end{cases}$；

等比中项（Geometric Mean）：若三个数 a、b、c 成等比数列，即 $b^2 = ac$，则称 b 为 a 和 c 的等比中项．

Example 11.1 Find a formula for the general term a_n of the sequence
$$\left\{ \frac{1}{2}, \frac{2}{4}, \frac{3}{8}, \frac{4}{16}, \frac{5}{32}, \cdots \right\}$$
if the pattern of the first terms continues.

Solution：
$$a_1 = \frac{1}{2}, a_2 = \frac{2}{4}, a_3 = \frac{3}{8}, a_4 = \frac{4}{16}, a_5 = \frac{5}{32}$$

Notice that the numerators of these fractions start with 1 and increase by 1 whenever we go to the next term.

The second term has numerator 2, the third term has numerator 3; in general, the

nth term will have numerator n.

The denominators are the powers of 2, so a_n has denominator 2^n. Therefore
$$a_n = \frac{n}{2^n}$$

11.1.2 数列的极限 Limits of Sequences

A sequence $\{a_n\}$ has the limit L and we write
$$\lim_{n\to\infty} a_n = L$$

If $\lim\limits_{n\to\infty} a_n$ exists, we say the sequence $\{a_n\}$ converges. Otherwise, we say the sequence $\{a_n\}$ diverges.

通过计算数列的极限，由极限存在与否，确定数列是收敛（Converges）或发散（Diverges）. 我们即将学习的无穷级数的敛散性是由数列的收敛或发散确定的.

Example 11.2 Does the sequence $a_n = \dfrac{n^3 + 3n^2 + 77}{n^5 - 11n^4 + 77}$ converge or diverge?

Solution:
$$\lim_{n\to\infty} a_n = \lim_{n\to\infty} \frac{n^3 + 3n^2 + 77}{n^5 - 11n^4 + 77} = 0$$

Consequently, the sequence $\{a_n\}$ converges to 0.

Example 11.3 Does the sequence $a_n = 3^n$ converge or diverge?

Solution:
$$\lim_{n\to\infty} a_n = \lim_{n\to\infty} 3^n = \infty$$

Consequently, the sequence $\{a_n\}$ diverge.

Example 11.4 Does the sequence $a_n = 3 \times (-1)^n$ converge or diverge?

Solution:

Note that the sequence $-3, 3, -3, \cdots$, the $\lim\limits_{n\to\infty} a_n = \lim\limits_{n\to\infty} 3 \times (-1)^n$ does not exist, the sequence $\{a_n\}$ diverge.

11.2 无穷级数 Infinite Series

理解无穷级数的概念，要注意从把握数列（Sequence）和无穷级数的联系和区别入手. AP 微积分 BC 主要考查无穷级数的收敛性.

11.2.1 无穷级数的定义 Definition of Infinite Series

Given a sequence of numbers $\{a_n\}$, an expression of the form
$$a_1 + a_2 + a_3 + \cdots + a_n + \cdots$$
is an infinite series. The number a_n is general term of the series. The sequence $\{S_n\}$ defined by
$$S_n = a_1 + a_2 + a_3 + \cdots + a_n = \sum_{i=1}^{n} a_i$$

is the sequence of partial sums of the series, the number S_n being the nth partial sum.

假设有数列 $\{a_n\}$，我们称 $\sum_{n=1}^{\infty} a_n$ 为无穷级数（Infinite Series），a_n 为级数的通项（General Term），$S_n = \sum_{i=1}^{n} a_i$ 为该级数的前 n 项部分和数列（Sequence of Partial Sums）。

我们从级数的部分和入手理解无穷级数。这是个小窍门，部分和 $\sum_{i=1}^{n} a_i$ 是数列的前 n 项和 S_n。

假设有一个数列 $\{a_n\}$，其通项 $a_n = \frac{1}{2^n}$，则级数 $\sum_{i=1}^{n} a_i$ 的前 n 项部分和数列 $\{S_n\}$ 如 Table 11.2.1 所示。

Table 11.2.1 级数的前 n 项部分和数列 $\{S_n\}$

项目	部分和（Sequence of Partial Sums）	值（Value）
第 1 项	$S_n = \frac{1}{2}$	$1 - \frac{1}{2}$
第 2 项	$S_n = \frac{1}{2} + \frac{1}{4}$	$1 - \frac{1}{4}$
第 3 项	$S_n = \frac{1}{2} + \frac{1}{4} + \frac{1}{8}$	$1 - \frac{1}{8}$
⋮		
第 n 项	$S_n = \frac{1}{2} + \frac{1}{4} + \frac{1}{8} + \cdots + \frac{1}{2^n}$	$1 - \left(\frac{1}{2}\right)^n$

我们可以这样理解：部分和组成一个新的数列，它的通项是：

$$S_n = 1 - \left(\frac{1}{2}\right)^n$$

根据前面讲述的数列收敛的定义，因为 $\lim_{n \to \infty} \left[1 - \left(\frac{1}{2}\right)^n\right] = 1$，数列 $\{S_n\}$ 收敛于 1。

即，无穷级数 $\frac{1}{2} + \frac{1}{4} + \frac{1}{8} + \cdots + \frac{1}{2^n} + \cdots = 1$。

上述过程，从"前 n 项的和"到"无穷多项的和"，从数列到无穷级数，帮助同学们理解无穷级数，为下面理解无穷级数的收敛性做准备。

Example 11.5 Suppose the series $\sum_{n=1}^{\infty} a_n = \sum_{n=1}^{\infty} \frac{2^n}{3}$, find S_n.

Solution: Since

$$\sum_{n=1}^{\infty} \frac{2^n}{3} = \frac{2}{3} + \frac{2}{3} \times 2 + \frac{2}{3} + 2^2 + \cdots + \frac{2}{3} \times 2^{n-1} + \cdots$$

Then,

$$S_n = \frac{2}{3} + \frac{2}{3} \times 2 + \frac{2}{3} + 2^2 + \cdots + \frac{2}{3} \times 2^{n-1}$$

Consequently,

$$S_n = \frac{\frac{2}{3} \times (1 - 2^n)}{1 - 2} = \frac{2}{3}(2^n - 1)$$

11.2.2 级数收敛或发散 Series Convergent or Divergent

Given a series $\sum_{n=1}^{\infty} a_n = a_1 + a_2 + a_3 + \cdots + a_n + \cdots$, S_n denote its nth partial sum:

$$S_n = \sum_{i=1}^{n} a_i = a_1 + a_2 + a_3 + \cdots + a_n$$

If the sequence $\{S_n\}$ is convergent and $\lim_{n\to\infty} S_n = L$ exists as a real number, then the series $\sum a_n$ is called convergent and we write

$$a_1 + a_2 + a_3 + \cdots + a_n + \cdots = L$$

The number L is called the sum of the series. If the sequence $\{S_n\}$ is divergent, then the series is called divergent.

在实际运用中，人们希望级数（Series）能够收敛（Convergent），因为收敛才有实用性。级数收敛是靠数列（Sequence）收敛定义的，随着学习的深入，我们将掌握许多级数收敛性的判别方法。

Example 11.6 Suppose we know that the sum of the first n terms of the series $\sum_{n=1}^{\infty} a_n$ is

$$S_n = a_1 + a_2 + a_3 + \cdots + a_n = \frac{3n}{5n+3}$$

Find the series $\sum_{n=1}^{\infty} a_n$.

Solution: The sum of the series is the limit of the sequence $\{S_n\}$:

$$\sum_{n=1}^{\infty} a_n = \lim_{n\to\infty} S_n$$

Then,

$$\sum_{n=1}^{\infty} a_n = \lim_{n\to\infty} \frac{3n}{5n+3} = \lim_{n\to\infty} \frac{3}{5+\frac{3}{n}} = \frac{3}{5}$$

Consequently, the series $\sum_{n=1}^{\infty} a_n$ converges to $\frac{3}{5}$.

11.2.3 级数收敛或发散定理 Theorems about Convergent or Divergent of Series

1) If the series $\sum_{n=1}^{\infty} a_n$ is convergent, then $\lim_{n\to\infty} a_n = 0$.

2) If $\lim_{n\to\infty} a_n$ does not exist or if $\lim_{n\to\infty} a_n \neq 0$, then the series $\sum_{n=1}^{\infty} a_n$ is divergent.

3) If the series $\sum_{n=1}^{\infty} a_n$ and the series $\sum_{n=1}^{\infty} b_n$ both converge, then $\sum_{n=1}^{\infty} (a_n + b_n)$ converge.

4) If the series $\sum_{n=1}^{\infty} a_n$ converge, but the series $\sum_{n=1}^{\infty} b_n$ diverge, then $\sum_{n=1}^{\infty}(a_n + b_n)$ diverge.

5) A finite number of terms may be added to or deleted from a series without affecting its convergence or divergence; thus $\sum_{n=1}^{\infty} a_n$ and $\sum_{n=k}^{\infty} a_n$ (k is any positive integer) both converge or both diverge.

6) The terms of a series may be multiplied by a nonzero constant without affecting the convergence or divergence; thus $\sum_{n=1}^{\infty} a_n$ and $\sum_{n=1}^{\infty} ca_n (c \neq 0)$ both converge or both diverge.

定理2）也称发散级数第 n 项判别法（The nth-Term Test for Divergence）. 无穷级数收敛（Converge）或发散（Diverge）最基本的判别方法是利用级数收敛或发散的定义及其基本定理进行判别.

Example 11.7 Show that the series $\sum_{n=1}^{\infty} \dfrac{n^3}{3n^3 + 5}$ diverges.

Solution:
$$\lim_{n \to \infty} a_n = \lim_{n \to \infty} \frac{n^3}{3n^3 + 5} = \lim_{n \to \infty} \frac{1}{3 + \dfrac{5}{n^3}} = \frac{1}{3} \neq 0$$

So, the series $\sum_{n=1}^{\infty} \dfrac{n^3}{3n^3 + 5}$ diverges.

Example 11.8 Find the sum of the series $\sum_{n=1}^{\infty} \left(\dfrac{3}{n(n+1)} + \dfrac{1}{2^n} \right)$.

Solution: Let
$$S_n = \sum_{n=1}^{n} \frac{3}{n(n+1)} = 3 \sum_{n=1}^{n} \frac{1}{n(n+1)} = 3 \sum_{n=1}^{n} \left(\frac{1}{n} - \frac{1}{n+1} \right)$$

Thus, we have
$$S_n = 3 \left[\left(1 - \frac{1}{2}\right) + \left(\frac{1}{2} - \frac{1}{3}\right) + \left(\frac{1}{3} - \frac{1}{4}\right) + \cdots + \left(\frac{1}{n} - \frac{1}{n+1}\right) \right] = 3 \left(1 - \frac{1}{n+1}\right)$$

And so,
$$\sum_{n=1}^{\infty} \frac{3}{n(n+1)} = \lim_{n \to \infty} S_n = \lim_{n \to \infty} 3 \left(1 - \frac{1}{n+1}\right) = 3$$

Therefore, the series $\sum_{n=1}^{\infty} \dfrac{3}{n(n+1)}$ converges to 3.

The series $\sum_{n=1}^{\infty} \dfrac{1}{2^n}$ is a geometric series with $a = \dfrac{1}{2}$ and $r = \dfrac{1}{2}$ (See 11.3.1), so
$$\sum_{n=1}^{\infty} \frac{1}{2^n} = \frac{\dfrac{1}{2}}{1 - \dfrac{1}{2}} = 1$$

Since $\sum_{n=1}^{\infty} \frac{3}{n(n+1)}$ and $\sum_{n=1}^{\infty} \frac{1}{2^n}$ both converge, then

$$\sum_{n=1}^{\infty} \left(\frac{3}{n(n+1)} + \frac{1}{2^n} \right) = \sum_{n=1}^{\infty} \frac{3}{n(n+1)} + \sum_{n=1}^{\infty} \frac{1}{2^n} = 3 + 1 = 4$$

【方法总结】判断某个级数是否收敛，通常先看 $\lim_{a \to \infty} a_n$ 是否为 0，如果不为 0，则发散；如果为 0，再继续讨论．因为级数收敛，必然 $\lim_{a \to \infty} a_n = 0$；如果 $\lim_{a \to \infty} a_n \neq 0$，级数必定发散．

11.3 四类重要级数 Four Important Series

学习了级数和级数收敛的定理之后，我们再来看看四类重要的级数．重点关注到四类重要级数收敛的条件．

11.3.1 几何级数 Geometric Series

Geometric series are series of the form

$$\sum_{n=1}^{\infty} a\, r^{n-1} = a + ar + a r^2 + \cdots + a r^{n-1} + \cdots$$

in which a and r are fixed real numbers and $a \neq 0$.

假如 $|r| < 1$，几何级数 $\sum_{n=1}^{\infty} a\, r^{n-1}$ 收敛（Converge），并且收敛于 $\frac{a}{1-r}$；假如 $|r| \geq 1$，几何级数 $\sum_{n=1}^{\infty} a\, r^{n-1}$ 发散（Diverge）．这称为几何级数判别法（The Geometric Series Test）．

Example 11.9 Show that the geometric series $\sum_{n=1}^{\infty} a\, r^{n-1}$, $|r| < 1$ converges to $\frac{a}{1-r}$.

Solution：

$$\sum_{n=1}^{\infty} a\, r^{n-1} = a + ar + a r^2 + \cdots + a r^n + \cdots$$
$$= \lim_{n \to \infty}(a + ar + a r^2 + \cdots + a r^n)$$

Let S represent the sum of the geometric series, then：

$$S = \lim_{n \to \infty}(a + ar + a r^2 + \cdots + a r^n)$$
$$rS = \lim_{n \to \infty}(ar + a r^2 + \cdots + a r^n + a r^{n+1})$$

Subtraction yields

$$(1-r)S = \lim_{n \to \infty}(a - a r^{n+1}) = a - \lim_{n \to \infty} a r^{n+1}$$

Since $|r| < 1$, $\lim_{n \to \infty} a r^{n+1} = 0$, then：

$$(1-r)S = a$$

$$S = \frac{a}{1-r}$$

Example 11.10 Does $1 - \frac{1}{3} + \frac{1}{9} - \frac{1}{27} + \cdots + \left(-\frac{1}{3}\right)^{n-1} + \cdots$ converge or diverge?

Solution:

$$\sum_{n=0}^{\infty} a_n = 1 - \frac{1}{3} + \frac{1}{9} - \frac{1}{27} + \cdots + \left(-\frac{1}{3}\right)^{n-1} + \cdots = \sum_{n=1}^{\infty} 1 \times \left(-\frac{1}{3}\right)^{n-1}$$

The series is geometric series with $a = 1$ and $r = -\frac{1}{3}$.

Since $|r| < 1$, the series converges, and its sum is

$$S = \frac{a}{1-r} = \frac{1}{1-\left(-\frac{1}{3}\right)} = \frac{3}{4}$$

Example 11.11 Does $\frac{\pi}{3} + \frac{\pi^2}{9} + \frac{\pi^3}{27} + \cdots + \left(\frac{\pi}{3}\right)^n + \cdots$ converge or diverge?

Solution:

$$\sum_{n=0}^{\infty} a_n = \frac{\pi}{3} + \frac{\pi^2}{9} + \frac{\pi^3}{27} + \cdots + \left(\frac{\pi}{3}\right)^n + \cdots = \sum_{n=1}^{\infty} \frac{\pi}{3} \times \left(\frac{\pi}{3}\right)^{n-1}$$

The series is geometric series with $a = \frac{\pi}{3}$ and $r = \frac{\pi}{3}$.

Since $|r| > 1$, the series diverge.

11.3.2 p 级数 p-Series

p-Series are series of the form

$$\sum_{n=1}^{\infty} \frac{1}{n^p} = \frac{1}{1^p} + \frac{1}{2^p} + \frac{1}{3^p} + \cdots + \frac{1}{n^p} + \cdots$$

in which p is fixed real numbers.

当 $p > 1$，p 级数 $\sum_{n=1}^{\infty} \frac{1}{n^p}$ 收敛 (Converge)；当 $p \leq 1$，p 级数 $\sum_{n=1}^{\infty} \frac{1}{n^p}$ 收敛发散 (Diverge)。这称为 p 级数判别法 (p-Series Test)。

Example 11.12 Does $1 + \frac{1}{8} + \frac{1}{27} + \frac{1}{64} + \cdots + \frac{1}{n^3} + \cdots$ converge or diverge?

Solution: Since

$$1 + \frac{1}{8} + \frac{1}{27} + \frac{1}{64} \cdots + \frac{1}{n^3} + \cdots = \sum_{n=1}^{\infty} \frac{1}{n^3}$$

It is a p-Series with $p = 3 > 1$. The series converges.

Example 11.13 Does $1 + \frac{1}{\sqrt{2}} + \frac{1}{\sqrt{3}} + \frac{1}{\sqrt{4}} + \cdots + \frac{1}{\sqrt{n}} + \cdots$ converge or diverge?

Solution: Since

$$1 + \frac{1}{\sqrt{2}} + \frac{1}{\sqrt{3}} + \frac{1}{\sqrt{4}} \cdots + \frac{1}{\sqrt{n}} + \cdots = \sum_{n=1}^{\infty} \frac{1}{n^{\frac{1}{2}}}$$

It is a p-Series with $p=\dfrac{1}{2}<1$. The series diverge.

11.3.3 调和级数 Harmonic Series

The p-Series with $p=1$ is the harmonic series.

$$\sum_{n=1}^{\infty}\dfrac{1}{n}=1+\dfrac{1}{2}+\dfrac{1}{3}+\cdots+\dfrac{1}{n}+\cdots$$

根据 p 级数判别法，调和级数是发散的（Diverge）。调和级数的发散或许让人感到惊讶，作为级数，调和级数形式简洁优美，充满魅力！

Example 11.14 Show that the harmonic series is divergent.

Solution：

$$\begin{aligned}\sum_{n=1}^{\infty}\dfrac{1}{n}&=1+\dfrac{1}{2}+\dfrac{1}{3}+\cdots+\dfrac{1}{n}+\cdots\\ &=1+\dfrac{1}{2}+\left(\dfrac{1}{3}+\dfrac{1}{4}\right)+\left(\dfrac{1}{5}+\dfrac{1}{6}+\dfrac{1}{7}+\dfrac{1}{8}\right)+\left(\dfrac{1}{9}+\dfrac{1}{10}+\cdots+\dfrac{1}{15}+\dfrac{1}{16}\right)+\cdots\\ &>1+\dfrac{1}{2}+2\times\dfrac{1}{4}+4\times\dfrac{1}{8}+8\times\dfrac{1}{16}+\cdots\\ &=1+\dfrac{1}{2}+\dfrac{1}{2}+\dfrac{1}{2}+\dfrac{1}{2}+\cdots\end{aligned}$$

Since the series $\sum_{n=1}^{\infty}\dfrac{1}{n}$ isn't bounded, the harmonic series is divergent.

11.3.4 交错级数 Alternating Series

Alternating series are series of the form

$$\sum_{n=1}^{\infty}(-1)^{n+1}a_n=a_1-a_2+a_3-a_4+\cdots+(-1)^{n+1}a_n+\cdots$$

in which $a_n>0$.

交错级数 $\sum_{n=1}^{\infty}(-1)^{n+1}a_n$，$a_n>0$ 如果同时满足下列两个条件，则收敛（Converge），否则发散（Diverge）：1) $a_n>a_{n+1}$；2) $\lim_{n\to\infty}a_n=0$.

这称为交错级数判别法（Alternating Series Test），也称为莱布尼兹定理（Leibniz's Theorem）。

同时，级数

$$\sum_{n=1}^{\infty}(-1)^{n+1}\dfrac{1}{n}=1-\dfrac{1}{2}+\dfrac{1}{3}-\dfrac{1}{4}+\cdots+(-1)^{n+1}\dfrac{1}{n}+\cdots$$

称为交错调和级数（Alternating Harmonic Series）。根据交错级数判别法，交错调和级数 $\sum_{n=1}^{\infty}(-1)^{n+1}\dfrac{1}{n}$ 收敛（Converge）.

Example 11.15 Show that the alternating harmonic series is convergent.

Solution：

The Alternating Harmonic Series is

$$1 - \frac{1}{2} + \frac{1}{3} - \frac{1}{4} + \cdots + (-1)^{n+1} \frac{1}{n} + \cdots = \sum_{n=1}^{\infty} (-1)^{n+1} \frac{1}{n}$$

Satisfies

1) $a_n > a_{n+1}$, because $\frac{1}{n} > \frac{1}{n+1}$;

2) $\lim_{n \to \infty} a_n = \lim_{n \to \infty} \frac{1}{n} = 0$.

So, the alternating harmonic series converge by the Alternating Series Test.

Example 11.16 Does $\frac{1}{1 \times 2} - \frac{1}{2 \times 3} + \frac{1}{3 \times 4} - \frac{1}{4 \times 5} + \cdots$ converge or diverge?

Solution:

$$\frac{1}{1 \times 2} - \frac{1}{2 \times 3} + \frac{1}{3 \times 4} - \frac{1}{4 \times 5} + \cdots = \sum_{n=1}^{\infty} (-1)^{n+1} \frac{1}{n(n+1)}$$

Since

$$a_n - a_{n+1} = \frac{1}{n(n+1)} - \frac{1}{(n+1)(n+2)}$$

$$= \frac{(n+2)-n}{n(n+1)(n+2)} = \frac{2}{n(n+1)(n+2)} > 0$$

Satisfies

1) $a_n > a_{n+1}$, because $a_n - a_{n+1} > 0$;

2) $\lim_{n \to \infty} a_n = \lim_{n \to \infty} \frac{1}{n(n+1)} = 0$;

So, the series converge by the alternating series test.

11.4 正项级数的四大判别法 Four Tests of Nonnegative Series

从级数通项（General Term）正负号来划分，AP 微积分中的级数可大致划分为正项级数和交错级数．四大判别法是判别级数收敛重要的方法．

11.4.1 正项级数 Nonnegative Series

If $a_n \geq 0$ for all n, then the series $\sum_{n=1}^{\infty} a_n$ is called a nonnegative series.

前面我们讲过的 p 级数（p-Series）和调和级数（Harmonic Series）就是典型的正项级数．

正项级数 $\sum_{n=1}^{\infty} a_n$ 收敛的充分必要条件是它的部分和数列 $\{S_n\}$（Sequence of Partial Sums）有上界，这个结论十分有用，它是正项级数收敛性的四大判别法的基础，有兴趣的同学可以拓展学习．

11.4.2 比值判别法 The Ratio Test

Let $\sum_{n=1}^{\infty} a_n$ be a series with positive terms and suppose that

$$\lim_{n\to\infty} \frac{a_{n+1}}{a_n} = \rho$$

Then, 1) the series converges if $\rho<1$; 2) the series diverge if $\rho>1$; 3) the test is inconclusive if $\rho=1$.

判别正项级数 $\sum_{n=1}^{\infty} a_n$ 收敛性，建议首先使用比值判别法（Ratio Test），因为比值判别法可以由 $\sum_{n=1}^{\infty} a_n$ 自身通过极限计算对其收敛性做出某种判断，而不需要类似于比较判别法那样去寻找一个比较级数．这是我们教学的一个经验．

Example 11.17 Does $\sum_{n=1}^{\infty} \frac{n}{2^n}$ converge or diverge?

Solution：

$$\frac{a_{n+1}}{a_n} = \frac{\frac{n+1}{2^{n+1}}}{\frac{n}{2^n}} = \frac{n+1}{n} \times \frac{2^n}{2^{n+1}} = \frac{n+1}{2n}$$

And

$$\lim_{n\to\infty} \frac{a_{n+1}}{a_n} = \lim_{n\to\infty} \frac{n+1}{2n} = \frac{1}{2}$$

Since, $\frac{1}{2}<1$, $\sum_{n=1}^{\infty} \frac{n}{2^n}$ converge by the Ratio Test.

Example 11.18 Use the ratio test to determine the series $\sum_{n=1}^{\infty} \frac{(2n)!}{n!\,n!}$ diverge.

Solution：

$$a_n = \frac{(2n)!}{n!\,n!} \text{ and } a_{n+1} = \frac{(2n+2)!}{(n+1)!\,(n+1)!}$$

$$\frac{a_{n+1}}{a_n} = \frac{\frac{(2n+2)!}{(n+1)!\,(n+1)!}}{\frac{(2n)!}{n!\,n!}} = \frac{(2n+2)(2n+1)}{(n+1)(n+1)} = \frac{4n+2}{n+1}$$

And

$$\lim_{n\to\infty} \frac{a_{n+1}}{a_n} = \lim_{n\to\infty} \frac{4n+2}{n+1} = 4$$

Since, $4>1$, $\sum_{n=1}^{\infty} \frac{(2n)!}{n!\,n!}$ diverge by the Ratio Test.

11.4.3 根值判别法 The nth Root Test

Let $\sum_{n=1}^{\infty} a_n$ be a series with positive terms and suppose that

$$\lim_{n \to \infty} \sqrt[n]{a_n} = \rho$$

Then, 1) the series converges if $\rho < 1$; 2) the series diverge if $\rho > 1$; 3) the test is inconclusive if $\rho = 1$.

当 a_n 是关于 n 的若干个因子相乘、相除或有阶乘出现时，侧重比值判别法（Ratio Test）；当 a_n 是关于 n 的若干个因子相乘、相除的 n 次幂时，用根值判别法（The nth Root Test）. 这是我们的一个经验.

Example 11.19 Does $\sum_{n=1}^{\infty} \left(\dfrac{n}{3n+1} \right)^n$ converge or diverge?

Solution:

$$a_n = \left(\frac{n}{3n+1} \right)^n$$

And

$$\lim_{n \to \infty} \sqrt[n]{a_n} = \lim_{n \to \infty} \sqrt[n]{\left(\frac{n}{3n+1} \right)^n} = \lim_{n \to \infty} \frac{n}{3n+1} = \lim_{n \to \infty} \frac{1}{3}$$

Since, $\dfrac{1}{3} < 1$, $\sum_{n=1}^{\infty} \left(\dfrac{n}{3n+1} \right)^n$ diverge by the nth Root Test.

11.4.4 比较判别法 The Comparison Test

Let $\sum_{n=1}^{\infty} a_n$ be a series with positive terms.

1) If $\sum_{n=1}^{\infty} a_n$ converges and $b_n \leqslant a_n$, then $\sum_{n=1}^{\infty} b_n$ converges;

2) If $\sum_{n=1}^{\infty} a_n$ diverges and $a_n \leqslant b_n$, then $\sum_{n=1}^{\infty} b_n$ diverges.

上述属于直接比较判别法（Comparison Test），其极限形式（Limit Comparison Test）：

Suppose that $a_n > 0$ and $b_n > 0$, if $\lim_{n \to \infty} \dfrac{a_n}{b_n} = L > 0$, then $\sum_{n=1}^{\infty} a_n$ and $\sum_{n=1}^{\infty} b_n$ both converge or both diverge.

当 $\lim_{n \to \infty} \dfrac{a_{n+1}}{a_n} = 1$ 或 $\lim_{n \to \infty} \sqrt[n]{a_n} = 1$，比值判别法或根值判别法失效，转向比较判别法. 对正项级数 $\sum_{n=1}^{\infty} a_n$，寻找合适的比较级数，建议 2 种方法：

1) 对 a_n 适当放大或缩小，若 $\sum_{n=1}^{\infty} b_n$ 收敛或发散，则 $\sum_{n=1}^{\infty} b_n$ 为所寻找的比较级数；

2) 确定 $n \to \infty$ 时 a_n 的等价无穷小，记为 b_n，则 $\sum\limits_{n=1}^{\infty} b_n$ 为所寻找的比较级数．

Example 11.20 Does $\sum\limits_{n=1}^{\infty} \dfrac{1}{1+2^n}$ converge or diverge?

Solution：

Let $\quad a_n = \dfrac{1}{1+2^n}$, $b_n = \dfrac{1}{2^n}$

The series $\sum\limits_{n=1}^{\infty} b_n$ is geometric series with $a = \dfrac{1}{2}$ and $r = \dfrac{1}{2}$.

Since $|r| < 1$, the series $\sum\limits_{n=1}^{\infty} b_n$ converges by the Geometric Series Test.

Since $a_n < b_n$ and $\sum\limits_{n=1}^{\infty} b_n$ converges, $\sum\limits_{n=1}^{\infty} \dfrac{1}{1+2^n}$ converge by the Comparison Test.

Example 11.21 Use the comparison test to determine the series $\sum\limits_{n=1}^{\infty} \dfrac{7}{3n^2+4n+5}$ converge.

Solution：

Let $a_n = \dfrac{7}{3n^2+4n+5}$, $b_n = \dfrac{7}{3n^2}$

We know that

$$\sum\limits_{n=1}^{\infty} b_n = \sum\limits_{n=1}^{\infty} \dfrac{7}{3n^2} = \dfrac{7}{3} \sum\limits_{n=1}^{\infty} \dfrac{1}{n^2}$$

is convergent because it's a constant time a p-Series with $p = 2 > 1$.

Since

$$\dfrac{7}{3n^2+4n+5} < \dfrac{7}{3n^2}$$

and $\sum\limits_{n=1}^{\infty} b_n$ converges, $\sum\limits_{n=1}^{\infty} \dfrac{7}{3n^2+4n+5}$ converge by the Comparison Test.

Example 11.22 Determine whether the series $\sum\limits_{n=1}^{\infty} \dfrac{1}{3n+1}$ converge or diverge?

Solution：

Let $a_n = \dfrac{1}{3n+1}$, $b_n = \dfrac{1}{n}$

$$\dfrac{a_n}{b_n} = \dfrac{\dfrac{1}{3n+1}}{\dfrac{1}{n}} = \dfrac{n}{3n+1}$$

And

$$\lim_{n \to \infty} \dfrac{a_n}{b_n} = \lim_{n \to \infty} \dfrac{n}{3n+1} = \dfrac{1}{3} > 0$$

Since $\sum\limits_{n=1}^{\infty} \dfrac{1}{n}$ diverges, $\sum\limits_{n=1}^{\infty} \dfrac{1}{3n+1}$ also diverges by the Limit Comparison Test.

Note：本题 $\dfrac{1}{3n+1} < \dfrac{1}{n}$ 且调和级数 $\sum\limits_{n=1}^{\infty} \dfrac{1}{n}$ 发散，直接使用比较判别法不能得到结论．

11.4.5　积分判别法 The Integral Test

Let $\{a_n\}$ be a sequence of positive terms. Suppose that $f(x)$ is a continuous, positive, decreasing function and $f(n)=a_n$, then $\sum\limits_{n=1}^{\infty} a_n$ converges if and only if the improper integral $\int_1^{\infty} f(x)\mathrm{d}x$ converges.

积分判别法以广义积分（Improper Integrals）为工具，用于各项递减的正项级数收敛性判别．广义积分收敛或发散的判别见 8.7.3 节．

Example 11.23　Use the integral test to determine the series $\sum\limits_{n=1}^{\infty} \dfrac{1}{n^2+1}$ converge.

Solution：

The function $f(x)=\dfrac{1}{x^2+1}$ is positive, continuous and decreasing for $x \geqslant 1$.

$$\int_1^{\infty} \dfrac{1}{x^2+1}\mathrm{d}x = \lim_{a \to \infty} \int_1^a \dfrac{1}{x^2+1}\mathrm{d}x = \lim_{a \to \infty} (\arctan x \big|_1^a)$$
$$= \lim_{a \to \infty} (\arctan a - \arctan 1)$$
$$= \dfrac{\pi}{2} - \dfrac{\pi}{4} = \dfrac{\pi}{4}$$

Therefore, the series $\sum\limits_{n=1}^{\infty} \dfrac{1}{n^2+1}$ converge by the Integral Test.

Example 11.24　Does $\sum\limits_{n=1}^{\infty} \dfrac{n}{\mathrm{e}^n}$ converge or diverge?

Solution：

The function $f(x)=\dfrac{x}{\mathrm{e}^x}$ is positive, continuous and decreasing for $x \geqslant 1$.

$$\int_1^{\infty} \dfrac{x}{\mathrm{e}^x}\mathrm{d}x = \lim_{a \to \infty} \int_1^a \dfrac{x}{\mathrm{e}^x}\mathrm{d}x$$

$$\int_1^a \dfrac{x}{\mathrm{e}^x}\mathrm{d}x = x(-\mathrm{e}^x) - \int(-\mathrm{e}^x)\mathrm{d}x = -\mathrm{e}^x(1+x)+C$$

Then,

$$\int_1^{\infty} \dfrac{x}{\mathrm{e}^x}\mathrm{d}x = \lim_{a \to \infty} \int_1^a \dfrac{x}{\mathrm{e}^x}\mathrm{d}x = \lim_{a \to \infty} -\mathrm{e}^x(1+x)\big|_1^a = -\lim_{a \to \infty}\left(\dfrac{1+a}{\mathrm{e}^a} - \dfrac{2}{\mathrm{e}}\right)$$

$$\lim_{a \to \infty} \dfrac{1+a}{\mathrm{e}^a} = \lim_{a \to \infty}\left(\dfrac{1+a}{\mathrm{e}^a}\right)' = \lim_{a \to \infty} \dfrac{1}{\mathrm{e}^a} = 0$$

Thus,

$$\int_1^{\infty} \dfrac{x}{\mathrm{e}^x}\mathrm{d}x = -\lim_{a \to \infty}\left(\dfrac{1+a}{\mathrm{e}^a} - \dfrac{2}{\mathrm{e}}\right) = \dfrac{2}{\mathrm{e}}$$

Therefore, the series $\sum\limits_{n=1}^{\infty} \dfrac{n}{\mathrm{e}^n}$ converge by the Integral Test.

11.5 绝对收敛和条件收敛 Absolute and Conditional Convergence

相对于正项级数（Nonnegative Series）有任意项级数，交错级数（Alternating Series）就是特殊的任意项级数．任意项级数是有无限多个正数项和无限多个负数项的级数．关于任意项级数，我们要学习绝对收敛和条件收敛．

11.5.1 绝对收敛 Absolute Convergence

A series $\sum_{n=1}^{\infty} a_n$ is absolutely convergent if the corresponding series of absolute values, $\sum_{n=1}^{\infty} |a_n|$, converges.

绝对收敛一定收敛．后续，我们将学习求幂级数的收敛半径（Radius of Convergence），掌握绝对收敛是它的基础．

Example 11.25 Determine whether $-\dfrac{1}{2} - \dfrac{1}{4} + \dfrac{1}{8} + \dfrac{1}{16} - \cdots = \sum_{n=1}^{\infty} (-1)^{\frac{n(n+1)}{2}} \dfrac{1}{2^n}$ converges absolutely, converges conditionally or diverges.

Solution：

$$-\frac{1}{2} - \frac{1}{4} + \frac{1}{8} + \frac{1}{16} - \cdots = \sum_{n=1}^{\infty} (-1)^{\frac{n(n+1)}{2}} \frac{1}{2^n}$$

We know that

$$\sum_{n=1}^{\infty} \left| (-1)^{\frac{n(n+1)}{2}} \frac{1}{2^n} \right| = \sum_{n=1}^{\infty} \frac{1}{2^n}$$

The series $\sum_{n=1}^{\infty} \dfrac{1}{2^n}$ is geometric series with $a = \dfrac{1}{2}$ and $r = \dfrac{1}{2}$.

Since $|r| < 1$, the series $\sum_{n=1}^{\infty} \dfrac{1}{2^n}$ converges by the Geometric Series Test.

Therefore, the series $\sum_{n=1}^{\infty} (-1)^{\frac{n(n+1)}{2}} \dfrac{1}{2^n}$ converges absolutely.

Note：$\sum_{n=1}^{\infty} (-1)^{\frac{n(n+1)}{2}} \dfrac{1}{2^n}$ 不是交错级数．

11.5.2 条件收敛 Conditional Convergence

A series $\sum_{n=1}^{\infty} a_n$ that converges but does not converge absolutely converge conditionally.

收敛但不是绝对收敛的级数条件收敛．交错调和级数（Alternating Harmonic Series）$\sum_{n=1}^{\infty} (-1)^{n+1} \dfrac{1}{n}$ 条件收敛．

Example 11.26 Show that the alternating harmonic series converges conditionally.

Solution:

The Alternating Harmonic Series is

$$1 - \frac{1}{2} + \frac{1}{3} - \frac{1}{4} + \cdots + (-1)^{n+1}\frac{1}{n} + \cdots = \sum_{n=1}^{\infty}(-1)^{n+1}\frac{1}{n}$$

Satisfies

1) $a_n > a_{n+1}$, because $\frac{1}{n} > \frac{1}{n+1}$;

2) $\lim_{n \to \infty} a_n = \lim_{n \to \infty} \frac{1}{n} = 0$.

So, the alternating harmonic series converge by the Alternating Series Test.

$$\sum_{n=1}^{\infty}\left|(-1)^{n+1}\frac{1}{n}\right| = \sum_{n=1}^{\infty}\frac{1}{n}$$

Which is the harmonic series and it is not absolutely convergent.

Therefore, the alternating harmonic series $\sum_{n=1}^{\infty}(-1)^{n+1}\frac{1}{n}$ converges conditionally.

11.5.3 交错级数的误差界 Alternating Series Error Bound

If $S = \sum_{n=1}^{\infty}(-1)^{n+1}a_n$ is the sum of an alternating series that satisfies $a_{n+1} \leqslant a_n$ and $\lim_{n \to \infty} a_n = 0$, then $|R_n| = |S - S_n| \leqslant a_{n+1}$.

一个交错级数的误差界（Error）$|R_n| = |S - S_n|$ 小于等于组成级数的下一项 a_{n+1}，注意此结论只对符合条件的交错级数有效.

Example 11.27 Find the sum of the series $\sum_{n=0}^{\infty}\frac{(-1)^n}{n!}$ correct to three decimal places.

Solution:

$$a_n = \frac{1}{n!}, \quad a_{n+1} = \frac{1}{(n+1)!} = \frac{1}{(n+1)n!}$$

$$a_{n+1} = \frac{1}{(n+1)n!} < \frac{1}{n!} = a_n, \text{ and } \lim_{n \to \infty} a_n = \lim_{n \to \infty} \frac{1}{n!} = 0$$

Let's write out the first few terms of the series:

$$S = \sum_{n=0}^{\infty}\frac{(-1)^n}{n!} = \frac{1}{0!} - \frac{1}{1!} + \frac{1}{2!} - \frac{1}{3!} + \frac{1}{4!} - \frac{1}{5!} + \frac{1}{6!} - \frac{1}{7!} + \cdots$$

$$= 1 - 1 + \frac{1}{2} - \frac{1}{6} + \frac{1}{24} - \frac{1}{120} + \frac{1}{720} - \frac{1}{5040} + \cdots$$

Note that

$$\frac{1}{720} \approx 0.001389 \text{ and } \frac{1}{5040} \approx 0.000198$$

By the Alternating Series Error Bound:
$$|R_n| = |S - S_n| \leq a_7 < 0.0002$$
Then,
$$S_6 = 1 - 1 + \frac{1}{2} - \frac{1}{6} + \frac{1}{24} - \frac{1}{120} + \frac{1}{720} \approx 0.368056$$

The error less than 0.000198 does not affect the third place, so we have $S \approx 0.368$ correct to three decimal places.

11.6　幂级数 Power Series

本节学习幂级数，讨论类似于 $\sum\limits_{n=0}^{\infty} x^n$ 这样的无穷多项式，解决用无穷多项式（幂级数）表示特定函数的问题．

11.6.1　幂级数的定义 Definition of Power Series

A power series about $x = 0$ is a series of the form
$$\sum_{n=0}^{\infty} a_n x^n = a_0 + a_1 x + a_2 x^2 + \cdots + a_n x^n + \cdots$$

A power series about $x = a$ is a series of the form
$$\sum_{n=0}^{\infty} a_n (x-a)^n = a_0 + a_1(x-a) + a_2(x-a)^2 + \cdots + a_n(x-a)^n + \cdots$$

in which the center a and the coefficients a_0, a_1, a_2, $\cdots$, a_n, $\cdots$ are constants.

前面我们学习了常数项级数（Constant Term Series）的敛散性，其中每一项都是实数．幂级数是最基本的函数项级数（Series with Function Terms），函数项级数的每一项都是函数．

令幂级数的系数（Coefficients）$a_0 = a_1 = a_2 = \cdots = a_n = \cdots = 1$，有：
$$\sum_{n=1}^{\infty} x^{n-1} = 1 + x + x^2 + \cdots + x^{n-1} + \cdots$$

它是一个 $a=1$，$r=x$ 的几何级数（Geometric Series），由我们前面学过的几何级数判别法（Geometric Series Test）可知：

当 $|x| < 1$ 时，此级数收敛（Converge）于 $S = \dfrac{a}{1-r} = \dfrac{1}{1-x}$：
$$1 + x + x^2 + \cdots + x^{n-1} + \cdots = \frac{1}{1-x}$$

当 $|x| \geq 1$ 时，此级数发散（Diverge）．

由此，我们见到：幂级数 $\sum\limits_{n=1}^{\infty} x^{n-1}$ 是中心（Center）在 $x=0$ 的幂级数，它在区间 $-1 < x < 1$ 收敛到 $\dfrac{1}{1-x}$，区间 $-1 < x < 1$ 也以 $x=0$ 为中心．下面，我们会看到这是典型情形，这个情形在一般的幂级数上也出现．

11.6.2 收敛半径和收敛区间 The Radius and Interval of Convergence

If power series $\sum_{n=0}^{\infty} a_n (x-a)^n$ converges when $|x-a| < R$ and diverges when $|x-a| > R$, then R is called the radius of convergence.

The set of all values of x for which power series $\sum_{n=0}^{\infty} a_n (x-a)^n$ converges is called its interval of convergence.

幂级数在收敛区间内部的每个点绝对收敛（Absolute Convergence）. 如果对所有 x 值绝对收敛，它的收敛半径是 ∞；如果仅在 $x=a$ 收敛，它的收敛半径是零.

求幂级数的收敛半径（Radius of Convergence）常用比值判别法（Ratio Test），即：

$$\lim_{n \to \infty} \left| \frac{a_{n+1}}{a_n} \right| = \rho$$

$$R = \begin{cases} \dfrac{1}{\rho}, & 0 < \rho < +\infty \\ +\infty, & \rho = 0 \\ 0, & \rho = +\infty \end{cases}$$

Example 11.28 Find what x does $\sum_{n=1}^{\infty} x^{n-1}$ converges?

Solution：

By the Ratio Test, the series converges if

$$\lim_{n \to \infty} \left| \frac{a_{n+1}}{a_n} \right| = \lim_{n \to \infty} \left| \frac{x^{n+1}}{x^n} \right| = \lim_{n \to \infty} |x| = |x| < 1$$

Thus, the radius of convergence is 1.

The endpoints must be tested separately since the Ratio Test fails when the limit equals 1.

When $x=1$, $\sum_{n=1}^{\infty} x^{n-1}$ becomes $1+1+1+\cdots$ and diverges；

When $x=-1$, $\sum_{n=1}^{\infty} x^{n-1}$ becomes $1-1+1-1+\cdots$ and diverges.

Thus the interval of convergence is $-1 < x < 1$.

Example 11.29 Find what values of x is the series $\sum_{n=1}^{\infty} n! \, x^n$ converges?

Solution：

Using the Ratio Test,

$$\lim_{n \to \infty} \left| \frac{a_{n+1}}{a_n} \right| = \lim_{n \to \infty} \left| \frac{(n+1)! \, x^{n+1}}{n! \, x^n} \right| = \lim_{n \to \infty} (n+1)|x| = \infty$$

By the Ratio Test, the series $\sum_{n=1}^{\infty} n! \, x^n$ diverges when $x \neq 0$. Thus the given series converges only when $x=0$.

Example 11.30 Find what x does $\sum_{n=1}^{\infty} \dfrac{x^n}{n!}$ converges?

Solution：

Using the Ratio Test,

$$\lim_{n\to\infty}\left|\dfrac{a_{n+1}}{a_n}\right|=\lim_{n\to\infty}\left|\dfrac{\dfrac{x^{n+1}}{(n+1)!}}{\dfrac{x^n}{n!}}\right|=\lim_{n\to\infty}\left|\dfrac{x^{n+1}}{(n+1)!}\times\dfrac{n!}{x^n}\right|=\lim_{n\to\infty}\dfrac{|x|}{n+1}=0$$

The series converges for all x. Its interval of convergence is $(-\infty,+\infty)$ and its radius of convergence is ∞.

【方法总结】求收敛区间（Interval of Convergence）一般有 3 个步骤：

1）用比值判别法求使得幂级数绝对收敛的区间，通常是一个开区间 $|x-a|<R$；

2）如果绝对收敛区间是有限的，检验每个端点收敛或发散，多用比较判别法（Comparison Test）、积分判别法（Integral Test）或交错级数判别法（Alternating Series Test）；

3）如果幂级数对 $|x-a|>R$ 发散，对这些 x 值第 n 项不趋于零．

11.6.3 幂级数表示函数 Representations of Functions as Power Series

Let the function $f(x)$ be defined by：

$$f(x)=\sum_{n=0}^{\infty}a_n(x-a)^n=a_0+a_1(x-a)+a_2(x-a)^2+\cdots+a_n(x-a)^n+\cdots$$

Its domain is the interval of convergence of the power series.

幂级数表示的函数 $f(x)$ 具有 3 个性质：

1）$f(x)$ 的定义域是幂级数的收敛区间，在定义域上，$f(x)$ 连续（Continuous）；

2）$f(x)$ 在定义域上具有各阶导数，对幂级数在收敛区间内的点逐项求导：

$$f'(x)=\sum_{n=1}^{\infty}na_n(x-a)^{n-1}=a_1+2a_2(x-a)+\cdots+na_n(x-a)^{n-1}+\cdots$$

求导得到的级数和原级数具有相同的收敛半径．

3）对 $f'(x)$ 求积分，得到的级数在原级数收敛区间上的每一点收敛：

$$\int_a^x f(t)\mathrm{d}t = a_0(x-a)+\dfrac{a_1(x-a)^2}{2}+\dfrac{a_2(x-a)^3}{3}+\cdots+\dfrac{a_n(x-a)^{n+1}}{n+1}+\cdots$$

$$\int_a^x f(t)\mathrm{d}t = \sum_{n=0}^{\infty}\dfrac{a_n(x-a)^{n+1}}{n+1}$$

Example 11.31 Let $f(x)=\sum_{n=1}^{\infty}\dfrac{x^n}{n}=x+\dfrac{x^2}{2}+\cdots+\dfrac{x^n}{n}+\cdots$ Find the intervals of convergence of the power series for $f(x)$ and $f'(x)$.

Solution：

Using the Ratio Test,

$$\lim_{n\to\infty}\left|\dfrac{a_{n+1}}{a_n}\right|=\lim_{n\to\infty}\left|\dfrac{\dfrac{x^{n+1}}{n+1}}{\dfrac{x^n}{n}}\right|=\lim_{n\to\infty}\left|\dfrac{x^{n+1}}{n+1}\times\dfrac{n}{x^n}\right|=\lim_{n\to\infty}\left|\dfrac{n}{n+1}\right|\times|x|=|x|$$

Also,
$$f(1)=1+\frac{1}{2}+\frac{1}{3}+\cdots+\frac{1}{n}+\cdots$$

And
$$f(-1)=-1+\frac{1}{2}-\frac{1}{3}+\cdots+(-1)^n\frac{1}{n}+\cdots$$

Hence, the power series for $f(x)$ converges if $-1 \leqslant x < 1$.

For the derivative
$$f'(x)=\sum_{n=1}^{\infty}x^{n-1}=1+x+x^2+\cdots+x^{n-1}+\cdots$$

Using the Ratio Test,
$$\lim_{n\to\infty}\left|\frac{a_{n+1}}{a_n}\right|=\lim_{n\to\infty}\left|\frac{x^n}{x^{n-1}}\right|=|x|$$

Also,
$$f(1)=1+1+1+\cdots+1+\cdots$$

And
$$f(-1)=1-1+1-\cdots+(-1)^{n-1}+\cdots$$

Hence, the power series for $f'(x)$ converges if $-1 < x < 1$.

Note：幂级数表示的函数，求导所得级数和原级数收敛半径相同，收敛区间不一定相同.

Example 11.32 Find a power series representation for $\frac{1}{x+3}$.

Solution：Since
$$\frac{1}{1-x}=1+x+x^2+\cdots+x^{n-1}+\cdots=\sum_{n=1}^{\infty}x^{n-1}, |x|<1$$

And
$$\frac{1}{x+3}=\frac{1}{3\left(1+\frac{x}{3}\right)}=\frac{1}{3}\times\frac{1}{1-\left(-\frac{x}{3}\right)}$$

Then,
$$\frac{1}{x+3}=\frac{1}{3}\sum_{n=1}^{\infty}\left(-\frac{x}{3}\right)^{n-1}=\sum_{n=1}^{\infty}(-1)^{n-1}\frac{x^{n-1}}{3^n}$$

The series converges when $\left|-\frac{x}{3}\right|<1$, that is, $|x|<3$. So the interval of convergence is $(-3,3)$.

11.7 泰勒级数和麦克劳林级数 Taylor and Maclaurin Series

如果函数 $f(x)$ 在一个区间上 n 阶可导，它能够表示成一个幂级数吗？如果能够，它的系数是什么？

11.7.1 泰勒级数 Taylor Series

Let $f(x)$ be a function with derivatives of all orders throughout some interval containing a as an interior point. Then the Taylor series generated by $f(x)$ at $x=a$ is

$$f(x)=f(a)+f'(a)(x-a)+\frac{f''(a)}{2!}(x-a)^2+\cdots+\frac{f^{(n)}(a)}{n!}(x-a)^n+\cdots$$

泰勒级数幂级数（Power Series）表示一个函数，相加项的系数（Coefficients）由函数在某一点的导数求得．泰勒级数在近似计算中有重要作用．

Example 11.33   Find the Taylor series for function $f(x)=\dfrac{1}{x}$ about $x=1$.

Solution：We arrange our computation in two columns as follows：

$f(x)=\dfrac{1}{x}$ $\qquad\qquad$ $f(1)=\dfrac{1}{1}=1$

$f'(x)=-\dfrac{1}{x^2}$ $\qquad\qquad$ $f'(1)=-1$

$f''(x)=\dfrac{2!}{x^3}$ $\qquad\qquad$ $f''(1)=2!$

$f^{(3)}(x)=-\dfrac{3!}{x^4}$ $\qquad\qquad$ $f^{(3)}(1)=-3!$

$f^{(4)}(x)=\dfrac{4!}{x^5}$ $\qquad\qquad$ $f^{(4)}(1)=4!$

$\cdots$ $\qquad\qquad$ $\cdots$

$f^{(n)}(x)=(-1)^n\dfrac{n!}{x^{n+1}}$ $\qquad\qquad$ $f^{(n)}(1)=(-1)^n n!$

Thus,

$$\frac{1}{x}=f(1)+f'(1)(x-1)+\frac{f''(1)}{2!}(x-1)^2+\cdots+\frac{f^{(n)}(1)}{n!}(x-1)^n+\cdots$$

$$=1-(x-1)+(x-1)^2+\cdots+(-1)^n(x-1)^n+\cdots$$

$$=\sum_{n=0}^{\infty}(-1)^n(x-1)^n$$

11.7.2 麦克劳林级数 Maclaurin Series

The Maclaurin series generated by $f(x)$ is the Taylor series generated by $f(x)$ at $x=0$：

$$f(x)=f(0)+f'(0)x+\frac{f''(0)}{2!}x^2+\cdots+\frac{f^{(n)}(0)}{n!}x^n+\cdots$$

麦克劳林级数是泰勒级数（Taylor Series）在 $x=0$ 处的特例．

Example 11.34   Find the Maclaurin series for function $f(x)=e^x$.

Solution：

$$f(x)=e^x, f'(x)=e^x, f''(x)=e^x, \cdots, f^{(n)}(x)=e^x, \cdots$$

And

$$f(0)=1, f'(0)=1, f''(0)=1, \cdots, f^{(n)}(0)=1, \cdots$$

Then,

$$e^x = f(0) + f'(0)x + \frac{f''(0)}{2!}x^2 + \cdots + \frac{f^{(n)}(0)}{n!}x^n + \cdots$$

$$= 1 + x + \frac{1}{2!}x^2 + \cdots + \frac{1}{n!}x^n + \cdots = \sum_{n=0}^{\infty} \frac{x^n}{n!}$$

Example 11.35 Find the Maclaurin series for function $f(x) = \sin x$.

Solution:

$f(x) = \sin x$ $f(0) = 0$
$f'(x) = \cos x$ $f'(0) = 1$
$f''(x) = -\sin x$ $f''(0) = 0$
$f^{(3)}(x) = -\cos x$ $f^{(3)}(0) = -1$
$f^{(4)}(x) = \sin x$ $f^{(4)}(0) = 0$
$\cdots$ $\cdots$

Thus,

$$\sin x = f(0) + f'(0)x + \frac{f''(0)}{2!}x^2 + \cdots + \frac{f^{(n)}(0)}{n!}x^n + \cdots$$

$$= 0 + x + 0 - \frac{x^3}{3!} + \cdots + (-1)^n \frac{x^{2n+1}}{(2n+1)!} + \cdots$$

$$= x - \frac{x^3}{3!} + \frac{x^5}{5!} - \cdots + (-1)^n \frac{x^{2n+1}}{(2n+1)!} + \cdots$$

$$= \sum_{n=0}^{\infty} (-1)^n \frac{x^{2n+1}}{(2n+1)!}$$

Example 11.36 Find the Maclaurin series for function $f(x) = \cos x$.

Solution:

$f(x) = \cos x$ $f(0) = 1$
$f'(x) = -\sin x$ $f'(0) = 0$
$f''(x) = -\cos x$ $f''(0) = -1$
$f^{(3)}(x) = \sin x$ $f^{(3)}(0) = 0$
$f^{(4)}(x) = \cos x$ $f^{(4)}(0) = 1$
$\cdots$ $\cdots$

Thus,

$$\cos x = 1 - \frac{x^2}{2!} + \frac{x^4}{4!} - \cdots + (-1)^n \frac{x^{2n}}{(2n)!} + \cdots = \sum_{n=0}^{\infty} (-1)^n \frac{x^{2n}}{(2n)!}$$

11.7.3 常用的麦克劳林级数 Common Maclaurin Series

$$e^x = \sum_{n=0}^{\infty} \frac{x^n}{n!} = 1 + x + \frac{x^2}{2!} + \frac{x^3}{3!} + \cdots + \frac{x^n}{n!} + \cdots, |x| < \infty$$

$$\sin x = \sum_{n=0}^{\infty} (-1)^n \frac{x^{2n+1}}{(2n+1)!} = x - \frac{x^3}{3!} + \frac{x^5}{5!} - \cdots + (-1)^n \frac{x^{2n+1}}{(2n+1)!} + \cdots, |x| < \infty$$

$$\cos x = \sum_{n=0}^{\infty} (-1)^n \frac{x^{2n}}{(2n)!} = 1 - \frac{x^2}{2!} + \frac{x^4}{4!} - \cdots + (-1)^n \frac{x^{2n}}{(2n)!} + \cdots, \ |x| < \infty$$

$$\ln(1+x) = \sum_{n=0}^{\infty} (-1)^{n-1} \frac{x^n}{n} = x - \frac{x^2}{2} + \frac{x^3}{3} - \cdots + (-1)^{n-1} \frac{x^n}{n} + \cdots, \ -1 < x \leqslant 1$$

$$\arctan x = \sum_{n=0}^{\infty} (-1)^n \frac{x^{2n+1}}{2n+1} = x - \frac{x^3}{3} + \frac{x^5}{5} - \cdots + (-1)^n \frac{x^{2n+1}}{2n+1} + \cdots, \ |x| \leqslant 1$$

利用常用的麦克劳林级数，通过适当运算能将更多函数展开成幂级数．

Example 11.37 Find the Maclaurin series for function $f(x) = e^{-x}$.

Solution：

$$e^x = 1 + x + \frac{x^2}{2!} + \frac{x^3}{3!} + \cdots + \frac{x^n}{n!} + \cdots, \ |x| < \infty$$

Let $-x = x$, then

$$e^{-x} = 1 - x + \frac{x^2}{2!} - \frac{x^3}{3!} + \cdots + (-1)^n \frac{x^n}{n!} + \cdots$$

$$= \sum_{n=0}^{\infty} (-1)^n \frac{x^n}{n!}, \ |x| < \infty$$

11.7.4　泰勒多项式 Taylor Polynomial

The Taylor polynomial $P_n(x)$ of order n generated by $f(x)$ at $x = a$ is the polynomial：

$$P_n(x) = f(a) + f'(a)(x-a) + \frac{f''(a)}{2!}(x-a)^2 + \cdots + \frac{f^{(n)}(a)}{n!}(x-a)^n$$

The Maclaurin polynomial of order n is

$$P_n(x) = f(0) + f'(0)x + \frac{f''(0)}{2!}x^2 + \cdots + \frac{f^{(n)}(0)}{n!}x^n$$

我们说 n 阶泰勒多项式，而不说 n 次，这是因为 $f^{(n)}(a)$ 可能为零．同时，我们注意到 n 阶泰勒多项式和泰勒级数（Taylor Series）的区别：n 阶泰勒多项式是有限的 n 项，而泰勒级数是无限项的．

泰勒多项式为函数 $f(x)$ 在点 $x = a$ 处提供了近似函数，即 $f(x) \approx P_n(x)$．微分应用中的线性估算（Linear Approximation）等同于一阶泰勒多项式，即：$f(x) \approx f(a) + f'(a)(x-a)$，见 6.6.3 节．高阶泰勒多项式提供相应阶的最佳多项式估算．

Example 11.38 Find the Taylor polynomial of order 4 at $x = 1$ for $f(x) = \ln x$. Use this to approximate $f(1.5)$.

Solution：The first four derivatives are

$$f(x) = \ln x, f'(x) = \frac{1}{x}, f''(x) = -\frac{1}{x^2}, f^{(3)}(x) = \frac{2}{x^3}, f^{(4)}(x) = -\frac{6}{x^4}$$

And,

$$f(1) = 0, f'(1) = 1, f''(1) = -1, f^{(3)}(1) = 2, f^{(4)}(1) = -6$$

The approximating Taylor polynomial of order 4 is therefore

$$\ln x \approx (x-1) - \frac{(x-1)^2}{2} + \frac{(x-1)^3}{3} - \frac{(x-1)^4}{4}$$

With $x=1.5$ we have
$$\ln 1.5 \approx (1.5-1) - \frac{(1.5-1)^2}{2} + \frac{(1.5-1)^3}{3} - \frac{(1.5-1)^4}{4} \approx 0.4010$$
This approximation of ln1.5 is correct to four places.

Example 11.39 Find the Taylor polynomials P_2, P_4, P_6 and P_8, at $x=0$ for $f(x)=\cos x$ and graph $f(x)$ and all four polynomials in $\left[-\frac{\pi}{2}, \frac{3\pi}{2}\right] \times [-2,2]$.

Solution:

The derivatives of $f(x)=\cos x$ at 0 are given by the following table:

Order of deriv	0	1	2	3	4	5	6	7	8
Deriv of $\cos x$	$\cos x$	$-\sin x$	$-\cos x$	$\sin x$	$\cos x$	$-\sin x$	$-\cos x$	$\sin x$	$\cos x$
Deriv of $\cos x$ at 0	1	0	-1	0	1	0	-1	0	1

From the table, we know that:
$$P_2(x) = 1 - \frac{x^2}{2!}$$
$$P_4(x) = 1 - \frac{x^2}{2!} + \frac{x^4}{4!}$$
$$P_6(x) = 1 - \frac{x^2}{2!} + \frac{x^4}{4!} - \frac{x^6}{6!}$$
$$P_8(x) = 1 - \frac{x^2}{2!} + \frac{x^4}{4!} - \frac{x^6}{6!} + \frac{x^8}{8!}$$

Figure 11.7.4(a) show the graphs of $\cos x$ and $P_2(x)$, Figure 11.7.4(b) show the graphs of $\cos x$ and $P_4(x)$, Figure 11.7.4(c) show the graphs of $\cos x$ and $P_6(x)$. Figure 11.7.4(d) show the graphs of $\cos x$ and $P_8(x)$.

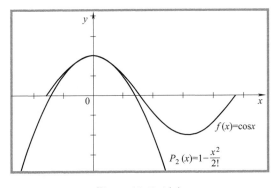

Figure 11.7.4(a)

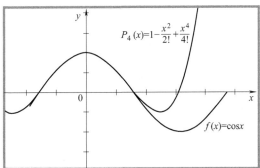

Figure 11.7.4(b)

11.7.5 拉格朗日误差界 Lagrange Error Bound

If a function $f(x)$ and its first $(n+1)$ derivatives are continuous on the interval $|x-a| < r$, then for each x in this interval

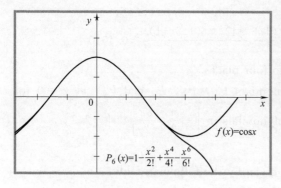
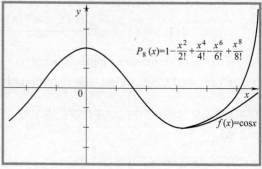

Figure 11.7.4(c)　　　　　　　　　Figure 11.7.4(d)

$$f(x)=f(a)+f'(a)(x-a)+\frac{f''(a)}{2!}(x-a)^2+\cdots+\frac{f^{(n)}(a)}{n!}(x-a)^n+R_n(x)$$

where

$$R_n(x)=\frac{f^{(n+1)}(c)(x-a)^{n+1}}{(n+1)!}$$

上述称为泰勒定理（Taylor's Theorem），$R_n(x)$ 为拉格朗日余项（Lagrange Remainder）。

对函数值 $f(x)$ 用其泰勒多项式 $P_n(x)$ 逼近的精确程度，可以使用拉格朗日余项 $R_n(x)$ 度量.

$$f(x)=P_n(x)+R_n(x)$$
准确值　近似值　余项

绝对值 $|R_n(x)|=|f(x)-P_n(x)|$ 称为相关逼近的误差（Error），并且，把下述称为拉格朗日误差界（Lagrange Error Bound），c 是介于 a 与 x 之间的实数：

$$R_n(x)<\max\left|\frac{f^{(n+1)}(c)(x-a)^{n+1}}{(n+1)!}\right|$$

Example 11.40 Find the third-degree Maclaurin polynomial for $f(x)=\sin\left(x+\frac{\pi}{4}\right)$, and determine the upper bound on the error in estimating $f(0.2)$.

Solution：

Make a table of the derivatives, evaluated at $x=0$ and giving us the coefficients：

n	$f^{(n)}(x)$	$f^{(n)}(0)$	$a_n=\dfrac{f^{(n)}(0)}{n!}$
0	$\sin\left(x+\dfrac{\pi}{4}\right)$	$\dfrac{\sqrt{2}}{2}$	$\dfrac{\sqrt{2}}{2}$
1	$\cos\left(x+\dfrac{\pi}{4}\right)$	$\dfrac{\sqrt{2}}{2}$	$\dfrac{\sqrt{2}}{2}$
2	$-\sin\left(x+\dfrac{\pi}{4}\right)$	$-\dfrac{\sqrt{2}}{2}$	$-\dfrac{\sqrt{2}}{2\times 2!}$
3	$-\cos\left(x+\dfrac{\pi}{4}\right)$	$-\dfrac{\sqrt{2}}{2}$	$-\dfrac{\sqrt{2}}{2\times 3!}$

Thus,

$$P_3(x) = \frac{\sqrt{2}}{2} + \frac{\sqrt{2}}{2}x - \frac{\sqrt{2}}{2\times 2!}x^2 - \frac{\sqrt{2}}{2\times 3!}x^3$$

Use the Lagrange error bound:

$$R_n(x) < \max\left|\frac{f^{(4)}(c)}{4!}x^4\right|, \text{ where } x = \frac{\pi}{2} \text{ and } 0 < c < 0.2$$

Since $f^{(4)}(c) = \sin\left(c + \frac{\pi}{4}\right)$ is increasing on the interval $0 < c < 0.2$, so its maximum value occurs at $c = 0.2$.

Consequently,

$$R_n(x) < \max\left|\frac{\sin\left(c + \frac{\pi}{4}\right)}{4!}x^4\right| = \frac{\sin\left(0.2 + \frac{\pi}{4}\right)}{4!} \times (0.2)^4 \approx 0.000056$$

11.8 幂级数的计算 Computations with Power Series

幂级数（Power Series）对于函数的逼近值、不确定形式极限的估计和定积分的估计都是非常有用，还能推导出数学上著名的欧拉公式（Euler's Formula）。

11.8.1 近似计算 Approximating Values

Example 11.41 Show how a series may be used to evaluate π.

Solution：

By the Geometric Series Test，we have that

$$\frac{1}{1-x} = 1 + x + x^2 + x^3 + \cdots + x^n + \cdots$$

Let $x = -t^2$, we obtain

$$\frac{1}{1+t^2} = 1 - t^2 + t^4 - t^6 + \cdots (-1)^n t^{2n} + \cdots$$

$$\int_0^x \frac{1}{1+t^2}dt = \int_0^x [1 - t^2 + t^4 - t^6 + \cdots (-1)^n t^{2n} + \cdots]dt$$

And

$$\int_0^x \frac{1}{1+t^2}dt = \arctan x$$

We get

$$\arctan x = x - \frac{x^3}{3} + \frac{x^5}{5} - \cdots + (-1)^n \frac{x^{2n+1}}{2n+1} + \cdots, \ |x| \leq 1$$

Since

$$\frac{\pi}{4} = \arctan 1$$

For $x = 1$, we have

$$\arctan 1 = 1 - \frac{1}{3} + \frac{1}{5} - \frac{1}{7} + \cdots$$

Then,

$$\frac{\pi}{4} = 1 - \frac{1}{3} + \frac{1}{5} - \frac{1}{7} + \cdots$$

And

$$\pi = 4 \times \left(1 - \frac{1}{3} + \frac{1}{5} - \frac{1}{7} + \cdots\right)$$

Here are some approximations for π using this series:

Number of terms	1	2	5	10	25	50	59	60
approximation	4	2.67	3.34	3.04	3.18	3.12	3.16	3.12

11.8.2 求函数的近似值 Approximating Values of Functions

Example 11.42 Approximate the function $f(x) = \sqrt[3]{x}$ by a Taylor polynomial of degree 2 at $a = 8$.

Solution:

Make a table of the derivatives, evaluated at $x = 8$ and giving us the coefficients:

n	$f^{(n)}(x)$	$f^{(n)}(8)$	$a_n = \dfrac{f^{(n)}(0)}{n!}$
0	$x^{\frac{1}{3}}$	2	2
1	$\frac{1}{3}x^{-\frac{2}{3}}$	$\frac{1}{12}$	$\frac{1}{12}$
2	$-\frac{2}{9}x^{-\frac{5}{3}}$	$-\frac{1}{144}$	$-\frac{1}{144 \times 2!}$

Thus,

$$P_2(x) = 2 + \frac{1}{12}(x-8) - \frac{1}{144 \times 2!}(x-8)^2$$

11.8.3 求不定形式极限 Evaluating Indeterminate Forms of Limits

Example 11.43 Use a series to evaluate $\lim\limits_{x \to 0}\left(\dfrac{1}{\sin x} - \dfrac{1}{x}\right)$

Solution:

$$\sin x = x - \frac{x^3}{3!} + \frac{x^5}{5!} - \cdots + (-1)^n \frac{x^{2n+1}}{(2n+1)!} + \cdots$$

Then,

$$\frac{1}{\sin x} - \frac{1}{x} = \frac{x - \sin x}{x \sin x} = \frac{x - \left(x - \dfrac{x^3}{3!} + \dfrac{x^5}{5!} - \cdots\right)}{x\left(x - \dfrac{x^3}{3!} + \dfrac{x^5}{5!} - \cdots\right)}$$

$$= \frac{x^3\left(\dfrac{1}{3!} - \dfrac{x^2}{5!} + \cdots\right)}{x^2\left(1 - \dfrac{x^2}{3!} + \dfrac{x^4}{5!} - \cdots\right)} = x \cdot \frac{\dfrac{1}{3!} - \dfrac{x^2}{5!} + \cdots}{1 - \dfrac{x^2}{3!} + \dfrac{x^4}{5!} - \cdots}$$

Thus,
$$\lim_{x\to 0}\left(\frac{1}{\sin x}-\frac{1}{x}\right)=\lim_{x\to 0}\left(x\,\frac{\frac{1}{3!}-\frac{x^2}{5!}+\cdots}{1-\frac{x^2}{3!}+\frac{x^4}{5!}-\cdots}\right)=0$$

11.8.4 估计定积分 Estimating Definite Integrals

Example 11.44 Use a series to evaluate $\int_0^{0.2} e^{-x^2}\,dx$ to four decimal places.

Solution:
$$e^x = 1 + x + \frac{x^2}{2!} + \frac{x^3}{3!} + \cdots + \frac{x^n}{n!} + \cdots$$

Let $x = -x^2$, then
$$e^{-x^2} = 1 - x^2 + \frac{x^4}{2!} - \frac{x^6}{3!} + \cdots + (-1)^n \frac{x^{2n}}{n!} + \cdots$$

So
$$\int_0^{0.2} e^{-x^2}\,dx = \int_0^{0.2}\left[1 - x^2 + \frac{x^4}{2!} - \frac{x^6}{3!} + \cdots + (-1)^n \frac{x^{2n}}{n!} + \cdots\right]dx$$
$$= \left(x - \frac{x^3}{3} + \frac{x^5}{5\times 2!} - \frac{x^7}{7\times 3!} + \cdots\right)\Big|_0^{0.2}$$
$$= 0.2 - \frac{0.008}{3} + \frac{0.00032}{10} - \frac{0.000013}{42} + \cdots$$
$$= 0.2 - 0.002667 + 0.000032 + R_6$$
$$= 0.20269 + R_6$$

Since this is a convergent alternating series, $|R_6| < 3\times 10^{-7}$, which will not affect the fourth decimal place. Then, correct to four decimal places,
$$\int_0^{0.2} e^{-x^2}\,dx = 0.2027$$

11.8.5 欧拉公式 Euler's Formula

Example 11.45 Show that the Euler's Formula $e^{i\pi} + 1 = 0$

Solution:
$$e^x = 1 + x + \frac{x^2}{2!} + \frac{x^3}{3!} + \cdots + \frac{x^n}{n!} + \cdots$$

Let $x = yi$
$$e^{yi} = 1 + yi + \frac{(yi)^2}{2!} + \frac{(yi)^3}{3!} + \frac{(yi)^4}{4!} + \cdots$$
$$= 1 + yi - \frac{y^2}{2!} - \frac{y^3 i}{3!} + \frac{y^4}{4!} + \cdots$$
$$= \left(1 - \frac{y^2}{2!} + \frac{y^4}{4!} + \cdots\right) + i\left(y - \frac{y^3}{3!} + \frac{y^5}{5!} + \cdots\right)$$

Since

$$\sin x = x - \frac{x^3}{3!} + \frac{x^5}{5!} - \cdots + (-1)^n \frac{x^{2n+1}}{(2n+1)!} + \cdots$$

And

$$\cos x = 1 - \frac{x^2}{2!} + \frac{x^4}{4!} - \cdots + (-1)^n \frac{x^{2n}}{(2n)!} + \cdots$$

Then

$$e^{yi} = \cos y + i \sin y$$

When $y = \pi$

$$e^{i\pi} + 1 = 0$$

Note：欧拉公式把数学中最重要的几个常数——自然对数的底数 e，圆周率 π，虚数单位 i，以及 0 和 1 联系到一起，令人着迷，以微积分展现了"数学之美"．

11.9 习题 Practice Exercises

(1) Find the sum of the series $\sum_{n=1}^{\infty} \frac{1}{n(n+1)}$.

(2) Does $1 - \frac{1}{2} + \frac{1}{4} - \frac{1}{8} + \cdots + \left(-\frac{1}{2}\right)^{n-1} + \cdots$ converge or diverge?

(3) Does $1 + \frac{1}{4} + \frac{1}{9} + \frac{1}{16} + \cdots + \frac{1}{n^2} + \cdots$ converge or diverge?

(4) Use the alternating series test to determine the series $\frac{1}{\sqrt{2}} - \frac{1}{\sqrt{3}} + \frac{1}{\sqrt{4}} - \frac{1}{\sqrt{5}} + \cdots$ converges.

(5) Use the ratio test to determine the series $\sum_{n=1}^{\infty} \frac{n+1}{n!}$ converges.

(6) Use the comparison test to determine the series $\sum_{n=1}^{\infty} \frac{2}{n^2+3}$ converges.

(7) Test the series $\sum_{n=1}^{\infty} (-1)^n \frac{n^2}{2^n}$ for absolute convergence.

(8) For what values of x does the series $\sum_{n=1}^{\infty} \frac{(x-3)^n}{n}$.

(9) Find the radius of convergence and interval of convergence of the series $\sum_{n=1}^{\infty} \frac{n(x+1)^n}{2^{n+1}}$.

(10) Find the Maclaurin series for function $f(x) = \ln(1+x)$.

(11) Find the Taylor polynomial of order 3 at $x = 1$ for e^x.

(12) Use a Maclaurin series to evaluate $e^{0.1}$ to three decimal places.

习题答案
Practice Answer

第 2 章

(1) 1) $\{x \mid x > -1, x \neq 1\}$ 2) $\{x \mid x < -1, 1 < x < 2\}$ 3) 全体实数

(2) The range of $f(g(x))$ is $\{y \mid y \leq \ln 25\}$

(3) 1) $f(-x) = (-x)^3 - (-x) = -x^3 + x = -(x^3 - x) = -f(x)$

Since $f(-x) = -f(x)$, $f(x)$ is an odd function.

2) $f(-x) = (-x)^6 - (-x)^2 + 7 = x^6 - x^2 + 7 = f(x)$

Since $f(-x) = f(x)$, $f(x)$ is an even function.

3) $f(-x) = (-x) + (-x)^2 = -x - x^2$

Since $f(-x) \neq f(x)$, and $f(-x) \neq -f(x)$, the function $f(x)$ is neither even nor odd.

(4) Express both sides with the same base, we get

$$2^{5x-7} = (2^3)^{x+1}$$
$$2^{5x-7} = 2^{3x+3}$$

Since exponential function are one-to-one, then

$$5x - 7 = 3x + 3$$
$$x = 5$$

(5) $\log_2(4+x) = 4$ change to exponential form

$$4 + x = 2^4$$

Solve for x

$$x = 12$$

(6) By Double-Angle formula,

$$f(x) = \frac{1}{2}\cos^2 x = \frac{1}{2} \times \frac{1+\cos 2x}{2} = \frac{1}{4} + \frac{1}{4}\cos 2x$$

The period T:

$$T = \frac{2\pi}{\omega} = \frac{2\pi}{2} = \pi$$

(7) We consider the equation

$$y = x^2 - 2$$

and solve for x, obtaining

$$x = \pm\sqrt{y+2}$$

Since x is nonnegative, we reject $x=-\sqrt{y+2}$ and let
$$f^{-1}(x)=\sqrt{x+2}$$

(8) Tangent and cotangent identities
$$\sin x(\tan x+\cot x)=\sin x\left(\frac{\sin x}{\cos x}+\frac{\cos x}{\sin x}\right)$$

and fractions
$$\sin x(\tan x+\cot x)=\sin x\left(\frac{\sin^2 x+\cos^2 x}{\sin x\cos x}\right)$$

Pythagorean identity
$$\sin x(\tan x+\cot x)=\sin x\left(\frac{1}{\sin x\cos x}\right)$$

Cancel $\sin x$
$$\sin x(\tan x+\cot x)=\frac{1}{\cos x}$$

Reciprocal identity
$$\sin x(\tan x+\cot x)=\sec x$$

(9) 1) $\arcsin\left(\sin\frac{\pi}{3}\right)=\arcsin\left(\frac{\sqrt{3}}{2}\right)=\frac{\pi}{3}$

2) $\arccos\left[\sin\left(-\frac{\pi}{3}\right)\right]=\arccos\left(-\frac{\sqrt{3}}{2}\right)=\frac{5\pi}{6}$

(10) Finding composite function,
$$f\circ g=f(g(x))=f(\sqrt{x})=(\sqrt{x})^2-3=x^2-3$$

the domain of $f\circ g$ is set of all x in $[0,+\infty]$.

(11) Using Trigonometric Identify $\sin^2 t+\cos^2 t=1$:
$$\cos t=\frac{x}{2} \text{ and } \sin t=\frac{y}{2}$$

Then,
$$\left(\frac{x}{2}\right)^2+\left(\frac{y}{2}\right)^2=1$$

Therefore,
$$x^2+y^2=4$$

This is a circle, centered at the origin, with radius 2.

第 3 章

(1) $\lim\limits_{x\to 0}\sqrt{\pi}=\sqrt{\pi}$

(2) $\lim\limits_{x\to -1}(3x^2+9)=\lim\limits_{x\to -1}[3(-1)^2+9]=12$

(3) $\lim\limits_{x\to 1}(2x^2-1)(\sqrt{8x^2+1})=\lim\limits_{x\to 1}(2\times 1^2-1)(\sqrt{8\times 1^2+1})=3$

(4) $\lim\limits_{x\to 2}\left(\frac{3x+1}{x-9}\right)=\lim\limits_{x\to 2}\left(\frac{3\times 2+1}{2-9}\right)=-1$

(5) $\lim\limits_{x\to\infty}\dfrac{x+1}{\sqrt{6x-1}}=\lim\limits_{x\to\infty}\dfrac{\frac{x+1}{\sqrt{x}}}{\frac{\sqrt{6x-1}}{\sqrt{x}}}=\lim\limits_{x\to\infty}\dfrac{\sqrt{x}+\frac{1}{\sqrt{x}}}{\sqrt{6-\frac{1}{x}}}=\infty$

(6) $\lim\limits_{x\to 0^-}\left(\dfrac{|x|}{x}\right)=-1$

(7) $\lim\limits_{x\to 0^+}\left(\dfrac{|x|}{x}\right)=1$

(8) The limit does Not Exist.

Since $\lim\limits_{x\to 4}\left(\dfrac{x+3}{x^2-x-12}\right)=\lim\limits_{x\to 4}\dfrac{x+3}{(x+3)(x-4)}=\lim\limits_{x\to 4}\dfrac{1}{x-4}$,

$\lim\limits_{x\to 4^-}\dfrac{1}{x-4}=-\infty$ and $\lim\limits_{x\to 4^+}\dfrac{1}{x-4}=+\infty$. So, the limit of $\lim\limits_{x\to 4}\left(\dfrac{x+3}{x^2-x-12}\right)$ does not exist.

(9) The limit does Not Exist.

(10) $\lim\limits_{x\to\infty}\dfrac{8x^4+64x+1}{x^5+4\sqrt{x}+9}=0$

(11) $\lim\limits_{x\to\infty}\dfrac{x^4+64x+5x^5}{x^5+x^2+9}=5$

(12) $\lim\limits_{x\to 0}\dfrac{\tan x}{x}=\lim\limits_{x\to 0}\dfrac{\sin x}{x\cos x}=\dfrac{\lim\limits_{x\to 0}\frac{\sin x}{x}}{\lim\limits_{x\to 0}\cos x}=1$

(13) $\lim\limits_{x\to 0}\dfrac{1-\cos x}{x^2}=\lim\limits_{x\to 0}\dfrac{2\sin^2\frac{x}{2}}{x^2}=\lim\limits_{x\to 0}\dfrac{2\sin^2\frac{x}{2}}{4\left(\frac{x}{2}\right)^2}=\dfrac{1}{2}\lim\limits_{x\to 0}\left(\dfrac{\sin\frac{x}{2}}{\frac{x}{2}}\right)^2=\dfrac{1}{2}\times 1^2=\dfrac{1}{2}$

(14) $\lim\limits_{x\to 0}\dfrac{(4+x)^2-16}{x}=\lim\limits_{x\to 0}\dfrac{16+8x+x^2-16}{x}=\lim\limits_{x\to 0}(8+x)=8$

(15) $\lim\limits_{x\to 0}\dfrac{\sqrt{x^2+16}-4}{x^2}=\lim\limits_{x\to 0}\dfrac{(\sqrt{x^2+16}-4)(\sqrt{x^2+16}+4)}{x^2(\sqrt{x^2+16}+4)}=\lim\limits_{x\to 0}\dfrac{1}{\sqrt{x^2+16}+4}=\dfrac{1}{8}$

(16) 1) $\lim\limits_{x\to 9^-}f(x)=3$; 2) $\lim\limits_{x\to 9^+}f(x)=3$; 3) $\lim\limits_{x\to 9}f(x)=3$

(17) 1) $\lim\limits_{x\to 9^-}f(x)=3$; 2) $\lim\limits_{x\to 9^+}f(x)=4$; 3) The limit does Not Exist.

(18) (B)

(19) (C)

(20) The graph of the function $f(x)=\arctan x$ has horizontal asymptotes at $y=\pm\dfrac{\pi}{2}$.

第 4 章

(1) Using the definition, we get

$$f(0)=-2 \text{ and } \lim\limits_{x\to 0^-}f(x)=\lim\limits_{x\to 0^+}f(x)=-2$$

Then
$$\lim_{x \to 0} f(x) = f(0) = -2$$

So, $f(x)$ is continuous at $x=0$.

(2) Since $\lim\limits_{x \to 0^-} f(x) = \lim\limits_{x \to 0^-} x^2 = 0$ and $\lim\limits_{x \to 0^+} f(x) = \lim\limits_{x \to 0^+} (x-2) = -2$

Then, $\lim\limits_{x \to 0} f(x)$ not exists.

So, $f(x)$ is not continuous at $x=0$.

Look at the graph of $f(x)$, see Figure 4.4.2.

(3) 1) $x=\pm 3$; 2) $x=3$; 3) $x=0$.

(4) $a=2$, $b=3$

(5) removable discontinuity: $x=4$; jump discontinuity: $x=0$, $x=2$

(6) No, it is discontinuous at the endpoints of the interval.

(7) $x \neq 1$.

Notice that $f(x)$ can be broken up as the composition of two continuous functions: $f(x) = g \circ h$, where $g(x) = \dfrac{1}{x}$ $h(x) = x-1$, so, $x-1 \neq 0$.

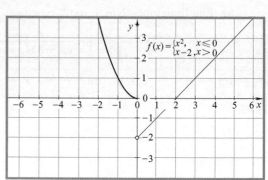

Figure 4.4.2

(8) Let $f(x) = x^5 - 3x$.

The function $f(x)$ is continuous on the closed interval $\left[\dfrac{1}{2}, 1\right]$.

take $a=1$, $b=2$ and $M=1$ in The Intermediate Value Theorem. We have
$$f(1) = (1)^5 - 3 \times 1 = -2$$
and
$$f(2) = (2)^5 - 2 \times 3 = 26$$

Thus $-2 < M < 26$. So, The Intermediate Value Theorem says there is a number ξ between 1 and 2 such that $f(\xi) = 1$.

In other words, the equation $x^5 - 3x = 1$ has at least one root ξ in the interval $(1, 2)$.

第5章

(1) 1) 注意 x^π 求导
$$f'(x) = 7x^6 + \pi x^{\pi-1}$$

2) 注意熟练 $\tan x$、$\cot x$、$\sec x$、$\csc x$ 等求导公式
$$f'(x) = \sec^2 x - \csc^2 x$$

3) 注意熟练 a^x、$\log_a x$ 等求导公式
$$f'(x) = \dfrac{1}{x \ln 2} + 5^x \ln 5$$

4）注意 π^x 求导
$$f'(x) = -\sin x + \pi^x \ln \pi$$

5) By the Sum Rule, we have
$$f'(x) = 3x^2 + 2x + 1$$

(2) A

(3) 1) Using the Product Rule, we have
$$y' = (x^3)' \sin x + x^3 (\sin x)' = 3x^2 \sin x + x^3 \cos x$$

2) By the Quotient Rule, we have
$$y' = \frac{(1-x)'(3x+1) - (1-x)(3x+1)'}{(3x+1)^2}$$
$$= \frac{-(3x+1) - 3(1-x)}{(3x+1)^2} = \frac{-4}{(3x+1)^2}$$

3) use the Quotient Rule,
$$y' = \frac{(\sqrt{x})' \sin x - \sqrt{x} (\sin x)'}{\sin^2 x}$$
$$= \frac{\frac{1}{2\sqrt{x}} \sin x - \sqrt{x} \cos x}{\sin^2 x} = \frac{\sin x - 2x \cos x}{2\sqrt{x} \sin^2 x}$$

4) use the Quotient Rule,
$$y' = \frac{(e^x - e^{-x})'(e^x + e^{-x}) - (e^x - e^{-x})(e^x + e^{-x})'}{(e^x + e^{-x})^2}$$
$$= \frac{(e^x + e^{-x})(e^x + e^{-x}) - (e^x - e^{-x})(e^x - e^{-x})}{(e^x + e^{-x})^2}$$
$$= \frac{(e^x + e^{-x})^2 - (e^x - e^{-x})^2}{(e^x + e^{-x})^2}$$
$$= \frac{(e^x + e^{-x} + e^x - e^{-x})(e^x + e^{-x} - e^x + e^{-x})}{(e^x + e^{-x})^2}$$
$$= \frac{2e^x \times 2e^{-x}}{(e^x + e^{-x})^2} = \frac{4}{(e^x + e^{-x})^2}$$

5) By the Difference Rule, we have
$$y' = (\sqrt{x})' - \left(\frac{1}{\sqrt{x}}\right)' = \frac{1}{2\sqrt{x}} + \frac{1}{2x\sqrt{x}}$$

(4) 1) Let $y = \cos u$, $u = \frac{1}{x}$
$$y' = \cos' u \cdot u' = -\sin u \left(-\frac{1}{x^2}\right) = \frac{1}{x^2} \sin\left(\frac{1}{x}\right)$$

2) Let $y = \ln u$, $u = \sqrt{v}$, $v = x^3 + 1$
$$y' = \ln' u \cdot u' = \frac{1}{u} \times \frac{1}{2\sqrt{v}} v' = \frac{1}{\sqrt{x^3+1}} \times \frac{1}{2\sqrt{x^3+1}} (x^3+1)' = \frac{3x^2}{2(x^3+1)}$$

3) Using the Chain Rule and the Product Rule,

$$y' = \ln^3 x + x \cdot 3\ln^2 x \cdot \frac{1}{x} = \ln^3 x + 3\ln^2 x$$

4) Let $y = 3^u$, $u = x^2$
$$y' = (3^u)'(x^2)' = 3^{x^2} \ln 3 \, (x^2)' = 3^{x^2} \ln 3 (2x)$$

5) 注意熟练 arcsin$x$、arccos$x$ 和arctan$x$ 等求导公式
$$y' = (\arcsin x)' - (\sqrt{1-x^2})' = \frac{1}{\sqrt{1-x^2}} - \frac{(1-x^2)'}{2\sqrt{1-x^2}} = \frac{1+x}{\sqrt{1-x^2}}$$

(5) Take the derivative of $f(x)$:
$$\frac{dy}{dx} = 2x + 1$$

Find the value of x when $y = 12$:
$$x^2 + x = 12, \quad x = 3$$

Using the derivative of an inverse function formula:
$$\frac{d}{dx} f^{-1}(y) \Big|_{y=12} = \frac{1}{(2x+1)\big|_{x=3}} = \frac{1}{7}$$

(6) Using implicit differentiation, we get:
$$3y^2 \frac{dy}{dx} + \left(3y + 3x \frac{dy}{dx}\right) - 2x = 0$$

Put all the terms containing $\frac{dy}{dx}$ on the left and all the other terms on the right:
$$3y^2 \frac{dy}{dx} + 3x \frac{dy}{dx} = 2x - 3y$$

Factor out $\frac{dy}{dx}$:
$$\frac{dy}{dx}(3y^2 + 3x) = 2x - 3y$$

Then, isolate $\frac{dy}{dx}$:
$$\frac{dy}{dx} = \frac{2x - 3y}{3y^2 + 3x}$$

(7) Using implicit differentiation, we get:
$$2y \frac{dy}{dx} = 2x$$

Simplify and solve for $\frac{dy}{dx}$:
$$\frac{dy}{dx} = \frac{2x}{2y} = \frac{x}{y}$$

Take derivative again:
$$\frac{d^2 y}{dx^2} = \frac{1 \times y - x \frac{dy}{dx}}{y^2} = \frac{y - x \frac{dy}{dx}}{y^2}$$

Substitute for $\dfrac{dy}{dx}$ and simplify:

$$\dfrac{d^2y}{dx^2} = \dfrac{y - x \dfrac{x}{y}}{y^2} = \dfrac{y^2 - x^2}{y^3}$$

(8) Take the log of the sides:

$$\ln y = \ln \dfrac{(x^2-3x)^2(5x^3+2x+1)^3}{(x^2+x)^2}$$

Using the logarithmic formula, we get

$$\ln y = \ln(x^2-3x)^2(5x^3+2x+1)^3 - \ln(x^2+x)^2$$
$$\ln y = \ln(x^2-3x)^2 + \ln(5x^3+2x+1)^3 - \ln(x^2+x)^2$$
$$\ln y = 2\ln(x^2-3x) + 3\ln(5x^3+2x+1) - 2\ln(x^2+x)$$

Take the derivative of both sides:

$$\dfrac{1}{y} \times \dfrac{dy}{dx} = 2 \times \dfrac{2x-3}{x^2-3x} + 3 \times \dfrac{15x^2+2}{5x^3+2x+1} - 2 \times \dfrac{2x+1}{x^2+x}$$

Multiply both sides by y:

$$\dfrac{dy}{dx} = y\left(2 \times \dfrac{2x-3}{x^2-3x} + 3 \times \dfrac{15x^2+2}{5x^3+2x+1} - 2 \times \dfrac{2x+1}{x^2+x}\right)$$

(9) Take the derivative of $x = x(t)$, $y = y(t)$ with respect to t:

$$\dfrac{dx}{dt} = 3t \text{ and } \dfrac{dy}{dt} = 12$$

Then, the derivative of $y = y(x)$ with respect to x:

$$\dfrac{dy}{dx} = \dfrac{\dfrac{dy}{dt}}{\dfrac{dx}{dt}} = \dfrac{12}{3t} = \dfrac{4}{t}$$

At $t = 3$,

$$\dfrac{dy}{dx}\bigg|_{t=3} = \dfrac{4}{3}, x = 31, y = 36$$

The equation of the tangent line:

$$y - 36 = \dfrac{4}{3}(x - 31)$$

(10)# The velocity vector is the derivative of the position vector, then, using the formula:

$$\overrightarrow{R'(t)} = \dfrac{d\overrightarrow{R}}{dt} = \left\langle 6\cos\dfrac{\pi}{6}t, 3\sin\dfrac{\pi}{3}t \right\rangle' = \left\langle -\pi\sin\dfrac{\pi}{6}t, \pi\cos\dfrac{\pi}{3}t \right\rangle$$

When $t = 3$, the velocity vector is

$$\overrightarrow{v(3)} = \overrightarrow{R'(3)} = \left\langle -\pi\sin\dfrac{\pi}{6} \times 3, \pi\cos\dfrac{\pi}{3} \times 3 \right\rangle = \langle -\pi, -\pi \rangle$$

(11)# Take the derivative $r(\theta)$ with respect to θ:

$$\dfrac{dr}{d\theta} = (2^\theta)' = 2^\theta \ln 2$$

Using the formula:

$$\frac{dy}{dx} = \frac{\frac{dr}{d\theta}\sin\theta + r\cos\theta}{\frac{dr}{d\theta}\cos\theta - r\sin\theta} = \frac{2^\theta \ln 2 \sin\theta + 2^\theta \cos\theta}{2^\theta \ln 2 \cos\theta - 2^\theta \sin\theta} = \frac{\ln 2 \sin\theta + \cos\theta}{\ln 2 \cos\theta - \sin\theta}$$

(12) Find $f'(x)$:

$$f'(x) = 2x\sin x + x^2 \cos x$$

Then,

$$dy = f'(x)dx = (2x\sin x + x^2 \cos x)dx$$

Substituting $x = \frac{\pi}{2}$ and $dx = 0.1$ in the expression for dy, we obtain

$$dy = \left(2 \times \frac{\pi}{2} \sin\frac{\pi}{2} + \left(\frac{\pi}{2}\right)^2 \cos\frac{\pi}{2}\right) \times 0.1 = (\pi + 0) \times 0.1 \approx 0.3142$$

(13) 1) Let $u = x^2$, then $y = \ln u$

$$dy = \frac{1}{u}du = \frac{1}{x^2}d(x^2) = \frac{1}{x^2} \times 2x\,dx = \frac{2}{x}dx$$

2) Using the Product Rule, we have

$$dy = \cos x\, d(e^{2x}) + e^{2x} d(\cos x)$$
$$= \cos x\, e^{2x} d(2x) + e^{2x}(-\sin x)dx$$
$$= 2\cos x\, e^{2x} dx - \sin x\, e^{2x} dx$$
$$= e^{2x}(2\cos x - \sin x)dx$$

3) Let $u = \frac{x}{3}$, then $y = \arctan u$

$$dy = \frac{1}{1+u^2}du = \frac{1}{1+\left(\frac{x}{3}\right)^2}d\left(\frac{x}{3}\right) = \frac{1}{3} \times \frac{1}{1+\left(\frac{x}{3}\right)^2}dx = \frac{3}{9+x^2}dx$$

4) Let $u = \sec x + \tan x$, then $y = \ln u$, and

$$du = d(\sec x + \tan x) = d(\sec x) + d(\tan x)$$
$$= \sec x \tan x\, dx + \sec^2 x\, dx = \sec x(\tan x + \sec x)dx$$
$$= \frac{1}{\cos x}\left(\frac{\sin x}{\cos x} + \frac{1}{\cos x}\right)dx = \frac{\sin x + 1}{\cos^2 x}dx$$

Thus,

$$dy = \frac{1}{u}du = \frac{1}{\sec x + \tan x} \times \frac{\sin x + 1}{\cos^2 x}dx$$
$$= \frac{\cos x}{1 + \sin x} \times \frac{\sin x + 1}{\cos^2 x}dx$$
$$= \frac{1}{\cos x}dx = \sec x\, dx$$

5) Let $u = x^3 + 2$, then $y = \sqrt{u}$

$$dy = \frac{1}{2\sqrt{u}}du = \frac{3x^2}{2\sqrt{x^3+2}}dx$$

(14) Since $A = \pi r^2$, the estimated increase is

$$dA = A'(10)dr = 2\pi \times 10 \times (0.01 \times 10) \approx 6.2832$$

(15) Take the derivative $r(\theta)$ with respect to θ:
$$\frac{dr}{d\theta} = (1+\cos\theta)' = -\sin\theta$$

Using the formula:
$$\frac{dy}{dx} = \frac{\dfrac{dr}{d\theta}\sin\theta + r\cos\theta}{\dfrac{dr}{d\theta}\cos\theta - r\sin\theta} = \frac{-\sin\theta\sin\theta + (1+\cos\theta)\cos\theta}{-\sin\theta\cos\theta - (1+\cos\theta)\sin\theta}$$

The slope of the tangent at the point where $\theta = \dfrac{\pi}{6}$
$$\left.\frac{dy}{dx}\right|_{\theta=\frac{\pi}{3}} = \frac{-\sin\dfrac{\pi}{6}\sin\dfrac{\pi}{6} + \left(1+\cos\dfrac{\pi}{6}\right)\cos\dfrac{\pi}{6}}{-\sin\dfrac{\pi}{6}\cos\dfrac{\pi}{6} - \left(1+\cos\dfrac{\pi}{6}\right)\sin\dfrac{\pi}{6}} = -1$$

Express the cardioid in parametric equation. We have
$$\begin{cases} x = r\cos\theta = (1+\cos\theta)\cos\theta \\ y = r\sin\theta = (1+\cos\theta)\sin\theta \end{cases}$$

Plug $\theta = \dfrac{\pi}{6}$ into the parametric equation
$$\begin{cases} x = \left(1+\cos\dfrac{\pi}{6}\right)\cos\dfrac{\pi}{6} = \dfrac{3+2\sqrt{3}}{4} \\ y = \left(1+\cos\dfrac{\pi}{6}\right)\sin\dfrac{\pi}{6} = \dfrac{2+\sqrt{3}}{4} \end{cases}$$

The equation of the tangent line is:
$$y - \frac{2+\sqrt{3}}{4} = -1 \times \left(x - \frac{3+2\sqrt{3}}{4}\right)$$
$$4x + 4y - (5 + 3\sqrt{3})$$

第 6 章

(1) Find the derivative of the function
$$y' = 4x + 3$$

Plug $x = 1$ into y':
$$y'|_{x=1} = 4 \times 1 + 2 = 6$$

The equation of the tangent line is:
$$y - 6 = 6(x - 1)$$
$$y = 6x$$

(2) Using implicit differentiation, we get:
$$2y\frac{dy}{dx} + 2x = y + x\frac{dy}{dx}$$

Simplify:
$$\frac{dy}{dx} = \frac{y-2x}{2y-x}$$

Since the tangent is parallel to the x-axis when $\frac{dy}{dx}=0$.

Then,
$$y=2x$$

substitute $y=2x$ in the equation of the curve, we get
$$(2x)^2+x^2=x(2x)+9$$

Thus $y=\pm 2\sqrt{3}$ and $x=\pm\sqrt{3}$.

The point, then, are $(\sqrt{3}, 2\sqrt{3})$ and $(-\sqrt{3}, -2\sqrt{3})$.

(3) Take the derivative of $x=x(t)$ and $y=y(t)$ with respect to t:
$$\frac{dx}{dt}=2\sin t \cos t, \frac{dy}{dt}=-\sin t$$

Then, the derivative of $y=y(x)$ with respect to x:
$$\frac{dy}{dx}=\frac{\frac{dy}{dt}}{\frac{dx}{dt}}=\frac{-\sin t}{2\sin t \cos t}=-\frac{1}{2\cos t}$$

Since $\begin{cases} x=\frac{3}{4} \\ y=\frac{1}{2} \end{cases}$ when $t=\frac{\pi}{3}$ and the slope at this point is:

$$-\frac{1}{\frac{dy}{dx}\big|_{t=\frac{\pi}{3}}} = -\left(-\frac{1}{2\times\frac{1}{2}}\right)=1$$

The equation of the normal line:
$$y-\frac{1}{2}=x-\frac{3}{4} \text{ or } 4x-4y-1=0$$

(4) The domain of function $f(x)$ is R and differentiable on the domain. Find the derivative of the function
$$f'(x)=6x^2+2x$$

Which is zero when $x=-\frac{1}{3}$ or $x=0$. We analyze the signs of $f'(x)$ in three intervals:

x	$\left(-\infty, -\frac{1}{3}\right)$	$\left(-\frac{1}{3}, 0\right)$	$(0, +\infty)$
$f'(x)$	+	−	+
$f(x)$	increasing	decreasing	increasing

Sketch the graph of $f(x)$, see Figure 6.8.4.

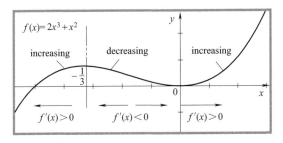

Figure 6.8.4

Thus,

$f(x)$ is increasing on $\left(-\infty, -\dfrac{1}{3}\right) \cup (0, +\infty)$ and decreasing on $\left(-\dfrac{1}{3}, 0\right)$.

(5) Take the derivative and set it equal to zero:

$$\dfrac{\mathrm{d}y}{\mathrm{d}x} = 3x^2 - 3 = 0$$

Solve for x:

$$x_1 = 1 \text{ and } x_2 = -1$$

We analyze the signs of $f'(x)$ in three intervals:

x	$[-4, -1)$	-1	$(-1, 1)$	1	$(1, 4]$
$f'(x)$	$+$	0	$-$	0	$+$
$f(x)$	↗	3	↘	-1	↗

Then, the local minimum values of $f(x)$ is -1 and the local maximum values of $f(x)$ is 3.

(6) Take the derivative and set it equal to zero:

$$\dfrac{\mathrm{d}y}{\mathrm{d}x} = 6x^2 - 1 = 0$$

Solve for x:

$$x_1 = \dfrac{\sqrt{6}}{6} \text{ and } x_2 = -\dfrac{\sqrt{6}}{6}$$

Take the second derivative of the function:

$$\dfrac{\mathrm{d}^2 y}{\mathrm{d}x^2} = 12x$$

Since $\dfrac{\mathrm{d}^2 y}{\mathrm{d}x^2}\bigg|_{x=\frac{\sqrt{6}}{6}} = 6 \times \dfrac{\sqrt{6}}{6} > 0$, $f(x)$ has a local minimum at $x = \dfrac{\sqrt{6}}{6}$:

$$f\left(\dfrac{\sqrt{6}}{6}\right) = 2 \times \left(\dfrac{\sqrt{6}}{6}\right)^3 - \dfrac{\sqrt{6}}{6} + 1 = 1 - \dfrac{\sqrt{6}}{9}$$

Since $\dfrac{\mathrm{d}^2 y}{\mathrm{d}x^2}\bigg|_{x=-\frac{\sqrt{6}}{6}} = 6 \times \left(-\dfrac{\sqrt{6}}{6}\right) < 0$, $f(x)$ has a local maximum at $x = -\dfrac{\sqrt{6}}{6}$:

$$f\left(-\dfrac{\sqrt{6}}{6}\right) = 2 \times \left(-\dfrac{\sqrt{6}}{6}\right)^3 - \left(-\dfrac{\sqrt{6}}{6}\right) + 1 = 1 + \dfrac{\sqrt{6}}{9}$$

Check the endpoint of the interval:

$$\text{At } x=-1, \ y=0 \text{ and at } x=1, \ y=2$$

Then, the absolute minimum values of $f(x)$ is 0 and the absolute maximum values of $f(x)$ is 2. See Figure 6.8.6.

(7) Let V be the volume of the balloon and let r be its radius. Relate V and r by the formula for the volume of a sphere:

$$V=\frac{4}{3}\pi r^3$$

Differentiate each side of this equation with respect to t:

$$\frac{dV}{dt}=4\pi r^2 \frac{dr}{dt}$$

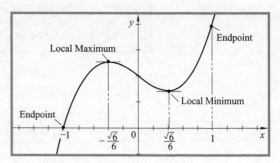

Figure 6.8.6

Put $r=4$ and $\frac{dV}{dt}=64\pi$ in this equation, we obtain

$$64\pi=4\pi \times 4^2 \times \frac{dr}{dt}$$

$$\frac{dr}{dt}=1$$

The radius of the balloon is increasing at the rate of 1in/sec.

(8) Take the derivative of $f(t)$:

$$v(t)=f'(t)=6t^2-30t+24=6(t-1)(t-4)$$

Take the derivative of $v(t)$:

$$a(t)=v'(t)=12t-30=6(2t-5)$$

Velocity $v(t)=0$ at $t=1$ and $t=4$, and:

t	[0, 1)	1	(1, 4)	4	(4, $+\infty$)
$v(t)$	+	0	−	0	+

Acceleration $a(t)=0$ at $t=\frac{5}{2}$, and:

t	$[0, \frac{5}{2})$	$\frac{5}{2}$	$(\frac{5}{2}, +\infty)$
$a(t)$	−	0	+

These signs of $v(t)$ and $a(t)$ immediately yield the answers, as follows:

1) The distance $f(t)$ decreases when t on $(1,4)$.

2) The velocity $v(t)$ increases when t on $\left(\frac{5}{2}, +\infty\right)$.

3) The speed $|v|$ is increasing when $v(t)$ and $a(t)$ are both positive, that is, for t

on $(4, +\infty)$, and when $v(t)$ and $a(t)$ are both negative, that is, for t on $\left(1, \dfrac{5}{2}\right)$.

4) P's motion can be indicated as show in Figure 6.8.8. P moves to the right if t on $[0, 1)$, reverses its

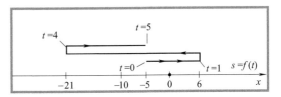

Figure 6.8.8

direction at $t=1$, moves to the left when t on $(1,4)$, reverses again at $t=4$, and continues to the right for all $t>4$. The position of P at certain time t are show in the following table:

t	0	1	4	5
$f(t)$	-5	6	-21	-10

Thus, P travels a total of 54 units between times $t=0$ and $t=5$.

(9) The function $f(x)$ is continuous on $[-\pi, 0]$ and is differentiable on $(-\pi, 0)$.
$$f(-\pi) = \sin(-\pi) = 0 \text{ and } f(0) = \sin(0) = 0$$
According to the Rolle's Theorem, we obtain:
$$f'(c) = 0$$
Find $f'(x)$:
$$f'(x) = \cos x$$
Thus,
$$f'(c) = \cos c = 0, \text{ and } c = -\dfrac{\pi}{2}$$

(10) 1) $[(1+2x)^6 - 1] \to 0$ and $x \to 0$ as $x \to 0$
$$\lim_{x \to 0} \dfrac{(1+2x)^6 - 1}{x} = \lim_{x \to 0} \dfrac{[(1+2x)^6 - 1]'}{x'} = \lim_{x \to 0} \dfrac{6(1+x)^5 \times 2}{1} = 12$$

2) $x^2 \to \infty$ and $e^{2x} \to \infty$ as $x \to \infty$
$$\lim_{x \to \infty} \dfrac{x^2}{e^{2x}} = \lim_{x \to \infty} \dfrac{(x^2)'}{(e^{2x})'} = \lim_{x \to \infty} \dfrac{2x}{2e^{2x}} = \lim_{x \to \infty} \dfrac{x}{e^{2x}} = \lim_{x \to \infty} \dfrac{1}{2e^{2x}} = 0$$

3) $[x - \sin x] \to 0$ and $x^2 \to 0$ as $x \to 0$
$$\lim_{x \to 0} \dfrac{x - \sin x}{x^2} = \lim_{x \to 0} \dfrac{1 - \cos x}{2x}$$
$[1 - \cos x] \to 0$ and $2x \to 0$ as $x \to 0$
$$\lim_{x \to 0} \dfrac{1 - \cos x}{2x} = \lim_{x \to 0} \dfrac{\sin x}{2} = 0$$

4) Let $y = x^{\frac{1}{1-x}}$, so that $\ln y = \ln x^{\frac{1}{1-x}} = \dfrac{1}{1-x} \ln x$, then,
$$\lim_{x \to 1} \ln y = \lim_{x \to 1} \left(\dfrac{1}{1-x} \ln x \right)$$
$(1-x) \to 0$ and $\ln x \to 0$ as $x \to 1$

$$\lim_{x\to 1}\frac{\ln x}{1-x}=\lim_{x\to 1}\frac{\frac{1}{x}}{-1}=-1$$

Since $\lim_{x\to 1}\ln y=-1$, $\lim_{x\to 1}x^{\frac{1}{1-x}}=\lim_{x\to 1}y=e^{-1}=\frac{1}{e}$

5) $\sin 2x \to 0$ and $3x \to 0$ as $x \to 0$

$$\lim_{x\to 0}\frac{\sin 2x}{3x}=\lim_{x\to 0}\frac{2\cos 2x}{3}=\frac{2}{3}$$

(11)
$$f'(1)\approx\frac{5^{1.001}-5^1}{1.001-1}=\frac{5.008-5}{0.001}=8$$

(12) Let $f(x)=\sqrt{x}$, $x_0=36$, $x=36.01$, the derivative of the function

$$f'(x)=\frac{1}{2\sqrt{x}}$$

Then,
$$f(36.01)\approx f(36)+f'(36)\times(36.01-36)$$
$$=\sqrt{36}+\frac{1}{2\sqrt{36}}\times(36.01-36)=6.00083$$

(13) We get,
$$x_0=0, \text{and } y_0=2, y'_0=x_0-y_0=0-2=-2$$

Use Euler's Method,
$$x_1=x_0+h=0+0.1=0.1$$
$$y_1=y_0+h\times y'_0=2+0.1\times(-2)=1.8$$
$$y'_1=x_1-y_1=0.1-1.8=-1.7$$

Next
$$x_2=x_1+h=0.1+0.1=0.2$$
$$y_2=y_1+hy'_1=1.8+0.1\times(-1.7)=1.63$$
$$y'_2=x_2-y_2=0.2-1.63=-1.43$$

Next
$$x_3=x_2+h=0.2+0.1=0.3$$
$$y_3=y_2+hy'_2=1.63+0.1\times(-1.43)=1.487$$

Thus,
$$y(0.3)\approx 1.487$$

第7章

(1) 1) Using the Power Rule
$$\int 3x\,dx=3\times\frac{x^2}{2}+C=\frac{3x^2}{2}+C$$

2) Using the Power Rule

$$\int 2\sqrt{x}\,\mathrm{d}x = \int 2x^{\frac{1}{2}}\,\mathrm{d}x = 2 \times \frac{x^{\frac{3}{2}}}{\frac{3}{2}} = \frac{4}{3}x^{\frac{3}{2}} + C$$

3) Using the Sum Rule
$$\int (3x^2 + 2x + 1)\,\mathrm{d}x = \int 3x^2\,\mathrm{d}x + \int 2x\,\mathrm{d}x + \int 1\,\mathrm{d}x$$
$$= 3 \times \frac{x^3}{3} + 2 \times \frac{x^2}{2} + x + C$$
$$= x^3 + x^2 + x + C$$

4) Expand the integrand:
$$\int (x+1)(x-1)\,\mathrm{d}x = \int (x^2 - 1)\,\mathrm{d}x$$
$$= \frac{x^3}{3} - x + C$$

5) Using the basic integration formulas
$$\int 3\sin x\,\mathrm{d}x = 3 \times (-\cos x) + C = -3\cos x + C$$

(2) The general antiderivative of $f'(x) = 2x$ is
$$f(x) = \int f'(x)\,\mathrm{d}x = \int 2x\,\mathrm{d}x = 2 \times \frac{x^2}{2} + C = x^2 + C$$

To determine C we use the fact that $f(1) = 2$:
$$f(1) = (1)^2 + C = 2$$

Solving for C, we get $C = 1$, so the solution is
$$f(x) = x^2 + 1$$

(3) Since $\ln x - 3$ is an antiderivative of the function $f(x)$, then
$$\int f(x)\,\mathrm{d}x = \ln x - 3$$

And
$$\left[\int f(x)\,\mathrm{d}x \right]' = f(x)$$
$$f(x) = \left[\int f(x)\,\mathrm{d}x \right]' = (\ln x - 3)' = \frac{1}{x}$$

Then,
$$f'(x) = \left(\frac{1}{x}\right)' = -\frac{1}{x^2} \text{ and } f'(\mathrm{e}^x) = -\frac{1}{(\mathrm{e}^x)^2} = -\frac{1}{\mathrm{e}^{2x}}$$

Thus,
$$\int \mathrm{e}^{2x} f'(\mathrm{e}^x)\,\mathrm{d}x = \int \mathrm{e}^{2x}\left(-\frac{1}{\mathrm{e}^{2x}}\right)\mathrm{d}x = \int (-1)\,\mathrm{d}x = -x + C$$

(4) 1) Let $u = 2x + 1$, then $\mathrm{d}u = 2\mathrm{d}x$
$$\int (2x+1)^5\,\mathrm{d}x = \int \frac{1}{2} u^5\,\mathrm{d}u$$

Integrate:

$$\int \frac{1}{2} u^5 \, du = \frac{1}{2} \times \frac{u^6}{6} + C$$

Substitute back for u:

$$\int (2x+1)^5 \, dx = \frac{(2x+1)^6}{12} + C$$

2) Let $u = x^2 + 3$, then $du = 2x \, dx$

$$\int x e^{x^2+3} \, dx = \int \frac{1}{2} e^u \, du$$

Integrate:

$$\int \frac{1}{2} e^u \, du = \frac{1}{2} e^u + C$$

Substitute back for u:

$$\int x e^{x^2+3} \, dx = \frac{1}{2} e^{x^2+3} + C$$

3) Let $u = 5x$, then $du = 5 \, dx$

$$\int \sin 5x \, dx = \int \frac{1}{5} \sin u \, du$$

Integrate:

$$\int \frac{1}{5} \sin u \, du = -\frac{1}{5} \cos u + C$$

Substitute back for u:

$$\int \sin 5x \, dx = -\frac{1}{5} \cos 5x + C$$

4) Let $u = 1 + 2x^2$, then $du = 4x \, dx$

$$\int \frac{x}{1+2x^2} \, dx = \int \frac{1}{4} \times \frac{1}{u} \, du$$

Integrate:

$$\int \frac{1}{4} \times \frac{1}{u} \, du = \frac{1}{4} \ln|u| + C$$

Substitute back for u:

$$\int \frac{x}{1+2x^2} \, dx = \frac{1}{4} \ln(1+2x^2) + C$$

5) Let $u = 1 + x^2$, then $du = 2x \, dx$

$$\int \frac{x}{(1+x^2)^2} \, dx = \int \frac{1}{2} \times \frac{1}{u^2} \, du$$

Integrate:

$$\int \frac{1}{2} \times \frac{1}{u^2} \, du = \frac{1}{2} \times \frac{u^{-1}}{-1} + C = -\frac{1}{2u} + C$$

Substitute back for u:

$$\int \frac{x}{(1+x^2)^2} \, dx = -\frac{1}{2(1+x^2)} + C$$

(5) 1) Let $u = \arctan x$, then $du = \frac{1}{1+x^2} \, dx$

$$\int \frac{\arctan x}{1+x^2} dx = \int u \, du$$

Integrate:
$$\int u \, du = \frac{u^2}{2} + C$$

Substitute back for u:
$$\int \frac{\arctan x}{1+x^2} dx = \frac{(\arctan x)^2}{2} + C$$

2) Use the trigonometric substitution $\tan^2 x = \sec^2 x - 1$:
$$\int \tan^2 2x \, dx = \int (\sec^2 2x - 1) \, dx$$

Break this up into two integrals:
$$\int (\sec^2 2x - 1) \, dx = \int \sec^2 2x \, dx - \int dx$$

Let $u = 2x$, then $du = 2dx$
$$\int \sec^2 2x \, dx - \int dx = \int \frac{1}{2} \sec^2 u \, du - \int dx$$

Integrate:
$$\int \frac{1}{2} \sec^2 u \, du - \int dx = \frac{1}{2} \tan u - x + C$$

Substitute back for u:
$$\int \tan^2 2x \, dx = \frac{1}{2} \tan 2x - x + C$$

3) Let $u = \cos x$, then $du = -\sin x \, dx$
$$\int \cos^2 x \sin x \, dx = \int -u^2 \, du$$

Integrate:
$$\int -u^2 \, du = -\frac{u^3}{3} + C$$

Substitute back for u:
$$\int \cos^2 x \sin x \, dx = -\frac{\cos^3 x}{3} + C$$

4) Since $\cos 2x = 2\cos^2 x - 1$, then
$$\cos^4 x = (\cos^2 x)^2 = \left(\frac{1+\cos 2x}{2}\right)^2 = \frac{1}{4}(1 + 2\cos 2x + \cos^2 2x)$$

And then
$$\cos^2 2x = \frac{1+\cos 4x}{2}$$

So,
$$\cos^4 x = \frac{1}{4}\left(1 + 2\cos 2x + \frac{1+\cos 4x}{2}\right) = \frac{1}{8}(3 + 4\cos 2x + \cos 4x)$$

Then,

$$\int \cos^4 x \, dx = \int \frac{1}{8}(3 + 4\cos 2x + \cos 4x) \, dx$$

Break this up into several integrals:

$$\int \frac{1}{8}(3 + 4\cos 2x + \cos 4x) \, dx = \frac{1}{8}\left(\int 3 \, dx + \int 4\cos 2x \, dx + \int \cos 4x \, dx\right)$$

Integrate:

$$\int \cos^4 x \, dx = \frac{3x}{8} + \frac{1}{4}\sin 2x + \frac{1}{32}\sin 4x + C$$

5) Let $u = \sin x$, then $du = \cos x \, dx$

$$\int \cos x \, e^{\sin x} \, dx = \int e^u \, du$$

Integrate:

$$\int e^u \, du = e^u + C$$

Substitute back for u:

$$\int \cos x \, e^{\sin x} \, dx = e^{\sin x} + C$$

(6) 1) Let $u = x$ and $dv = \cos x \, dx$, then $du = dx$ and $v = \sin x$

$$\int x \cos x \, dx = x \sin x - \int \sin x \, dx$$
$$= x \sin x + \cos x + C$$

2) Let $u = \ln x$ and $dv = x^3 \, dx$, then $du = \frac{1}{x} dx$ and $v = \frac{x^4}{4}$

$$\int x^3 \ln x \, dx = \ln x \, \frac{x^4}{4} - \int \frac{x^4}{4} \times \frac{1}{x} dx = \ln x \, \frac{x^4}{4} - \frac{x^4}{16} + C$$

3) Let $u = \arcsin x$ and $dv = dx$, then $du = \frac{1}{\sqrt{1-x^2}} dx$ and $v = x$

$$\int \arcsin x \, dx = x \arcsin x - \int x \, \frac{1}{\sqrt{1-x^2}} dx$$

Let $u = 1 - x^2$, then $du = -2x \, dx$

$$\int x \, \frac{1}{\sqrt{1-x^2}} dx = \int -\frac{1}{2} \times \frac{1}{\sqrt{u}} du$$
$$= -\sqrt{u} + C = -\sqrt{1-x^2} + C$$

Hence,

$$\int \arcsin x \, dx = x \arcsin x - \int x \, \frac{1}{\sqrt{1-x^2}} dx$$
$$= x \arcsin x + \sqrt{1-x^2} + C$$

4) Let $u = x^2$ and $dv = e^{-x} dx$, then $du = 2x \, dx$ and $v = -e^{-x}$

$$\int x^2 e^{-x} \, dx = -x^2 e^{-x} + 2\int x e^{-x} \, dx$$

Integrate by parts again with $u = x$ and $dv = e^{-x} dx$, then $du = dx$ and $v = -e^{-x}$

$$\int x e^{-x} \, dx = -x e^{-x} + \int e^{-x} \, dx = -x e^{-x} - e^{-x} + C = -e^x(x+1) + C$$

Hence,
$$\int x^2 e^{-x} dx = -x^2 e^{-x} + 2\int x e^{-x} dx = -e^{-x}(x^2+2x+2)+C$$

5) Let $u=e^x$ and $dv=\cos x\, dx$, then $du=e^x dx$ and $v=\sin x$
$$\int e^x \cos x\, dx = e^x \sin x - \int \sin x\, e^x dx$$

Again, let $u=e^x$, $dv=\sin x\, dx$, then $du=e^x dx$ and $v=-\cos x$
$$\int \sin x\, e^x dx = -e^x \cos x + \int e^x \cos x\, dx$$

Hence,
$$\int e^x \cos x\, dx = e^x \sin x + e^x \cos x - \int e^x \sin x\, dx$$

Adding the integral to both sides and adding the constant of integration gives,
$$2\int e^x \cos x\, dx = e^x \sin x + e^x \cos x + C_1$$

Dividing by 2 and renaming the constant of integration gives
$$\int e^x \cos x\, dx = \frac{1}{2} e^x (\sin x + \cos x) + C$$

(7)#. 1) The partial fraction decomposition has the from
$$\frac{7x-12}{(x-1)(x-2)} = \frac{A}{x-1} + \frac{B}{x-2}$$

Clear fractions and get,
$$7x-12 = A(x-2) + B(x-1)$$

Simplify and group the terms:
$$7x-12 = (A+B)x - 2A - B$$

Equate coefficients of like powers of x obtaining,
$$\begin{cases} A+B=7 \\ -2A-B=-12 \end{cases}$$

Solve the equations, we get
$$A=5, B=2$$

Hence,
$$\int \frac{7x-12}{(x-1)(x-2)} dx = \int \frac{5}{x-1} dx + \int \frac{2}{x-2} dx$$
$$= 5\ln|x-1| + 2\ln|x-2| + C$$

2) The partial fraction decomposition has the from
$$\frac{x+9}{(x+2)^2} = \frac{A}{x+2} + \frac{B}{(x+2)^2}$$

Clear fractions and get,
$$x+9 = Ax + 2A + B$$

Equate coefficients of like powers of x obtaining,
$$\begin{cases} A=1 \\ 2A+B=9 \end{cases}$$

Solve the equations, we get
$$A=1, B=7$$
That is,
$$\frac{x+9}{(x+2)^2}=\frac{1}{x+2}+\frac{7}{(x+2)^2}$$
Hence,
$$\int \frac{x+9}{(x+2)^2}dx = \int \left[\frac{1}{x+2}+\frac{7}{(x+2)^2}\right]dx$$
$$= \int \frac{1}{x+2}dx + \int \frac{7}{(x+2)^2}dx$$
$$= \ln|x+2| - \frac{7}{x+2} + C$$

3) The partial fraction decomposition has the from
$$\frac{x+2}{x(x+1)^2}=\frac{A}{x}+\frac{Bx+C}{(x+1)^2}$$
Clear fractions and get,
$$x+2=A(x+1)^2+x(Bx+C)$$
Simplify and group the terms:
$$x+2=(A+B)x^2+(2A+C)x+A$$
Equate coefficients of like powers of x obtaining,
$$\begin{cases} A+B=0 \\ 2A+C=1 \\ A=2 \end{cases}$$
Solve the equations, we get
$$A=2, B=-2, C=-3$$
That is,
$$\frac{3x+2}{x(x+1)^2}=\frac{2}{x}-\frac{2x+3}{(x+1)^2}$$
And let,
$$\frac{2x+3}{(x+1)^2}=\frac{D}{x+1}+\frac{E}{(x+1)^2}$$
Then,
$$D=2, E=1$$
So
$$\frac{x+2}{x(x+1)^2}=\frac{2}{x}-\frac{2}{x+1}-\frac{1}{(x+1)^2}$$
Hence,
$$\int \frac{x+2}{x(x+1)^2}dx = \int \left[\frac{2}{x}-\frac{2}{x+1}-\frac{1}{(x+1)^2}\right]dx$$
$$= \int \frac{2}{x}dx - \int \frac{2}{x+1}dx - \int \frac{1}{(x+1)^2}dx$$

$$= 2\ln|x| - 2\ln|x+1| + \frac{1}{x+1} + C$$

(8) Since $v(t) = \frac{ds}{dt} = 9.8t - 5$, Then,
$$s = \int (9.8t - 5)dt = 4.9t^2 - 5t + C$$

Since $s(0) = 20$, $4.9 \times 0^2 - 5 \times 0 + C = 20$, then
$$C = 20$$

And that the position function is
$$s = 4.9t^2 - 5t + 20$$

Plug in $t = 5$ to find the position at that time:
$$s(3) = 4.9 \times 5^2 - 5 \times 5 + 20 = 117.5$$

第 8 章

(1) Draw four rectangles that look like Figure 8.8.1:

The width of each rectangle is
$$\frac{b-a}{n} = \frac{5-1}{4} = 1$$

The heights of the rectangle are:
$y_1 = f(1) = 1, y_2 = f(2) = 2, y_3 = f(3)$
$\quad = 5$ and $y_4 = f(4) = 10$

Therefore, the area is:
$L(4) = 1 \times 1 + 1 \times 2 + 1 \times 5 + 1 \times 10 = 18$

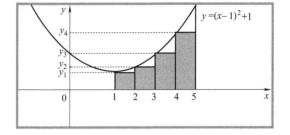

Figure 8.8.1

(2) Since $f(x)$ increase on $[a, b]$ and is concave down, the inequality is
$$R(n) \leqslant T(n) \leqslant \int_a^b f(x)dx \leqslant M(n) \leqslant L(n)$$

(3) 1)
$$\int_2^5 (x^3 + x^2 + x)dx = \left(\frac{x^4}{4} + \frac{x^3}{3} + \frac{x^2}{2}\right)\bigg|_2^5$$
$$= \left(\frac{5^4}{4} + \frac{5^3}{3} + \frac{5^2}{2}\right) - \left(\frac{2^4}{4} + \frac{2^3}{3} + \frac{2^2}{2}\right)$$
$$= \frac{807}{4}$$

2)
$$\int_0^{\frac{\pi}{2}} \sin x\, dx = -\cos x \bigg|_0^{\frac{\pi}{2}} = -\cos\left(\frac{\pi}{2}\right) + \cos 0 = 1$$

3)
$$\int_2^3 \frac{1}{3x-2}dx = \frac{1}{3}\ln|3x-2|\bigg|_2^3 = \frac{1}{3}\ln 7 - \frac{1}{3}\ln 4 = \frac{1}{3}\ln\frac{7}{4}$$

4)
$$\int_{-4}^{4} \frac{1}{16+x^2} dx = \int_{-4}^{4} \frac{1}{16} \times \frac{1}{1+\frac{x^2}{16}} dx$$

$$= \frac{1}{16} \int_{-4}^{4} \frac{1}{1+\left(\frac{x}{4}\right)^2} dx = \frac{1}{4} \arctan \frac{x}{4} \Big|_{-4}^{4} = \frac{\pi}{8}$$

5) Let $u = \ln x$, then $du = \frac{1}{x} dx$

$$\int \frac{\ln x}{x} dx = \int u\, du = \frac{u^2}{2} + C = \frac{(\ln x)^2}{2} + C$$

Then,
$$\int_{1}^{e} \frac{\ln x}{x} dx = \frac{(\ln x)^2}{2} \Big|_{1}^{e} = \frac{1}{2}$$

(4) By the second fundamental theorem of calculus,
$$\frac{d}{dx} \int_{5}^{x} (5t^6 - 2t^3) dt = 5x^6 - 2x^3$$

(5) Let $u = x^2$, then $du = 2x\, dx$ and $dx = \frac{du}{2x}$

$$\frac{d}{dx} \int_{5}^{x^2} (t^3 - 1) dt = 2x \frac{d}{du} \int_{5}^{u} (t^3 - 1) dt$$

By the second fundamental theorem of calculus,
$$\frac{d}{du} \int_{5}^{u} (t^3 - 1) dt = u^3 - 1$$

Therefor,
$$\frac{d}{dx} \int_{5}^{x^2} (t^3 - 1) dt = 2x [(x^2)^3 - 1] = 2x^7 - 2x$$

(6) Find the derivative of the function
$$F'(x) = \frac{d}{dx} \int_{1}^{x} (t^2 - 2t - 3) dt = x^2 - 2x - 3$$

Set the derivative equal to zero and solve for x:
$$x^2 - 2x - 3 = 0, x_1 = -1 \text{ and } x_2 = 3$$

Take the second derivative of the function:
$$F''(x) = (x^2 - 2x - 3)' = 2x - 2$$

Since $F''(-1) = -4 < 0$ and $F''(3) = 4 > 0$,

then, the function $F(x)$ attain a relative minimum at $x = 3$.

(7) The function $f(x) = 2x^2 + x + 2$ is continuous on $[1,4]$ and the average value of $f(x)$ is

$$f(\xi) = \frac{1}{b-a} \int_{a}^{b} f(x) dx = \frac{1}{4-1} \int_{1}^{4} (2x^2 + x + 2) dx$$

$$= \frac{1}{3} \left(\frac{2}{3} x^3 + \frac{x^2}{2} + 2x \right) \Big|_{1}^{4} = \frac{111}{6}$$

(8)#.1)
$$\int_1^{+\infty} \frac{1}{x^3} dx = \lim_{b \to +\infty} \int_1^b \frac{1}{x^3} dx = \lim_{b \to +\infty} \left(-\frac{1}{2x^2}\right) \Big|_1^b$$
$$= \lim_{b \to +\infty} \left(-\frac{1}{2b^2} + \frac{1}{2}\right) = \frac{1}{2}$$

2) Let $u = x^2$, then $du = 2x\,dx$
$$\int x e^{x^2} dx = \frac{1}{2} \int e^u du = \frac{e^u}{2} + C = \frac{e^{x^2}}{2} + C$$
$$\int_{-\infty}^0 x e^{x^2} dx = \lim_{a \to -\infty} \left(\frac{e^{x^2}}{2} \Big|_a^0\right) = \lim_{a \to -\infty} \left(\frac{1}{2} - \frac{e^{a^2}}{2}\right) = -\infty$$

3) $\int_0^2 \frac{1}{x-1} dx = \int_0^1 \frac{1}{x-1} dx + \int_1^2 \frac{1}{x-1} dx$

Where
$$\int_0^1 \frac{1}{x-1} dx = \lim_{c \to 1^-} \int_0^c \frac{1}{x-1} dx = \lim_{c \to 1^-} (\ln|x-1|) \Big|_0^c$$
$$= \lim_{c \to 1^-} (\ln|c-1| - \ln 1) = -\infty$$

Thus, $\int_0^1 \frac{1}{x-1} dx$ is divergent. So, $\int_0^2 \frac{1}{x-1} dx$ is divergent.

第9章

(1) Make a sketch of the region, see Figure 9.5.1:

The area between the curves is
$$\int_{-2}^1 (x^2 + 1) dx$$

Then, evaluate it:
$$\int_{-2}^1 (x^2 + 1) dx = \left(\frac{x^3}{3} + x\right) \Big|_{-2}^1 = 6$$

(2) Make a sketch of the region, see Figure 9.5.2:

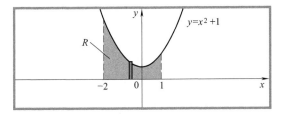

Figure 9.5.1

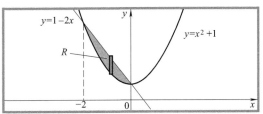

Figure 9.5.2

Find the points of intersection:
$$x^2 + 1 = 1 - 2x$$

Solve for x:
$$x = -2, \quad x = 0$$

The limits of integration $a = -2$, $b = 0$. The area between the curves is

$$\int_{-2}^{0}[(1-2x)-(x^2+1)]dx = \int_{-2}^{0}(-x^2-2x)dx$$
$$=\left(-\frac{x^3}{3}-x^2\right)\Big|_{-2}^{0} = \frac{4}{3}$$

(3) Make a sketch of the region, see Figure 9.5.3:

Find the points of intersection:
$$y^3-3y=y$$

Solve for y:
$$y=0, y=-2, y=2$$

Notice that $x=y^3-3y$ is to the right of $x=y$ from $y=-2$ and $y=0$, then $x=y^3-3y$ is to the left of $x=y$ from $y=0$ and $y=2$. The area between the curves is

$$\int_{-2}^{0}[(y^3-3y)-y]dy + \int_{0}^{2}[y-(y^3-3y)]dy$$
$$=\int_{-2}^{0}(y^3-4y)dy + \int_{0}^{2}(2y-y^3)dy$$
$$=\left(\frac{y^4}{4}-4\times\frac{y^2}{2}\right)\Big|_{-2}^{0} + \left(4\times\frac{y^2}{2}-\frac{y^4}{4}\right)\Big|_{0}^{2} = 4+4=8$$

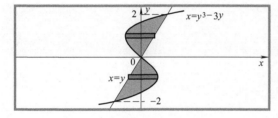

Figure 9.5.3   Figure 9.5.4

(4) Make a sketch of the region, see Figure 9.5.4:

Find the points of intersection:
$$\sin x = \cos x$$

Solve for y:
$$x=\frac{\pi}{4}$$

The area between the curves is

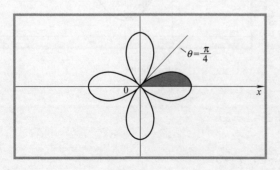

Figure 9.5.5

$$\int_{0}^{\frac{\pi}{4}}\sin x\, dx + \int_{\frac{\pi}{4}}^{\frac{\pi}{2}}\cos x\, dx = 2-\sqrt{2}$$

(5) Sketch the graph using graphing calculator, see Figure 9.5.5, where one eighth of the required area is shaded.

The shaded area of the four-leaved rose swept out by a radius as θ varies from 0 to $\frac{\pi}{4}$:

$$\int_\alpha^\beta \frac{1}{2} r^2 \mathrm{d}\theta = \int_0^{\frac{\pi}{4}} \frac{1}{2} \cos^2\theta \mathrm{d}\theta \approx 0.1963$$

Thus,
$$A = 8 \times 0.1963 = 1.5708$$

(6) Make a sketch of the region, a typical radius, and the generated solid, see Figure 9.5.6:

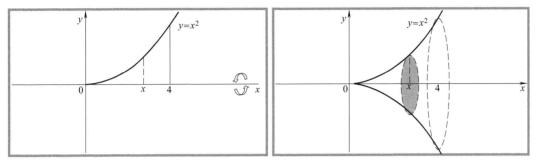

Figure 9.5.6

The volume is
$$\int_0^4 \pi (x^2)^2 \mathrm{d}x = \int_0^4 \pi x^4 \mathrm{d}x = \pi \times \frac{x^5}{5}\bigg|_0^4 = \frac{1024}{5}\pi$$

(7) Make a sketch of the region and the generated solid, see Figure 9.5.7:

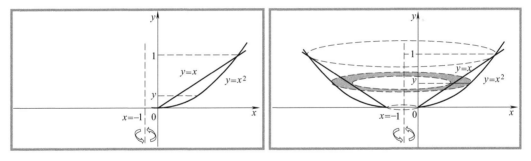

Figure 9.5.7

The curve $y = x^2$ on the outside and the curve $y = x$ on the inside, then the volume is:
$$V = \int_0^1 \pi [(\sqrt{y}+1)^2 - (y+1)^2] \mathrm{d}y = \pi \int_0^1 (2y^{\frac{1}{2}} - y^2 - y) \mathrm{d}y$$
$$= \pi \left(\frac{4}{3} y^{\frac{3}{2}} - \frac{y^3}{3} - \frac{y^2}{2}\right)\bigg|_0^1 = \frac{\pi}{2}$$

(8) Make a sketch of the region and the generated solid, see Figure 9.5.8:
Find the points of intersection:
$$\sqrt{x} = 2x - 1$$

Solve for x:
$$x = 1$$

The volume is:

$$V = \int_0^1 2\pi x[\sqrt{x} - (2x-1)]\,dx = 2\pi \int_0^1 x(\sqrt{x} - 2x + 1)\,dx = \frac{7\pi}{15}$$

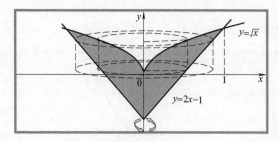
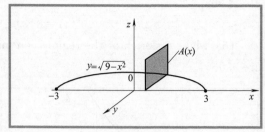

Figure 9.5.8　　　　　　　　　　　Figure 9.5.9

(9) Make a sketch of the region, see Figure 9.5.9:

The side of the squares is $\sqrt{9-x^2}$. The area of a cross-section is
$$A(x) = (\sqrt{9-x^2})^2 = 9 - x^2$$

The volume is:
$$V = \int_{-3}^3 (9-x^2)\,dx = \left(9x - \frac{x^3}{3}\right)\bigg|_{-3}^3 = 36$$

(10) Find the derivative of the function:
$$\frac{dy}{dx} = \frac{x^2}{4} - \frac{1}{x^2}$$

The length of the curve from $x=1$ to $x=3$ is

$$L = \int_1^3 \sqrt{1 + \left(\frac{x^2}{4} - \frac{1}{x^2}\right)^2}\,dx = \int_1^3 \sqrt{1 + \left(\frac{x^4 - 4}{4x^2}\right)^2}\,dx$$

$$= \int_1^3 \sqrt{1 + \left(\frac{x^8 - 8x^4 + 16}{16x^4}\right)}\,dx = \int_1^3 \sqrt{\frac{x^8 + 8x^4 + 16}{16x^4}}\,dx$$

$$= \int_1^3 \sqrt{\left(\frac{x^4 + 4}{4x^2}\right)^2}\,dx = \int_1^3 \frac{x^4 + 4}{4x^2}\,dx = \int_1^3 \left(\frac{x^2}{4} + \frac{1}{x^2}\right)dx$$

$$= \left(\frac{x^3}{12} - \frac{1}{x}\right)\bigg|_1^3 = \frac{17}{6}$$

第 10 章

(1) We rewrite the equation as
$$dy = (2x+1)\,dx$$

Integrate both side of the equation:
$$\int dy = \int (2x+1)\,dx$$

$$y = 2 \times \frac{x^2}{2} + x + C = x^2 + x + C$$

Since $y(0) = 3$, we get: $C = 3$, the particular solution is
$$y = x^2 + x + 3$$

(2) Separate variables,
$$\frac{1}{y}dy = \frac{1}{\sqrt{x}}dx$$

Integrate both side of the equation:
$$\int \frac{1}{y}dy = \int \frac{1}{\sqrt{x}}dx$$
$$\ln|y| = 2\sqrt{x} + C$$

The initial point yields
$$\ln 1 = 2\sqrt{4} + C$$

Hence
$$C = -4$$

With $y > 0$, the particular solution is
$$\ln y = 2\sqrt{x} - 4 \text{ or } y = e^{2\sqrt{x}-4}$$

(3) Integrate both side of the equation:
$$\int f''(x)dx = \int (x-1)dx$$
$$f'(x) = \frac{x^2}{2} - x + C$$

Since $f'(1) = 0$, we get: $C = \frac{1}{2}$,
$$f'(x) = \frac{x^2}{2} - x + \frac{1}{2}$$

Integrate both side of the equation:
$$\int f'(x)dx = \int \left(\frac{x^2}{2} - x + \frac{1}{2}\right)dx$$
$$f(x) = \frac{x^3}{6} - \frac{x^2}{2} + \frac{x}{2} + C'$$

Since $f(1) = 4$, we get: $C' = \frac{23}{6}$,
$$f(x) = \frac{x^3}{6} - \frac{x^2}{2} + \frac{x}{2} + \frac{23}{6}$$

(4) Since $a(t) = 4t^2 - 6t$, then
$$v(t) = \int a(t)dt = \int (3t^2 - 6t)dt = t^3 - 3t^2 + C_1$$

Similarly,
$$s(t) = \int v(t)dt = \int (t^3 - 3t^2 + C_1)dt = \frac{1}{4}t^4 - t^3 + C_1 x + C_2$$

Since $s(-1) = 3$ and $s(1) = 5$, then
$$\begin{cases} s(-1) = \frac{1}{4} + 1 - C_1 + C_2 = 3 \\ s(1) = \frac{1}{4} - 1 + C_1 + C_2 = 5 \end{cases}$$

So,
$$C_1 = 2$$
Therefore,
$$v(t) = t^3 - 3t^2 + 2$$

(5) Set up a table of values for $\dfrac{dy}{dx}$ selected values of x and y.

y \ $x+y$ \ x	-3	-2	-1	0	1	2	3
-3	-6	-5	-4	-3	-2	-1	0
-2	-5	-4	-3	-2	-1	0	1
-1	-4	-3	-2	-1	0	1	2
0	-3	-2	-1	0	1	2	3
1	-2	-1	0	1	2	3	4
2	-1	0	1	2	3	4	5
3	0	1	2	3	4	5	6

Draw short tangent line segments with the given slopes at various points. See Figure 10.7.5

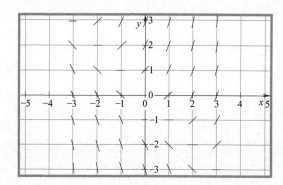

Figure 10.7.5

(6) We let M be the number at time t and M_0 the number initially.

Then,
$$M = M_0 e^{kt}$$

Since $M_0 = 821$, $k = 0.04$ when $t = 10$, then:
$$M(10) = 821 e^{0.04 \cdot 10} = 821 e^{0.4} \approx 1224.79$$

(7) $\dfrac{dy}{dx} \sin x + y \cos x = 0 \Rightarrow \dfrac{dy}{dx} = -\dfrac{y \cos x}{\sin x} = -y \cot x$

(8) If $P(t)$ is the number of people aware of the product at time t, then
$$\dfrac{dP(t)}{dt} = k[20000 - P(t)] \text{ and } P(t) = 20000 - C_0 e^{-kt}$$

Since $P(0) = 0$, then $C_0 = 20000$ and
$$P(t) = 20000(1 - e^{-kt})$$

Also, $P(10) = 2000$, so
$$2000 = 20000(1 - e^{-10k})$$
$$e^{-10k} = 0.9 \text{ and } k = \dfrac{\ln 0.9}{-10} = 0.0105$$

We now seek t when $P(t) = 5000$
$$5000 = 20000(1 - e^{-0.0105t})$$

$$t \approx 27.4 \text{ days}$$

(9)
$$\frac{dP}{dt} = 0.08P\left(1 - \frac{P}{1000}\right) \Rightarrow \frac{dP}{dt} = \frac{0.08}{1000}P(1000 - P)$$

The differential equation is a logistic equation with $k = \frac{0.08}{1000}$, carrying capacity $A = 1000$, and initial population $P_0 = 100$:

$$P_0 = \frac{A}{1+C_0} = \frac{1000}{1+C_0} = 100, C_0 = 9$$

Thus

$$P(t) = \frac{A}{1+C_0 e^{-Akt}} = \frac{1000}{1+9e^{-0.08t}}$$

So, the population sizes when $t = 50$,

$$P(50) = \frac{1000}{1+9e^{-0.08 \times 50}} \approx 858.4$$

第 11 章

(1) Let

$$S_n = \sum_{n=1}^{n} \frac{1}{n(n+1)} = \sum_{n=1}^{n} \left(\frac{1}{n} - \frac{1}{n+1}\right)$$

Thus, we have

$$S_n = \left(1 - \frac{1}{2}\right) + \left(\frac{1}{2} - \frac{1}{3}\right) + \left(\frac{1}{3} - \frac{1}{4}\right) + \cdots + \left(\frac{1}{n} - \frac{1}{n+1}\right) = 1 - \frac{1}{n+1}$$

And so,

$$\sum_{n=1}^{\infty} \frac{1}{n(n+1)} = \lim_{n \to \infty} S_n = \lim_{n \to \infty} \left(1 - \frac{1}{n+1}\right) = 1$$

Therefore, the series $\sum_{n=1}^{\infty} \frac{1}{n(n+1)}$ converges to 1.

(2) We have that

$$\sum_{n=0}^{\infty} a_n = 1 - \frac{1}{2} + \frac{1}{4} - \frac{1}{8} + \cdots + \left(-\frac{1}{2}\right)^{n-1} + \cdots = \sum_{n=1}^{\infty} 1 \times \left(-\frac{1}{2}\right)^{n-1}$$

The series is geometric series with $a = 1$ and $r = -\frac{1}{2}$.

Since $|r| < 1$, the series converges, and its sum is

$$S = \frac{a}{1-r} = \frac{1}{1-\left(-\frac{1}{2}\right)} = \frac{2}{3}$$

(3) Since

$$1 + \frac{1}{4} + \frac{1}{9} + \frac{1}{16} + \cdots + \frac{1}{n^2} + \cdots = \sum_{n=1}^{\infty} \frac{1}{n^2}$$

It is a p-Series with $p = 2 > 1$. The series converges.

(4) Since

$$\frac{1}{\sqrt{2}}-\frac{1}{\sqrt{3}}+\frac{1}{\sqrt{4}}-\frac{1}{\sqrt{5}}+\cdots=\sum_{n=1}^{\infty}(-1)^{n+1}\frac{1}{\sqrt{n+1}}$$

Then,

$$a_n-a_{n+1}=\frac{1}{\sqrt{n+1}}-\frac{1}{\sqrt{n+2}}=\frac{\sqrt{n+2}-\sqrt{n+1}}{\sqrt{n+1}\times\sqrt{n+2}}>0$$

Satisfies

1) $a_n > a_{n+1}$, because $a_n - a_{n+1} > 0$;

2) $\lim\limits_{n\to\infty} a_n = \lim\limits_{n\to\infty}\frac{1}{\sqrt{n+1}}=0$

So, the series converge by the alternating series test.

(5) Since,

$$a_n=\frac{n+1}{n!} \text{ and } a_{n+1}=\frac{n+2}{(n+1)!}$$

$$\frac{a_{n+1}}{a_n}=\frac{\frac{n+2}{(n+1)!}}{\frac{n+1}{n!}}=\frac{n+2}{(n+1)(n+1)}=\frac{n+2}{n^2+2n+1}$$

And

$$\lim_{n\to\infty}\frac{a_{n+1}}{a_n}=\lim_{n\to\infty}\frac{n+2}{n^2+2n+1}=0$$

Thus, $0<1$, $\sum\limits_{n=1}^{\infty}\frac{n+1}{n!}$ converge by the Ratio Test.

(6) Let $a_n=\frac{2}{n^2+3}$, $b_n=\frac{2}{n^2}$

We know that

$$\sum_{n=1}^{\infty}b_n=\sum_{n=1}^{\infty}\frac{2}{n^2}=2\sum_{n=1}^{\infty}\frac{1}{n^2}$$

is convergent because it's a constant time a p-Series with $p=2>1$.
Since

$$\frac{2}{n^2+3}<\frac{2}{n^2}$$

and $\sum\limits_{n=1}^{\infty}b_n$ converges, $\sum\limits_{n=1}^{\infty}\frac{1}{n^2+3}$ converge by the Comparison Test.

(7) We know that

$$\sum_{n=1}^{\infty}\left|(-1)^n\frac{n^2}{2^n}\right|=\sum_{n=1}^{\infty}\frac{n^2}{2^n}$$

$$\lim_{n\to\infty}\frac{a_{n+1}}{a_n}=\lim_{n\to\infty}\frac{\frac{(n+1)^2}{2^{n+1}}}{\frac{n^2}{2^n}}=\lim_{n\to\infty}\frac{(n+1)^2}{2n^2}=\frac{1}{2}$$

Thus, by the Ratio Test, the given series $\sum\limits_{n=1}^{\infty}(-1)^n\dfrac{n^2}{2^n}$ is absolutely convergent.

(8) By the Ratio Test,

$$\lim_{n\to\infty}\left|\dfrac{a_{n+1}}{a_n}\right|=\lim_{n\to\infty}\left|\dfrac{\dfrac{(x-3)^{n+1}}{n+1}}{\dfrac{(x-3)^n}{n}}\right|=\lim_{n\to\infty}\left|\dfrac{n(x-3)}{n+1}\right|$$

$$=\lim_{n\to\infty}\left|\dfrac{1}{1+\dfrac{1}{n}}(x-3)\right|=|x-3|$$

The given series is absolutely convergent, when $|x-3|<1$, then
$$2<x<4$$

When $x=2$, the series is $\sum\limits_{n=1}^{\infty}\dfrac{(-1)^n}{n}$, which converges by the Alternating Series Test.

When $x=4$, the series is $\sum\limits_{n=1}^{\infty}\dfrac{1}{n}$, the Harmonic Series, which is divergent.

Thus, the given series converges for $2\leqslant x<4$.

(9) By the Ratio Test,

$$a_n=\dfrac{n(x+1)^n}{2^{n+1}} \text{ and } a_{n+1}=\dfrac{(n+1)(x+1)^{n+1}}{2^{n+2}}$$

$$\lim_{n\to\infty}\left|\dfrac{a_{n+1}}{a_n}\right|=\lim_{n\to\infty}\left|\dfrac{\dfrac{(n+1)(x+1)^{n+1}}{2^{n+2}}}{\dfrac{n(x+1)^n}{2^{n+1}}}\right|$$

$$=\lim_{n\to\infty}\left|\dfrac{(n+1)(x+1)}{2n}\right|=\dfrac{|x+1|}{2}$$

The given series is convergent if $\dfrac{|x+1|}{2}<1$ and it diverges if $\dfrac{|x+1|}{2}>1$.

So, it converges if $|x+1|<2$ and diverges $|x+1|>2$. Thus, the radius of convergence is $R=2$.

The $|x+1|<2$ can be written as: $-3<x<1$.

When $x=-3$, the series is $\sum\limits_{n=1}^{\infty}\dfrac{n(-2)^n}{2^{n+1}}=\dfrac{1}{2}\sum\limits_{n=1}^{\infty}(-1)^n n$ which diverges by $\lim\limits_{n\to\infty}(-1)^n n\neq 0$.

When $x=1$, the series is $\sum\limits_{n=1}^{\infty}\dfrac{n(2)^n}{2^{n+1}}=\dfrac{1}{2}\sum\limits_{n=1}^{\infty}n$ which also diverges by $\lim\limits_{n\to\infty}n\neq 0$.

Thus, the interval of convergence is $(-3,1)$.

(10) The derivatives of $f(x)=\ln(1+x)$ at 0 are given by the following table:

$f(x)=\ln(1+x)$ $f(0)=0$

$f'(x)=\dfrac{1}{x+1}$ $f'(0)=1$

$$f''(x) = -(x+1)^{-2} \qquad f''(0) = -1$$
$$f^{(3)}(x) = 2(x+1)^{-3} \qquad f^{(3)}(0) = 2$$
$$\cdots \qquad \cdots$$
$$f^{(n)}(x) = (-1)^{n-1}\frac{(n-1)!}{(x+1)^n} \qquad f^{(n)}(0) = (-1)^{n-1}(n-1)!$$

Thus,
$$\ln(1+x) = f(0) + f'(0)x + \frac{f''(0)}{2!}x^2 + \cdots + \frac{f^{(n)}(0)}{n!}x^n + \cdots$$
$$= 0 + x - \frac{x^2}{2} + \frac{x^3}{3} - \cdots + (-1)^{n-1}\frac{x^n}{n} + \cdots$$
$$= \sum_{n=0}^{\infty}(-1)^{n-1}\frac{x^n}{n}$$

(11) The first three derivatives are
$$f(x) = e^x, f'(x) = e^x, f''(x) = e^x, f^{(3)}(x) = e^x$$

And
$$f(1) = e, f'(1) = e, f''(1) = e, f^{(3)}(1) = e$$

The approximating Taylor polynomial of order 3 is therefore
$$e^x \approx e + e(x-1) + \frac{e(x-1)^2}{2!} + \frac{e(x-1)^3}{3!}$$

(12) Since,
$$e^x = 1 + x + \frac{x^2}{2!} + \frac{x^3}{3!} + \cdots + \frac{x^n}{n!} + \cdots$$

Let $x = -x$, then
$$e^{-x} = 1 - x + \frac{x^2}{2!} - \frac{x^3}{3!} + \cdots + (-1)^n\frac{x^n}{n!} + \cdots$$

So
$$e^{0.1} = 1 - 0.1 + \frac{0.1^2}{2!} - \frac{0.1^3}{3!} + \cdots = 1 - 0.1 + 0.005 + R_2$$

Since this is a convergent alternating series, $|R_2| < \dfrac{0.1^3}{3!} < 0.0002$, which will not affect the third decimal place. Then, correct to three decimal places,
$$e^{0.1} \approx 0.905$$

附录
Appendix

A.1 常用公式和定理 Common Formulas and Theorems

A1.1 代数学 Algebra

A1.1.1 求根公式 The Roots of The Quadratic Equation

If $a \neq 0$, the roots of $ax^2 + bx + c = 0$ are $x = \dfrac{-b \pm \sqrt{b^2 - 4ac}}{2a}$

A1.1.2 特殊的乘法和因式分解公式 Special Product/Factoring Formulas

Special Product Formulas	Special Factoring Formulas
$(x+y)(x-y) = x^2 - y^2$	$x^2 - y^2 = (x+y)(x-y)$
$(x+y)^2 = x^2 + 2xy + y^2$	$x^2 + 2xy + y^2 = (x+y)^2$
$(x-y)^2 = x^2 - 2xy + y^2$	$x^2 - 2xy + y^2 = (x-y)^2$
$(x+y)^3 = x^3 + 3x^2y + 3xy^2 + y^3$	$x^3 - y^3 = (x-y)(x^2 + xy + y^2)$
$(x-y)^3 = x^3 - 3x^2y + 3xy^2 - y^3$	$x^3 + y^3 = (x+y)(x^2 - xy + y^2)$

A1.1.3 指数和根数 Exponents and Radicals

Exponents	Radicals
$a^m a^n = a^{m+n}$	$a^{1/n} = \sqrt[n]{a}$
$(a^m)^n = a^{mn}$	$a^{m/n} = \sqrt[n]{a^m}$
$(ab)^n = a^n b^n$	$a^{m/n} = (\sqrt[n]{a})^m$
$\left(\dfrac{a}{b}\right)^n = \dfrac{a^n}{b^n}$	$\sqrt[n]{ab} = \sqrt[n]{a} \times \sqrt[n]{b}$
$\dfrac{a^m}{a^n} = a^{m-n}$	$\sqrt[n]{\dfrac{a}{b}} = \dfrac{\sqrt[n]{a}}{\sqrt[n]{b}}$
$a^{-n} = \dfrac{1}{a^n}$	$\sqrt[m]{\sqrt[n]{a}} = \sqrt[mn]{a}$

A1.1.4 指数和对数 Exponential and Logarithm

$y = \log_a x$ means $a^y = x$	
$\log_a xy = \log_a x + \log_a y$	$\log_a \dfrac{x}{y} = \log_a x - \log_a y$
$\log_a x^r = r \log_a x$	$\ln x^r = r \ln x$
$a^{\log_a x} = x$	$\log_a a^x = x$
$\log_a 1 = 0$	$\log_a a = 1$
$\log x = \log_{10} x$	$\ln x = \log_e x$
$\log_b u = \dfrac{\log_a u}{\log_a b}$	$\log_b u = \dfrac{\ln u}{\ln b}$

A1.1.5 二项式定理 Binomial Theorem

$$(x+y)^n = x^n + \binom{n}{1} x^{n-1} y + \binom{n}{2} x^{n-2} y^2 + \cdots + \binom{n}{k} x^{n-k} y^k + \cdots + y^n$$

where

$$\binom{n}{k} = \dfrac{n!}{k!(n-k)!}$$

A1.1.6 不等式和绝对值 Inequalities and Absolute Value

If $a > b$ and $b > c$, then $a > c$;	If $a > b$, then $a + c > b + c$;
If $a > b$ and $c > 0$, then $ac > bc$;	If $a > b$ and $c < 0$, then $ac < bc$;
$\|x\| < d \, (d > 0)$ if and only if $-d < x < d$	
$\|x\| > d \, (d > 0)$ if and only if either $x > d$ or $x < -d$	

A1.2 几何学 Geometry

1)	Pythagorean Theorem	In a right triangle: $c^2 = a^2 + b^2$	
2)	Triangle	$A = \dfrac{1}{2} bh$	
3)	Parallelogram	$A = bh$	
4)	Trapezoid	$A = \dfrac{1}{2}(a+b)h$	
5)	Circle	$A = \pi r^2$	$C = 2\pi r$
6)	Circular Sector	$A = \dfrac{1}{2} r^2 \theta$	$l = r\theta$
7)	Circular Ring	$A = \pi(R^2 - r^2)$	
8)	Sphere	$V = \dfrac{4}{3} \pi r^3$	$S = 4\pi r^2$

续表

9)	Cylinder	$V = \pi r^2 h$	$S(\text{Lateral}) = 2\pi r h$
10)	Cone	$V = \dfrac{1}{3}\pi r^2 h$	$S(\text{Lateral}) = \pi r \sqrt{r^2 + h^2}$

面积(Area)A,表面积(Surface Area)S,体积(Volume)V,圆周(Circumference)C,底边(Base)a、b,高(Height)h,半径(Radius)r,圆心角(Central Angle)θ,弧长(Arc Length)l

A1.3 解析几何学 Analytic Geometry

A1.3.1 距离 Distance

$P_1(x_1, y_1)$, $P_2(x_2, y_2)$, distance d between P_1 and the line $Ax + By + C = 0$

$$|P_1 P_2| = \sqrt{(x_1 - x_2)^2 - (y_1 - y_2)^2} \qquad d = \dfrac{|Ax_1 + By_1 + C|}{\sqrt{A^2 + B^2}}$$

A1.3.2 直线方程 Equations of The Straight Line

$AB \neq 0$, $P_1(x_1, y_1)$, x-intercepts a, y-intercepts b and slope m

General Form	$Ax + By + C = 0$
Point-Slope Form	$y - y_1 = m(x - x_1)$
Slope-Intercept Form	$y = mx + b$
Two-Point Form	$\dfrac{y - y_1}{x - x_1} = \dfrac{y_2 - y_1}{x_2 - x_1}$
Intercept Form	$\dfrac{x}{a} + \dfrac{y}{b} = 1$

A1.3.3 圆锥曲线 Conical Section

Circle	$(x - a)^2 + (y - b)^2 = r^2$ center (a, b) and radius r
Ellipse	$\dfrac{x^2}{a^2} + \dfrac{y^2}{b^2} = 1$ with $b^2 + c^2 = a^2$
Hyperbola	$\dfrac{x^2}{a^2} - \dfrac{y^2}{b^2} = 1$ with $a^2 + b^2 = c^2$
Parabola	$x^2 = 4py$ focus at $F(0, p)$

A1.4 三角学 Trigonometry

A1.4.1 基本恒等式 Fundamental Identities

平方(Square)关系	商数(Quotient)关系	倒数(Reciprocal)关系
$\sin^2 x + \cos^2 x = 1$	$\tan x = \dfrac{\sin x}{\cos x}$	$\tan x = \dfrac{1}{\cot x}$
$1 + \tan^2 x = \sec^2 x$	$\cot x = \dfrac{\cos x}{\sin x}$	$\sec x = \dfrac{1}{\cos x}$
$1 + \cot^2 x = \csc^2 x$		$\csc x = \dfrac{1}{\sin x}$

A1.4.2 和差角公式 Addition and Subtraction Formulas

和角(Sum)公式	差角(Difference)公式
$\sin(x+y) = \sin x \cos y + \cos x \sin y$	$\sin(x-y) = \sin x \cos y - \cos x \sin y$
$\cos(x+y) = \cos x \cos y - \sin x \sin y$	$\cos(x-y) = \cos x \cos y + \sin x \sin y$
$\tan(x+y) = \dfrac{\tan x + \tan y}{1 - \tan x \tan y}$	$\tan(x-y) = \dfrac{\tan x - \tan y}{1 + \tan x \tan y}$

A1.4.3 诱导公式 Reduction Formulas

1)	$\sin(2k\pi + x) = \sin x$	$\cos(2k\pi + x) = \cos x$	$\tan(2k\pi + x) = \tan x$
2)	$\sin(\pi + x) = -\sin x$	$\cos(\pi + x) = -\cos x$	$\tan(\pi + x) = \tan x$
3)	$\sin(\pi - x) = \sin x$	$\cos(\pi - x) = -\cos x$	$\tan(\pi - x) = -\tan x$
4)	$\sin(-x) = -\sin x$	$\cos(-x) = \cos x$	$\tan(-x) = -\tan x$
5)	$\sin\left(\dfrac{\pi}{2} + x\right) = \cos x$	$\cos\left(\dfrac{\pi}{2} + x\right) = -\sin x$	$\tan\left(\dfrac{\pi}{2} + x\right) = -\cot x$
6)	$\sin\left(\dfrac{\pi}{2} - x\right) = \cos x$	$\cos\left(\dfrac{\pi}{2} - x\right) = \sin x$	$\tan\left(\dfrac{\pi}{2} - x\right) = \cot x$

A1.4.4 倍角公式 Double-Angle Formulas

$\sin x = 2\sin x \cos x$	$\tan 2x = \dfrac{2\tan x}{1 - \tan^2 x}$
$\cos 2x = \cos^2 x - \sin^2 x = 2\cos^2 x - 1 = 1 - 2\sin^2 x$	

A1.4.5 半角公式 Half-Angle Formulas

$\sin\dfrac{x}{2} = \pm\sqrt{\dfrac{1-\cos x}{2}}$	$\cos\dfrac{x}{2} = \pm\sqrt{\dfrac{1+\cos x}{2}}$	$\tan\dfrac{x}{2} = \dfrac{1-\cos x}{\sin x} = \dfrac{\sin x}{1+\cos x}$
$\sin^2 x = \dfrac{1-\cos 2x}{2}$	$\cos^2 x = \dfrac{1+\cos 2x}{2}$	$\tan^2 x = \dfrac{1-\cos 2x}{1+\cos 2x}$

A1.4.6 和差化积 & 积化和差 Sum-to-Product & Product-to-Sum

和差化积(Sum-to-Product)	积化和差(Product-to-Sum)
$\sin x + \sin y = 2\sin\dfrac{x+y}{2}\cos\dfrac{x-y}{2}$	$\sin x \cos y = \dfrac{1}{2}[\sin(x+y) + \sin(x-y)]$
$\sin x - \sin y = 2\cos\dfrac{x+y}{2}\sin\dfrac{x-y}{2}$	$\cos x \sin y = \dfrac{1}{2}[\sin(x+y) - \sin(x-y)]$
$\cos x + \cos y = 2\cos\dfrac{x+y}{2}\cos\dfrac{x-y}{2}$	$\cos x \cos y = \dfrac{1}{2}[\cos(x+y) + \cos(x-y)]$
$\cos x - \cos y = -2\sin\dfrac{x+y}{2}\sin\dfrac{x-y}{2}$	$\sin x \sin y = -\dfrac{1}{2}[\cos(x+y) - \cos(x-y)]$

A1.4.7 正余弦定理 Law of Sines/Cosines

If a, b, c are the sides of triangle ABC, and A, B, C are respectively the opposite interior angles, then:

Law of Sines	$\dfrac{a}{\sin A} = \dfrac{b}{\sin B} = \dfrac{c}{\sin C}$
Law of Cosines	$c^2 = a^2 + b^2 - 2ab\cos C$
The Area	$A = \dfrac{1}{2}ab\sin C$

A1.5 极坐标 Polar Coordinates

A1.5.1 与直角坐标系的转化 Relations with Rectangular Coordinates

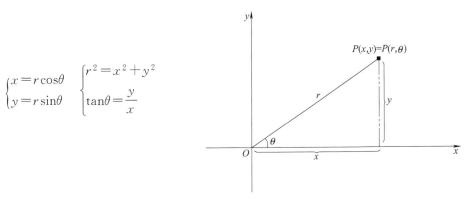

$\begin{cases} x = r\cos\theta \\ y = r\sin\theta \end{cases}$ $\begin{cases} r^2 = x^2 + y^2 \\ \tan\theta = \dfrac{y}{x} \end{cases}$

A1.5.2 常用极坐标方程 Common Polar Equations

Line	$x = a$	$r = a\sec\theta$
	$y = b$	$r = b\csc\theta$
Circle	$r = a$, center at pole, radius a	
	$r = 2a\cos\theta$, center at $(a, 0)$, radius a	
	$r = 2a\sin\theta$, center at $(0, a)$, radius a	
Roses (four leaves)	$r = \cos 2\theta$	
	$r = \sin 2\theta$	
cardioid	$r = a(1 \pm \cos\theta)$	
Lemniscate	$r^2 = \cos 2\theta$	

A.2 AP 微积分公式总结 Summary AP Calculus Formula

总结 AP 微积分考试中常用的重要公式，供 AP 微积分考生参考复习．

A2.1 极限和连续 Limits and Continuity

A2.1.1 极限的运算法则 Rules of Limits

1)	$\lim\limits_{x\to x_0} kf(x)=k\lim\limits_{x\to x_0} f(x)=kL$
2)	$\lim\limits_{x\to x_0}[f(x)\pm g(x)]=\lim\limits_{x\to x_0} f(x)\pm\lim\limits_{x\to x_0} g(x)=L\pm M$
3)	$\lim\limits_{x\to x_0}[f(x)g(x)]=\lim\limits_{x\to x_0} f(x)\lim\limits_{x\to x_0} g(x)=LM$
4)	$\lim\limits_{x\to x_0}\dfrac{f(x)}{g(x)}=\dfrac{\lim\limits_{x\to x_0} f(x)}{\lim\limits_{x\to x_0} g(x)}=\dfrac{L}{M}(M\neq 0)$
5)	$\lim\limits_{x\to x_0}[f(x)]^n=[\lim\limits_{x\to x_0} f(x)]^n$
\multicolumn{2}{l}{L,M,x_0 and k are real numbers, and $\lim\limits_{x\to x_0} f(x)=L$, $\lim\limits_{x\to x_0} g(x)=M$}

A2.1.2 两个重要极限 Two Important Limits

$$\lim_{x\to 0}\frac{\sin x}{x}=1 \qquad \lim_{x\to\infty}\left(1+\frac{1}{x}\right)^x=e$$

A2.1.3 有理函数的极限 Limit of a Rational Function

If $P(x)$ and $Q(x)$ are polynomials and $Q(x_0)\neq 0$, $a_0 b_0\neq 0$

$$\lim_{x\to\infty}\frac{P_n(x)}{Q_m(x)}=\lim_{x\to\infty}\frac{a_0 x^n+a_1 x^{n-1}+\cdots+a_{n-1}x+a_n}{b_0 x^m+b_1 x^{m-1}+\cdots+b_{m-1}x+b_m}$$

then,

$$\lim_{x\to\infty}\frac{P_n(x)}{Q_m(x)}=\begin{cases}\dfrac{a_0}{b_0}, & n=m \\ \infty, & n>m \\ 0, & n<m\end{cases}$$

A2.1.4 点连续 Continuity at a Point

A function $f(x)$ to be continuous at a point $x=x_0$ must fulfill all three of the following conditions:

1) $f(x_0)$ exists; 2) $\lim\limits_{x\to x_0} f(x)$ exists; 3) $\lim\limits_{x\to x_0} f(x)=f(x_0)$

A2.2 导数和微分 Derivative and Differentials

A2.2.1 导数的定义 Definition of the Derivative

$$f'(x_0)=\lim_{\Delta x\to 0}\frac{f(x_0+\Delta x)-f(x_0)}{\Delta x}=\lim_{x\to x_0}\frac{f(x)-f(x_0)}{x-x_0}=\lim_{\Delta x\to 0}\frac{\Delta y}{\Delta x}$$

$$f'_-(x_0)=\lim_{\Delta x\to 0^-}\frac{f(x_0+\Delta x)-f(x_0)}{\Delta x} \qquad f'_+(x_0)=\lim_{\Delta x\to 0^+}\frac{f(x_0+\Delta x)-f(x_0)}{\Delta x}$$

A2.2.2 导数的基本公式 Basic Differentiation Formulas

1)	$(C)' = 0$, C is constants	
2)	$(x^n)' = nx^{n-1}$	
3)	$(a^x)' = a^x \ln a$	$(e^x)' = e^x$
4)	$(\log_a x)' = \dfrac{1}{x \ln a}$	$(\ln x)' = \dfrac{1}{x}$
5)	$(\sin x)' = \cos x$	$(\cos x)' = -\sin x$
	$(\tan x)' = \sec^2 x$	$(\cot x)' = -\csc^2 x$
	$(\sec x)' = \sec x \tan x$	$(\csc x)' = -\csc x \cot x$
6)	$(\arcsin x)' = \dfrac{1}{\sqrt{1-x^2}}$	$(\arccos x)' = -\dfrac{1}{\sqrt{1-x^2}}$
	$(\arctan x)' = \dfrac{1}{1+x^2}$	$(\operatorname{arccot} x)' = -\dfrac{1}{1+x^2}$
	$(\operatorname{arcsec} x)' = \dfrac{1}{|x|\sqrt{x^2-1}}$	$(\operatorname{arccsc} x)' = -\dfrac{1}{|x|\sqrt{x^2-1}}$

A2.2.3 加减乘除法则 The Sum, Difference, Product and Quotient Rules

1)	$[ku(x)]' = k[u(x)]'$, k is constants
2)	$[u(x) + v(x)]' = u'(x) + v'(x)$
3)	$[u(x) - v(x)]' = u'(x) - v'(x)$
4)	$[u(x)v(x)]' = u'(x)v(x) + u(x)v'(x)$
5)	$\left[\dfrac{u(x)}{v(x)}\right]' = \dfrac{u'(x)v(x) - u(x)v'(x)}{v^2(x)}$, $v(x) \neq 0$

A2.2.4 微分学中的其他主题 Other Topics in Differential Calculus

The Chain Rule	$y = f[g(x)] \Rightarrow \dfrac{dy}{dx} = \dfrac{dy}{du} \times \dfrac{du}{dx}$
Inverse Function	$\dfrac{d}{dx} f^{-1}(x)\Big|_{x=c} = \dfrac{1}{\left[\dfrac{d}{dy} f(y)\right]_{y=a}}$
Parametric Equations	$\begin{cases} x = x(t) \\ y = y(t) \end{cases} \Rightarrow \dfrac{dy}{dx} = \dfrac{\dfrac{dy}{dt}}{\dfrac{dx}{dt}}$
Polar Functions	$\dfrac{dy}{dx} = \dfrac{\dfrac{dy}{d\theta}}{\dfrac{dx}{d\theta}} = \dfrac{\dfrac{dr}{d\theta}\sin\theta + r\cos\theta}{\dfrac{dr}{d\theta}\cos\theta - r\sin\theta}$

A2.3 积分和微分方程 Integral and Differential Equation

A2.3.1 基本积分公式 Basic Integration Formulas

1)	$\int k\,dx = kx + C$, C is constants	
2)	$\int x^n\,dx = \dfrac{x^{n+1}}{n+1} + C\,(n \neq -1)$	
3)	$\int a^x\,dx = \dfrac{a^x}{\ln a} + C$	$\int e^x\,dx = e^x + C$
4)	$\int \dfrac{1}{x}\,dx = \ln|x| + C$	
5)	$\int \sin x\,dx = -\cos x + C$	$\int \cos x\,dx = \sin x + C$
	$\int \tan x\,dx = -\ln|\cos x| + C$	$\int \cot x\,dx = \ln|\sin x| + C$
	$\int \sec x\,dx = \ln|\sec x + \tan x| + C$	$\int \csc x\,dx = \ln|\csc x - \cot x| + C$
	$\int \sec^2 x\,dx = \tan x + C$	$\int \csc^2 x\,dx = -\cot x + C$
	$\int \sec x \tan x\,dx = \sec x + C$	$\int \csc x \cot x\,dx = -\csc x + C$
6)	$\int \dfrac{dx}{\sqrt{1-x^2}} = \arcsin x + C$	
	$\int \dfrac{dx}{1+x^2} = \arctan x + C$	
	$\int \dfrac{dx}{x\sqrt{x^2-1}} = \text{arcsec}\,x + C$	

A2.3.2 不定积分运算法则 Rules for Finding Indefinite Integral

1)	$\int kf(x)\,dx = k\int f(x)\,dx$	
2)	$\int [f(x) + g(x)]\,dx = \int f(x)\,dx + \int g(x)\,dx$	
	$\int [f(x) - g(x)]\,dx = \int f(x)\,dx - \int g(x)\,dx$	
3)	$\left[\int f(x)\,dx\right]' = f(x)$	$d\int f(x)\,dx = f(x)\,dx$
4)	$\int f'(x)\,dx = f(x) + C$	$\int df(x) = f(x) + C$

A2.3.3 不定积分的其他主题 Other Topics in Indefinite Integral

Each Other to Reverse	
U-Substitution	$\int f[g(x)]g'(x)\,dx = \int f(u)\,du$
Integration by Parts	$\int u\,dv = uv - \int v\,du$

A2.3.4 黎曼和与梯形法则 Riemann Sums and Trapezoid Rule

Left Riemann Sums	$L(n) = \sum_{i=1}^{n} \left(\dfrac{b-a}{n}\right) y_i$
Right Riemann Sums	$R(n) = \sum_{i=1}^{n} \left(\dfrac{b-a}{n}\right) y_{i+1}$
Midpoint Riemann Sums	$M(n) = \sum_{i=1}^{n} \left(\dfrac{b-a}{n}\right) y_{\frac{2i-1}{2}}$
Trapezoidal Rule	$T(n) = \left(\dfrac{b-a}{2n}\right)(y_0 + 2y_1 + \cdots + 2y_{n-1} + y_n)$

A2.3.5 定积分基本定理 The Fundamental Theorem of Definite Integral

1)	$\int_a^b f(x)\mathrm{d}x = F(b) - F(a)$
2)	$\dfrac{\mathrm{d}F}{\mathrm{d}x} = \dfrac{\mathrm{d}}{\mathrm{d}x}\int_a^x f(t)\mathrm{d}t = f(x)$
3)	$\int_a^a f(x)\mathrm{d}x = 0$
4)	$\int_a^b kf(x)\mathrm{d}x = k\int_a^b f(x)\mathrm{d}x$
5)	$\int_a^b [f(x) \pm g(x)]\mathrm{d}x = \int_a^b f(x)\mathrm{d}x \pm \int_a^b g(x)\mathrm{d}x$
6)	$\int_a^b f(x)\mathrm{d}x = \int_a^c f(x)\mathrm{d}x + \int_c^b f(x)\mathrm{d}x, a < c < b$
7)	$\int_a^b f(x)\mathrm{d}x = -\int_b^a f(x)\mathrm{d}x$
8)	$f(\xi) = \dfrac{1}{b-a}\int_a^b f(x)\mathrm{d}x, a < \xi < b$
9)	$\int_a^b f[\varphi(x)]\varphi'(x)\mathrm{d}x = \int_{\varphi(a)}^{\varphi(b)} f(u)\mathrm{d}u$
10)	$\int_a^b u\mathrm{d}v = uv\Big|_a^b - \int_a^b v\mathrm{d}u$

A2.3.6 广义积分 Improper Integrals

Integration on an Infinite Interval	
$\int_a^{+\infty} f(x)\mathrm{d}x = \lim\limits_{b \to +\infty} \int_a^b f(x)\mathrm{d}x$	$\int_{-\infty}^b f(x)\mathrm{d}x = \lim\limits_{a \to -\infty} \int_a^b f(x)\mathrm{d}x$
$\int_{-\infty}^{+\infty} f(x)\mathrm{d}x = \int_{-\infty}^c f(x)\mathrm{d}x + \int_c^{+\infty} f(x)\mathrm{d}x$	
Integrands with Infinite Discontinuities	
$\int_a^b f(x)\mathrm{d}x = \lim\limits_{c \to a+} \int_c^b f(x)\mathrm{d}x$	$\int_a^b f(x)\mathrm{d}x = \lim\limits_{c \to b-} \int_a^c f(x)\mathrm{d}x$
$\int_a^b f(x)\mathrm{d}x = \int_a^c f(x)\mathrm{d}x + \int_c^b f(x)\mathrm{d}x$	

A2.3.7 弧长 Arc Length

| $L = \int_a^b \sqrt{1 + \left(\dfrac{dy}{dx}\right)^2}\,dx$ | $L = \int_c^d \sqrt{1 + \left(\dfrac{dx}{dy}\right)^2}\,dy$ | $L = \int_a^b \sqrt{\left(\dfrac{dx}{dt}\right)^2 + \left(\dfrac{dy}{dt}\right)^2}\,dt$ |

A2.3.8 常用微分方程 Common Differential Equations

指数增长与衰变 (Exponential Growth and Decay)	
$\dfrac{dy}{dt} = ky$	$y = C_0 e^{kt}$
约束增长与衰减 (Restricted Growth and Decay)	
$\dfrac{dy}{dt} = k(A - y)$	$y = A - C_0 e^{-kt}$
$\dfrac{dy}{dt} = -k(y - A)$	$y = A + C_0 e^{-kt}$
Logistic 微分方程	
$\dfrac{dy}{dt} = ky(A - y)$	$y = \dfrac{A}{1 + C_0 e^{-Akt}}$

A2.4 无穷级数 Infinite Series

A2.4.1 无穷级数的定义 Definition of Infinite Series

Infinite Series	$\sum\limits_{i=1}^{\infty} a_i = a_1 + a_2 + a_3 + \cdots + a_n + \cdots$
Partial Sum	$S_n = \sum\limits_{i=1}^{n} a_i = a_1 + a_2 + a_3 + \cdots + a_n$
Convergent	$\lim\limits_{n \to \infty} S_n = L$ exists;
Divergent	$\lim\limits_{n \to \infty} S_n$ does not exists;

A2.4.2 四类重要级数 Four Important Series

Geometric Series	$\sum\limits_{n=1}^{\infty} a r^{n-1} = a + ar + ar^2 + \cdots + ar^{n-1} + \cdots$
p-Series	$\sum\limits_{n=1}^{\infty} \dfrac{1}{n^p} = \dfrac{1}{1^p} + \dfrac{1}{2^p} + \dfrac{1}{3^p} + \cdots + \dfrac{1}{n^p} + \cdots$
Harmonic Series	$\sum\limits_{n=1}^{\infty} \dfrac{1}{n} = 1 + \dfrac{1}{2} + \dfrac{1}{3} + \cdots + \dfrac{1}{n} + \cdots$
Alternating Series	$\sum\limits_{n=1}^{\infty} (-1)^{n+1} a_n = a_1 - a_2 + \cdots + (-1)^{n+1} a_n + \cdots$

A2.4.3 级数判别法 Series Tests

Geometric Series	If $|r|<1$, Converge, $\sum_{n=1}^{\infty} a\, r^{n-1} = \dfrac{a}{1-r}$	
	If $|r| \geqslant 1$, Diverge	
p-Series	If $p>1$, Converge	
	If $p \leqslant 1$, Diverge	
Harmonic Series	Diverge	
Alternating Series	If $a_n > a_{n+1}$ and $\lim\limits_{n \to \infty} a_n = 0$, Converge	
Ratio Test	$\lim\limits_{n \to \infty} \dfrac{a_{n+1}}{a_n} = \rho$	converges if $\rho < 1$
		diverge if $\rho > 1$
		inconclusive if $\rho = 1$
Root Test	$\lim\limits_{n \to \infty} \sqrt[n]{a_n} = \rho$	converges if $\rho < 1$
		diverge if $\rho > 1$
		inconclusive if $\rho = 1$
Comparison Test ($0 \leqslant a_n \leqslant b_n$)	If $\sum_{n=1}^{\infty} b_n$ converges, then $\sum_{n=1}^{\infty} a_n$ converges	
	If $\sum_{n=1}^{\infty} a_n$ diverge, then $\sum_{n=1}^{\infty} b_n$ diverge	
Integral Test	If $f(n) = a_n \geqslant 0$ is continuous and decreasing for $x \geqslant 1$, then $\sum_{n=1}^{\infty} a_n$ and $\int_{1}^{\infty} f(x)\,\mathrm{d}x$ both converges or diverge	

A2.4.4 幂级数和泰勒级数 Power Series and Taylor Series

Power Series:

$$\sum_{n=0}^{\infty} a_n x^n = a_0 + a_1 x + a_2 x^2 + \cdots + a_n x^n + \cdots$$

$$\sum_{n=0}^{\infty} a_n (x-a)^n = a_0 + a_1(x-a) + a_2(x-a)^2 + \cdots + a_n(x-a)^n + \cdots$$

Taylor Series:

$$f(x) = f(a) + f'(a)(x-a) + \frac{f''(a)}{2!}(x-a)^2 + \cdots + \frac{f^{(n)}(a)}{n!}(x-a)^n + \cdots$$

Maclaurin Series:

$$f(x) = f(0) + f'(0)x + \frac{f''(0)}{2!}x^2 + \cdots + \frac{f^{(n)}(0)}{n!}x^n + \cdots$$

A2.4.5 常用的麦克劳林级数 Common Maclaurin Series

$$\mathrm{e}^x = \sum_{n=0}^{\infty} \frac{x^n}{n!} = 1 + x + \frac{x^2}{2!} + \frac{x^3}{3!} + \cdots + \frac{x^n}{n!} + \cdots, \quad |x| < \infty$$

$$\sin x = \sum_{n=0}^{\infty} (-1)^n \frac{x^{2n+1}}{(2n+1)!} = x - \frac{x^3}{3!} + \frac{x^5}{5!} - \cdots + (-1)^n \frac{x^{2n+1}}{(2n+1)!} + \cdots, \ |x| < \infty$$

$$\cos x = \sum_{n=0}^{\infty} (-1)^n \frac{x^{2n}}{(2n)!} = 1 - \frac{x^2}{2!} + \frac{x^4}{4!} - \cdots + (-1)^n \frac{x^{2n}}{(2n)!} + \cdots, \ |x| < \infty$$

$$\ln(1+x) = \sum_{n=1}^{\infty} (-1)^{n-1} \frac{x^n}{n} = x - \frac{x^2}{2} + \frac{x^3}{3} - \cdots + (-1)^{n-1} \frac{x^n}{n} + \cdots, \ -1 < x \leqslant 1$$

$$\arctan x = \sum_{n=0}^{\infty} (-1)^n \frac{x^{2n+1}}{2n+1} = x - \frac{x^3}{3} + \frac{x^5}{5} - \cdots + (-1)^n \frac{x^{2n+1}}{2n+1} + \cdots, \ |x| \leqslant 1$$

A.3 VIP 服务及网站

VIP 体系服务于报名学员，有保障的 5 分（满分）训练体系，欢迎报名．

官网：http://calculus.apexams.net【AP 微积分·考试网】

网校：http://wx.apexams.net

训练体系：《AP 微积分真题导练》

笔者个人微信号：709645945

公众号：AP 微积分（apcalculus）

参考文献 References

[1] 欧拉. 无穷分析引论 [M]. 张延伦译. 山西教育出版社，1997.
[2] 温惠林，王学敏. 微积分 [M]. 北京邮电大学出版社，2001.
[3] David S Kahn. Cracking the AP Calculus AB & BC Exams [M]. the Princeton Review Inc. 2012.
[4] Lopez R J. Implicit Differentiation [M]. Maple via Calculus. Birkhäuser Boston，1994.
[5] J Stewart. Single Variable Calculus [M]. Brooks/Cole Publishing Company，2012.
[6] Thomas. Thomas' Calculus [M]. Pearson Education，2005.
[7] David Bock，等. Barron's AP Calculus 11th Edition [M]. 北京：世界图书出版社，2012.
[8] 同济大学数学系. 高等数学：第六版 [M]. 北京：高等教育出版社，2007.
[9] 胡振媛. 对"微积分基本定理"的认识和理解 [J]. 教育与教学研究，2000（3）.
[10] 华东师范大学数学系. 数学分析 [M]. 北京：高等教育出版社，2012.
[11] 林群. 微积分快餐 [M]. 北京：科学出版社，2010.
[12] Massaza C，Terracini L，Valabrega P. On the Intermediate Value Theorem over a Valued Field [J]. Mathematics，2015.
[13] John A. Gubner. The Intermediate-Value Theorem [J]. Department of Electrical and Computer Engineering，University of Wisconsin – Madison.
[14] Earl W，Swokowski and Jeffery A，Cole. PreCalculus：Functions and Graphs，11e [M]. Thomson Higher Education，2008.
[15] 陈启浩. 微积分精讲精练 [M]. 北京：北京师范大学出版社，2009.
[16] 张景中. 不用极限的微积分 [M]. 北京：中国少年儿童出版社，2012.